SYSTEM AND BAYESIAN RELIABILITY

SERIES IN QUALITY, RELIABILITY & ENGINEERING STATISTICS

Series Editors: *M. Xie* (*National University of Singapore*)
T. Bendell (*Nottingham Polytechnic*)
A. P. Basu (*University of Missouri*)

Published

Vol. 1: Software Reliability Modelling
M. Xie

Vol. 2: Recent Advances in Reliability and Quality Engineering
H. Pham

Vol. 3: Contributions to Hardware and Software Reliability
P. K. Kapur, R. B. Garg & S. Kumar

Vol. 4: Frontiers in Reliability
A. P. Basu, S. K. Basu & S. Mukhopadhyay

Forthcoming title

Reliability Optimization & Design of Fault Tolerant Systems
H. Pham

VOL 5 Series on Quality, Reliability & Engineering Statistics

SYSTEM AND BAYESIAN RELIABILITY

Essays in Honor of Professor Richard E. Barlow on His 70th Birthday

Editors

Yu Hayakawa

Victoria University of Wellington, New Zealand

Telba Irony

Food and Drug Administration, USA

Min Xie

National University of Singapore, Singapore

World Scientific
New Jersey • London • Singapore • Hong Kong

Published by

World Scientific Publishing Co. Pte. Ltd.

P O Box 128, Farrer Road, Singapore 912805

USA office: Suite 1B, 1060 Main Street, River Edge, NJ 07661

UK office: 57 Shelton Street, Covent Garden, London WC2H 9HE

British Library Cataloguing-in-Publication Data
A catalogue record for this book is available from the British Library.

SYSTEM AND BAYESIAN RELIABILITY

ISBN 981-02-4865-2

Printed in Singapore by World Scientific Printers

Foreword

In the modern world we usually take for granted the reliability of the equipment that we use: the freezer will continue to preserve our food whilst we are away, the airplane engine will keep operating when we are in the air, even the computer will continue functioning, though the software may occasionally play us tricks. Much of this reliability is due to the skill of engineers in the design of the equipment and in the use of suitable materials in their construction; the failures, when they do rarely occur, are often due to human error. Engineering skill is often unappreciated and taken for granted, so that when an engineering failure does occur, as recently happened in the construction of a foot bridge over the Thames in London, we express astonishment at the disaster.

There is a body of opinion which holds that if the engineering is done properly and the equipment sensibly used, then failures need never arise. This is false; uncertainty is an integral part of all aspects of life and is openly recognized in, for example, quantum physics and Mendelian genetics, and engineering is no exception. Failure is an uncertain phenomenon whose occurrence cannot be predicted nor entirely prevented. Even a bridge can fail. Once uncertainty enters the picture, it is essential to use the tools of probability because, as de Finetti and others have shown, probability is the only satisfactory language in which to speak about uncertainty and, in particular, the uncertainty that even the skill of the engineer cannot entirely avoid. There has therefore grown up a discipline which studies, by probabilistic methods, the manner in which breakdowns occur, how they can be reduced in number, and how experiments on new products can be designed so that there is a trustworthy guide to how they will behave in practice. The subject is usually called Engineering Reliability and Richard Barlow chose it as the title of his most recent book, although, as he says, "it is really about statistics". Barlow has been a leader in this field and progress in it owes much to the ideas he has developed in the course of his career.

The first encounter I can recall between us arose when he used a statistical method that, in my view, was unsound and, never being shy of declaring what appeared to be an error, I said so. Two interesting things then happened; first, my view was sensible, second that view was listened to, neither event being as frequent as I might wish. As a result, he obtained a grant, an activity at which he was very successful, his success being indicative of the high regard in which his work was held by his peers, that took me to visit him in Berkeley. This was to be the first of many such visits he organized, making my formal retirement from London both enjoyable and rewarding. Many hours we have spent together in the Café Espresso, just down the road from Etcheverry Hall and opposite the campus of the University of California at Berkeley, discussing questions of technical interest to us both. Neither Dick nor I are great on small talk and we would concentrate on such vital issues as the correct interpretation of probability, its role in engineering reliability, the proper analysis of data and why, with a few honourable exceptions, members of the department of statistics over the way were, in our view, wrong. Reliability theorists and reliability engineers tend to deal with abstract issues, whereas engineers are better at material matters, and they tend to work in isolation, so that the gathering together over coffee, or tea in England, serves not just a social function, but can be integral to the development of sound research.

Barlow obtained a master's degree in mathematics at Eugene, Oregon in 1955 and a doctorate at Stanford in 1960. There he worked with Sam Karlin and began the interaction with Frank Proschan that continued for many years afterwards. After a brief spell outside academia, Barlow obtained a post at Berkeley in 1963, where he remained until his recent retirement. As others have discovered, once settled in Berkeley why should one leave? Here is one of the world's leading universities, in a place with an excellent climate, in a society which is more sensible, that is, left-wing, than most in the United States. Only the threat of an earthquake could disturb the idyll, and after all that is a topic worthy of study by a reliability engineer.

Problems in probability divide into two types; those of direct probability and those of inverse probability. The division has been recognized ever since the earliest serious studies of the subject, and first arose after the introduction of the binomial distribution. If the probability of success in each of a number of trials is known as p, then in n trials, judged independent given p, r the number of successes will be binomial. This is a direct problem, passing from a known probability p to uncertainty regarding the data, r out of n. The corresponding inverse problem is, having observed r and n, what can be said about p, an early solution being due to Bayes. Nowadays it is common to think of the direct problems as being in the field of probability, whereas the inverse ones belong to statistics.

Both types arise in engineering reliability. For a single component, the reliability is commonly described by its failure rate, the equivalent of p in the binomial example above, and different forms of this function lead directly to different observed patterns of failure, the analogue of r and n. The problems become more interesting, and more realistic, when many components are put together in the form of a coherent structure, when it is required to assess the behaviour of the structure in terms of those of its components and their interactions. An interesting early result obtained by Barlow, jointly with Marshall and Proschan, was that components with failure rate increasing with time formed, under convolution, a structure with the same increasing property. In contrast, even in systems with parallel components, increasing failure rates of components does not imply the property for the system. The concept of increase on average has been found more useful, as have realistic bounds on failure rates. Thus invariance does not always obtain with coherent structures and Barlow has been responsible for many of our advances in understanding the failure patterns of systems; for example, in fault-tree analysis. A major contribution has been made with Marshall on obtaining inequalities and crossing properties of survival functions with increasing failure rate, enabling useful bounds to be put on the occurrence of failures.

These are all direct problems but equally he has been responsible for influential studies of statistical, inverse questions; indeed, his last book is mainly devoted to this field. An important concept here is the total time on test and how it changes with time. A more recent enthusiasm has been the use of influence diagrams which enable the structure of a system to be more easily appreciated. Some of his work has been in co-operation with the Lawrence Livermore Laboratory, where he has analysed reliability data from experiments conducted there, including accelerated life tests where significant contributions have been made to the relationship between the reliability in the field and that in the stressed, laboratory environment.

A topic that embraces both the direct and inverse concepts of probability is that of decision analysis, which impinges on reliability in the development of maintenance systems. When should equipment be withdrawn from service for maintenance? How extensive should be repair be? When is it sensible to replace a component? These questions need both a statistical analysis of experience in the field, and development of a model, in order to construct sensible strategies.

A common, basic assumption in the literature is that, conditional on parameters, observations are independent and identically distributed. Recent work Barlow has done with Mendel escapes from this assumption and develops finite tools more relevant to practice. This depends on the earlier work of de Finetti and makes reliability move even further into the subjective appreciation of probability and what is nowadays called the Bayesian viewpoint.

The subject of reliability engineering today is very different from the form it took in the early days of operational research and much of this change has been due to the work of Richard Barlow. In this volume several of his colleagues and friends, who appreciate his considerable contributions, recognize their value by writing papers that build on the work he has done over the last forty years. Many are former students of his, which reminds us to recognize the significant effect Barlow has had on reliability studies through the effort and enthusiasm he has put into teaching many people who have gone on to do important work in the field. In writing the Foreword to this important volume, I would like to express my thanks to a person who has flattered me by not only listening to what Savage, Ramsey, de Finetti and the other great contributors to our proper understanding of probability, as elaborated by me in the coffee shop, but who has gone on to incorporate their ideas into engineering with such important consequences. May you have a very happy seventieth birthday Dick and see in this volume the respect with which you and your work is held.

London, August 2001

Dennis V. Lindley

Foreword

Richard Barlow is a professor emeritus in the College of Engineering at the University of California, Berkeley. He has had a long and distinguished career. Since 1963 he has been a professor jointly in the Department of Industrial Engineering and Operations Research and in the Department of Statistics at UC, Berkeley and in addition a research engineer in UC, Berkeley's Engineering Systems Research Center. He served as a consultant to the U. S. Department of Defense and for nearly 30 years he also acted as a consultant to Lawrence Livermore National Laboratory. In 1991 he was awarded the von Neumann prize, (together with Frank Proschan).

His book Mathematical Theory of Reliability (published in 1965 and re-issued in the SIAM classic series in 1996) written jointly with Frank Proschan and their 1975 book Statistical Theory and Reliability and Life Testing are major contributions in the field. They are probably the most widely referred literature in reliability theory worldwide. The latter book has been translated into German, Russian and Chinese. So many Barlow and Proschan joint works have been cited that Frank Proschan once introduced himself: "I am Proschan. Many people think my first name is Barlow."

Richard earned a B.A. and an M.A. in mathematics from Knox College and the University of Oregon respectively. Two years of graduate study in statistics at the University of Washington was followed by a Ph.D. in statistics from Stanford University in 1960. It was during his time at Stanford that he began his collaboration with Frank Proschan which resulted in the two books previously mentioned and numerous papers in many areas of reliability. Richard also has an extensive bibliography of papers on his own and with other co-authors which have been published in the leading statistical journals. He and his associates introduced a number of key ideas in modern reliability theory. Among these are:

system efficiency and reliability,
classes of distributions based on aging,
replacement policies,
inference for restricted families,
reliability growth,
"burn-in" procedures,
fault tree analysis,
total time on test procedures,
influence diagrams,
combining expert opinions,
stress-rupture life of Kevler/epoxy spherical pressure vessels,
group decision making,
Bayesian analysis of reliability problems.

In addition to influencing the direction of academic research, Richard has made major contributions to government and industry. He has served on the committee on Applied and Theoretical Statistics of the National Research Council and has been associate editor of several professional journals.

In the 50's Milton Sobel and Benjamin Epstein were making the most important contributions in reliability. They inspired me and many of my contemporaries to enter the field of reliability. It was Milton Sobel who "introduced" me to Richard Barlow while I was a graduate student at the University of Minnesota doing my Ph.D. thesis under Professor Sobel. Sobel asked me to review for the journal Econometrica the then just published book, Mathematical Theory of Reliability by Barlow and Proschan, pointing out that the book contained some interesting new ideas. I have been "associated" with Barlow and Proschan ever since.

During the late 70's, on my invitation, Richard graciously agreed to deliver a series of 10 lectures at the University of Missouri. Both Barlow and Proschan served as members of the advisory committees of four international research conferences on reliability held at the University of Missouri-Columbia in 1984, 1986, 1988 and 1991. On and off Richard has stressed to me the importance of developing Bayesian reliability theory. During the 1991 conference Richard served on a panel on the topic "Future Directions of Reliability Research." He impressed upon the audience, in no uncertain terms, that the future direction lies in carrying out further research in Bayesian reliability theory. His most recent book Engineering Reliability, published by SIAM in 1998, emphasizes this strongly held belief.

Richard has exerted a strong and worldwide influence in the areas of system reliability and Bayesian reliability for a period of more than forty years. This influence is reflected in the papers comprising this present volume. On a personal level, Richard has been a supportive mentor to his students and a source of inspiration to other young researchers and to his colleagues throughout the statistical world.

Columbia, Missouri, August 2001 — Asit P. Basu

Preface

Professor Richard E. Barlow is famous for his pioneering research in reliability theory. He is the author of several important books and numerous leading articles in reliability theory and Bayesian reliability. His "Mathematical Theory of Reliability", co-authored with Frank Proschan, which was published originally by Wiley in 1965 was reprinted as one of the SIAM Classics in 1996. It was translated into several languages and was a stimulus to research in reliability theory worldwide. To honor Professor Barlow's contribution, his former PhD students and other reliability researchers worldwide have cooperated in the preparation of this volume.

After receiving his PhD in Statistics from Stanford University in 1960, Dick Barlow joined the Institute of Defense Analysis, and then moved to General Telephone Laboratories. At that time he gained substantial practical insight into reliability and started active research which lead to breakthroughs and new ideas in reliability theory for practical applications.

Professor Barlow moved to the Department of Industrial Engineering and Operations Research of the University of California at Berkeley in 1963. He was a faculty member there until his retirement in July of 1999 when he became Professor Emeritus. He received many honors, including the Von Neumann Prize Award presented by ORSA-TIMS (jointly with Frank Proschan in 1991). He is a Fellow of the Institute of Mathematical Statistics, a Fellow of the American Statistical Association and an elected member of the International Statistical Institute.

Professor Barlow has an impressive list of published articles on Reliability and Bayesian Analysis. The following are some of his books, several of which have had very significant impact in the reliability and Bayesian research communities.

Barlow, R. E. and Proschan, F. (1965). *Mathematical Theory of Reliability.* Wiley, New York. (reprinted in 1996 as SIAM Classics)

Barlow, R. E., Bartholomew, D. J., Bremner, J. M. and Brunk, H. D. (1972). *Statistical Inference under Order Restrictions: The Theory and Application of Isotonic Regression.* Wiley, New York.

Barlow, R. E. and Proschan, F. (1975). *Statistical Theory of Reliability and Life Testing: Probability Models.* Holt, Reinhart and Winston, New York. (reprinted by To Begin With, Silver Spring, MD, 1981)

Barlow, R. E., Fussell, J. B. and Singpurwalla, N. D. (Editors) (1975). *Reliability and Fault Tree Analysis.* SIAM, Philadelphia.

Serra, A. and Barlow, R. E. (Editors) (1986). *Theory of Reliability.* North-Holland, Amsterdam.

Barlow, R. E., Clarotti, C. A. and Spizzichino, F. (Editors) (1993). *Reliability and Decision Making.* Chapman & Hall, London.

Barlow, R. E. (1998). *Engineering Reliability.* ASA and SIAM, New York.

This volume contains contributions from active researchers in reliability and Bayesian analysis worldwide. Many are also Professor Barlow's former students or close friends. All have responded to our call for contributions enthusiastically.

The contributions are in three groups, each reflecting an area in which Professor Barlow has had significant influence with his books, research articles and philosophy.

The first group deals with system reliability analysis. Professor Barlow's two books with Frank Proschan are the classics in this field, and his stimulus has been important to many researchers in the area. Professor Barlow's contribution to system reliability ranges from coherent systems and reliability bounds, to replacement models and fault tree analysis. Due to his pioneering work, system reliability is now an active research topic worldwide.

The second group of articles deals with aging properties which are important in reliability related analysis and decision making. The research carried out by Professors Barlow and Proschan in the sixties and seventies has led to numerous research and applications in reliability and maintenance engineering. Professor Barlow contributed actively in the research of aging concepts, closure properties and characterization. Research in this area is still ongoing and several papers in this volume deal with this topic.

Finally, the third group deals with Bayesian analysis. Professor Barlow is a leading figure in Bayesian methods, particularly, of course, in Bayesian reliability. His recent book "Engineering Reliability" provides a good summary of his views and sets out a number of useful Bayesian methods. It is already having a significant effect on research as can be seen from the articles presented here.

The articles in this volume are, in the main, self-contained and can therefore be read separately but each group also gives an excellent overview of the current research in its area. Some articles present new insights or results and some include stimulating topic ideas for both students and researchers.

We hope that this book can bring together reliability and Bayesian researchers to further contribute to the development of research and applications in these important and challenging fields.

Yu Hayakawa
Wellington, New Zealand

Telba Irony
Washington, D. C., U.S.A.

Min Xie
Kent Ridge, Singapore

Acknowledgements

Following our initial call for contribution in 2000, we received an overwhelming number of responses from potential contributors worldwide. We would like to thank all authors of this volume for their time and effort. We would like to thank especially Professors Dennis V. Lindley and Asit P. Basu for their kind forewords to this volume.

We would also like to thank World Scientific Publisher and their staff who have helped us with the preparation of this volume.

The support and assistance of our host institutions that enabled us to work on this volume is greatly appreciated.

Finally we are grateful to our families, colleagues and friends for their encouragement and support in one way or another.

Contents

PART 1

SYSTEM RELIABILITY ANALYSIS

CHAPTER 1

ON REGULAR RELIABILITY MODELS

Jen-Chun Chang
Department of Information Management
Ming Hsin Institute of Technology
Hsin-Fong, Hsin-Chu, Taiwan 304
E-mail: jcchang@csie.nctu.edu.tw

Rong-Jaye Chen
Department of Computer Science & Information Engineering
National Chiao-Tung University
Hsin-Chu, Taiwan 300
E-mail: rjchen@csie.nctu.edu.tw

Frank K. Hwang
Department of Applied Mathematics
National Chiao-Tung University
Hsin-Chu, Taiwan 300
E-mail: fhwang@math.nctu.edu.tw

This paper proposes the regular reliability model, a tool to specify and analyze various systems. When analyzing the reliability of a system, we first specify the system structure with the regular reliability model, apply the automata theory to derive a minimal state heterogeneous Markov chain, and then an efficient reliability algorithm can be obtained by implementing the Markov chain approach with the sparse matrix data structure. For most systems, the reliability algorithms derived from the regular reliability model are more efficient than the best published ones. For other systems, the reliability algorithms are at least as efficient as the best Markov chain approaches since the number of states are minimized.

1. Introduction

Barlow and Proschan[12] gave a section to describe the *coherent* system in their book. There are three intuitive hypotheses for coherent systems. The first is that if all the components function, the system functions. The next hypothesis states that if all components fail, the system fails. The last hypothesis is that functioning components do not interfere with the functioning of the system.

This paper proposes the regular reliability model. It is a useful tool to describe and analyze various system structures. In addition to coherent systems, many non-coherent systems can also be described and analyzed with the regular reliability model. For example, the regular model can be used to analyze the n-component system which functions when the number of functioning components in it is even. This system is not coherent since none of the three hypotheses given in Barlow and Proschan's book are satisfied when n is odd.

Before introducing the notion of the regular reliability model, we need a few basic definitions.

Definition 1: A *system* is a finite sequence of components where each component is either in the working state (state 1) or the failed state (state 0) with certain probabilities. All components are statistically independent.

Let $\Pr\{\text{component } i \text{ is in state } 1\} = p_i$ and $\Pr\{\text{component } i \text{ is in state } 0\} = 1 - p_i$.

Definition 2: A *system instance* is a binary string to represent the status of the system, where the ith bit of the string is 1 if and only if component i is in state 1.

Definition 3: A *reliability model*, or abbreviated as *model*, is a (maybe infinite) set of binary strings. Under a reliability model M, a system instance s is working if and only if $s \in M$; otherwise, s is failed under model M.

The automaton [1] is a mathematical model of a system, with discrete inputs and outputs. The system can be in any one of a finite number of of internal configurations or "states." The state of a system summarizes the information concerning past inputs that is needed to determine the behavior of the system on subsequent inputs. In computer science the theory of automata is a useful design tool for many finite state systems, such as switching circuits, text editors and lexical analyzers, etc. In automata theory, the languages (string sets) accepted by automata are easily described by simple expressions called "regular expressions". A *regular expression* is an expression to denote a set of binary strings with the following components and operations:

(1) ϕ denotes the empty set of strings,
(2) ε denotes $\{\varepsilon\}$, where ε is the empty string,
(3) 0 denotes $\{0\}$, and 1 denotes $\{1\}$,
(4) If r denotes R, s denotes S, then

(4.1) $r+s$ denotes $R \cup S$,
(4.2) rs denotes $RS = \{xy \mid x \in R,\ y \in S\}$,
(4.3) r^* denotes $R^* = \bigcup_{i=0}^{\infty} R^i$, where $R^0 = \{\varepsilon\},\ R^i = R^{i-1}R$, for $i \geq 1$.

Definition 4: ϕ, ε, 0, and 1 are *regular expressions*; if r and s are *regular expressions*, then $r+s$, rs, and r^* also are.

Definition 5: A reliability model is *regular* if it can be denoted by a regular expression.

Applying automata theory [1] to the definitions given above, we have the following results, where 0,1* is the set of all binary strings, including the empty string.

Theorem 6: If M is a regular model, then $\overline{M} = \{0,1\}^* - M$ is also a regular model.

Proof: Let r be the regular expression to denote M. By automata theory, from r we can construct an automaton A which accepts the strings in M and rejects the strings not in M. Then we can convert A to $\overline{A}$, which accepts the strings not in M and rejects the strings in M, or equivalently, $\overline{A}$ accepts the strings in $\overline{M}$ and rejects the strings not in $\overline{M}$. Next, from $\overline{A}$ we can find a regular expression $\overline{r}$ to denote $\overline{M}$. Therefore $\overline{M}$ is also regular. □

2. *F* Reliability Models and *G* Reliability Models

Various reliability models under the general name of consecutive-k systems have been proposed in the literature. The definitions of these models (and their variants) are summarized as follows.

The consecutive-k: F model [3,4]: Under this model, a system instance fails if and only if it contains k consecutive components all in state 0. The model can be expressed as the set {binary strings which do not contain k consecutive 0's}.

The consecutive-k: G model [5]: Under this model, a system instance works if and only if it contains k consecutive components all in state 1. The model can be expressed as the set {binary strings which contain k consecutive 1's}.

The f-or-consecutive-k: F model [7,8]: Under this model, a system instance fails if and only if it contains f 0-state components or k consecutive 0-state components. The model can be expressed as the set {binary strings which contain neither f 0's nor k consecutive 0's}.

The g-or-consecutive-k: G model: Under this model, a system instance works if and only if it contains g 1-state components or k consecutive 1-state components. The model can be expressed as the set {binary strings which contain g 1's or k consecutive 1's}.

The f-within-consecutive-k: F model [6]: Under this model, a system instance fails if and only if it contains f 0-state components within k consecutive components. The model can be expressed as the set {binary strings which do not contain f 0's within a segment of k consecutive digits}.

The g-within-consecutive-k: G model: Under this model, a system instance works if and only if it contains g 1-state components within k consecutive components. The model can be expressed as the set {binary strings which contain g 1's within a segment of k consecutive digits}.

After a short observation on above definitions, it is easy to find that these models are basically defined by "good" or "bad" patterns such that a system instance works (or fails) if and only if it contains a good (or bad) pattern. Moreover, the patterns in published reliability models are usually expressible by simple regular expressions. Therefore, we can define the F and G models as follows.

Definition 7: M is a *G reliability model* if and only if M can be expressed by a regular expression of the form $(0+1)^*r(0+1)^*$, where r is a regular expression.

Definition 8: M is a *F reliability model* if and only if $\overline{M}$ can be expressed by a regular expression of the form $(0+1)^*r(0+1)^*$, where r is a regular expression.

In order to explain the above definitions, we give some examples here.

Example 9: M_1: a consecutive-4: G model

$$\overline{M}_1 = (0+1)^*1111(0+1)^*.$$

Example 10: M_2: a consecutive-4: F model

$$\overline{M}_2 = (0+1)^*0000(0+1)^*,$$
$$M_2 = \{0,1\}^* - \overline{M}_2$$

Example 11: M_3: a 5-or-consecutive-3: G model

$$\begin{aligned} M_3 &= (0+1)^*(10^*10^*10^*10^*1)(0+1)^* + (0+1)^*111(0+1)^* \\ &= (0+1)^*(0^10^*10^*10^*10^*1)(0+1)^* + (0+1)^*111(0+1)^* \\ &= (0+1)^*[(0^*1)5 + 111](0+1)^* \end{aligned}$$

Example 12: M_4: a 2-within-consecutive-4: G model

$$\begin{aligned} M_4 = (0+1)^*(&1100+1010+1001+0110+0101 \\ &+0011+0111+1011+1101+1110+1111)(0+1)^* \end{aligned}$$

The following corollaries give important properties of the F models and G models.

Corollary 13: *If M is a G reliability model and $b \in M$, then for all $x, y \in \{0,1\}^*$, $xby \in M$.*

Corollary 14: *If M is a F reliability model and $b \notin M$, then for all $x, y \in \{0,1\}^*$, $xby \notin M$.*

The following results pinpoint the relations among the F models, the G models and the regular models.

Corollary 15: *If M is a G or F reliability model, then M is regular.*

Lemma 16: *If M is both a F model and a G model, then M is either ϕ or $\{0,1\}*$.*

Proof: Consider the empty string ε. If $\varepsilon \in M$, then $M = \{0,1\}^*$ because M is a G model. If $\varepsilon \notin M$, then $\varepsilon \notin \overline{M}$ and $\overline{M} = 0, 1^*$ because M is a F model. Therefore M must be either ϕ or $\{0,1\}^*$. □

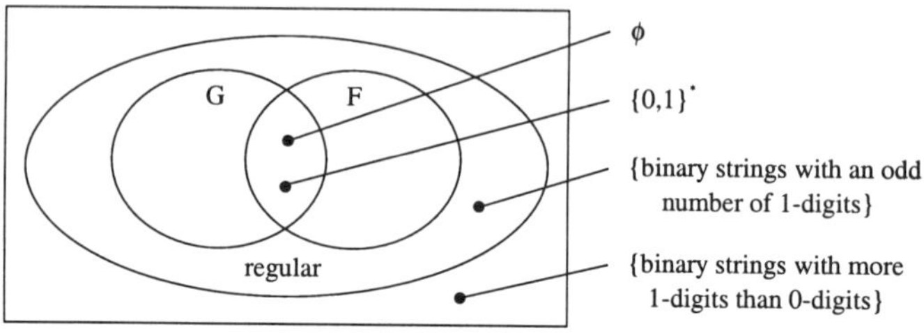

Fig. 1. The classification of models.

Lemma 17: *The union set of F and G models is a proper subset of the set of regular models.*

Proof: Consider a model $M = \{$binary strings with an odd number of 1-digits$\}$. Obviously, the model M can be denoted by the regular expression $0^*1(0^*10^*1)^*0^*$, so it is regular. But M is neither a F model nor a G model because $1 \in M,\ 11 \notin M,\ 111 \in M$. □

Lemma 18: *The set of regular models is a proper subset of the set of all models.*

Proof: Consider a model $M = \{$binary strings which contain more 1-digits than 0-digits$\}$. By the "pumping lemma"[2], M can not be denoted by any regular expression. □

Based on these results, the relations among the F models, the G models and the regular models are given in Figure 1.

3. Efficient Reliability Algorithms for Regular Models

In literature, reliabilities of the various consecutive-k systems are computed case by case (see [9] for a summary and references). Here we propose a general approach to evaluate the reliability under any regular model. Our approach is:

(1) For a given regular model M, find a regular expression r to denote M.
(2) Construct an automaton A corresponding to r, that is, A accepts exactly all the strings in M.

(3) Reduce the number of states in A by automata theory until a minimal state automaton A_m is found.
(4) Use A_m in the reliability computation as follows.

Let the state set of A_m be $Q = \{1, 2, ..., m\}$, where state 1 is the initial state. Let $\delta : Q \times \{0, 1\} \to Q$ be the state transition function of A_m defined as $\delta(i, x) = j$ if state i goes into state j when inputting a digit $x \in \{0, 1\}$. In addition, we define

$$d(i) = \begin{cases} 1 \text{ , if state } i \text{ is an acception state} \\ 0 \text{ , if state } i \text{ is a rejecting state} \end{cases}$$

Then the reliability of an n-component system can be evaluated as

$$\underbrace{[100\ldots0]}_{m}\left(\prod_{t=1}^{n}\Lambda_t\right)[d(1)d(2)\ldots d(m)]^T, \tag{1}$$

where

$$\Lambda_t = [a_{i,j}^t]_{m\times m},$$

$$a_{i,j}^t = \begin{cases} p_t & \text{, if } \delta(i,1) = j \\ 1 - p_t & \text{, if } \delta(i,0) = j \\ 0 & \text{, otherwise.} \end{cases}$$

The computation of Equation (1) takes only $O(mn)$ time, where m is the number of states in A_m. Therefore the design of efficient reliability algorithms is reduced to the search of automata with less states. By automata theory, once the minimal state automata are found, very efficient algorithms can be derived easily.

4. Applications

There are many applications on the regular models. And the reliability algorithms for the regular models are also helpful in model analysis. Here we list some examples.

4.1. *The f-or-consecutive-k: F Model*

The f-or-consecutive-k: F model has been studied in many published papers [7,8] with practical applications. For a f-or-consecutive-k: F model M with $f > k > 1$, we can write

$$\overline{M} = (0+1)^*[(1^*0)^f + 0^k](0+1)^*.$$

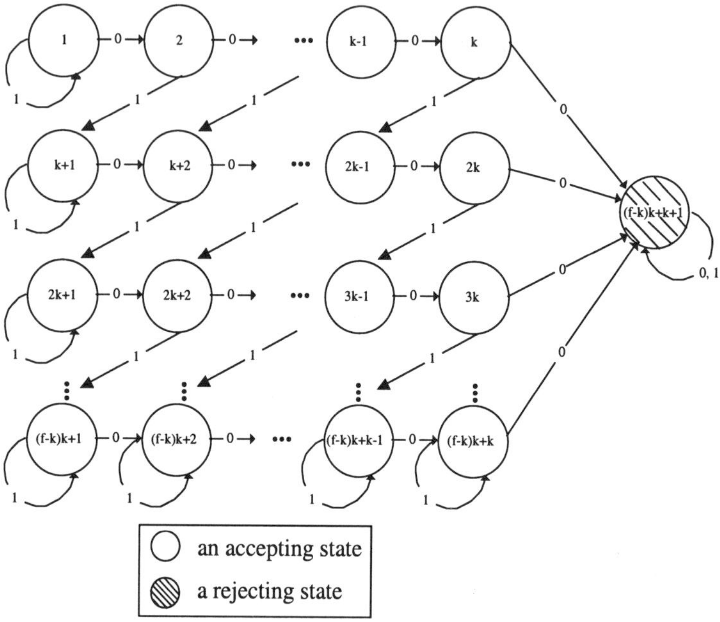

Fig. 2. The state transition diagram of the minimal automaton.

The state transition diagram given in Figure 2 explains the structure of the minimal state automaton accepting exactly the strings in M.

Because the number of states is $(f-k+1)k+1$, the reliability of an n-component system can be computed in $O((f-k)kn)$ time [10], which is more efficient than all other published algorithms.

4.2. *The f-within-consecutive-k: F Model*

In literature the f-within-consecutive-k: F model has been shown to be able to model some real systems [6]. Many reliability algorithms were published for this model, but they all came with an unbearable time complexity.

Now let M be an f-within-consecutive-k: F model, where $k > f > 1$. Then $\overline{M}$ can be expressed as

$$\overline{M} = (0+1)^*[\sum_{\substack{x_1,x_2,\ldots,x_k\in\{0,1\}\\ x_1+x_2+\ldots+x_k\leq k-f}} x_1x_2\ldots x_k](0+1)^*.$$

In our previous paper [11], the minimal state automaton accepting exactly the strings in M has the following set of states Q. We label the states with k-bit binary strings.

$$\begin{aligned}
Q &= Q_0 \cup Q_1 \cup \cdots \cup Q_f, \text{ where} \\
Q_0 &= \{\underbrace{1\cdots 1}_{k-f}\underbrace{0\cdots 0}_{f}\}, \\
Q_1 &= \{b_1 b_2 \cdots b_k |\, b_1 = 1,\ \sum_{i=2}^{k} b_i = k - f\}, \\
Q_2 &= \{b_1 b_2 \cdots b_k |\, b_1 = b_2 = 1,\ \sum_{i=3}^{k} b_i = k - f\}, \\
&\cdots \\
Q_f &= \{b_1 b_2 \cdots b_k |\, b_1 = b_2 = \cdots = b_f = 1,\ \sum_{i=f+1}^{k} b_i = k - f\} = \{\underbrace{1\cdots 1}_{k}\}.
\end{aligned}$$

The only one state in Q_f is the initial state. The only one state in Q_0 is a rejecting state; all other states (including the initial state) are accepting states. The state transition function $\delta : Q \times \{0,1\} \to Q$ is defined as:

$$\delta(b_1 \cdots b_k, 0) = \begin{cases} \underbrace{1\cdots 1}_{k-f}\underbrace{0\cdots 0}_{f}, & \text{if } \sum_{i=2}^{k} b_i = k - f \\ b_2 \cdots b_k 0 & \text{, otherwise} \end{cases}$$

$$\delta(b_1 \cdots b_k, 1) = \begin{cases} \underbrace{1\cdots 1}_{k-f}\underbrace{0\cdots 0}_{f} & \text{, if } \sum_{i=1}^{k} b_i = k - f \\ \underbrace{1\cdots 1}_{t-2} b_t \cdots b_k 1 & \text{, otherwise, where} \\ & t = \min\{x |\, \sum_{i=x}^{k} b_i = k - f - 1\}. \end{cases}$$

Thus in the minimal state automaton,

$$\begin{aligned}
|Q| &= 1 + \binom{k-1}{k-f} + \binom{k-2}{k-f} + \cdots + \binom{k-f}{k-f} \\
&= 1 + \binom{k-1}{f-f} + \binom{k-2}{f-2} + \cdots + \binom{k-f}{f-f} \\
&= 1 + \binom{k}{f-1}
\end{aligned}$$

Therefore, the reliability of an n-component system can be computed in $O(\binom{k}{f-1}n)$ time, which is far more efficient than other published algorithms.

4.3. *The k-mod-q Model*

The k-mod-q model can be used to predict the probability distribution of a random binary number, where each bit has its different probability to be one. A k-mod-q model M is defined as $M=\{$binary strings, when interpreted as binary numbers, whose values are k modulo $q\}$.

Obviously, M is neither an F nor a G model. But it is a regular model. Though its regular expression is difficult to find directly, we can construct an automaton for it and then transform the automaton to a regular expression.

The set of states in the minimal automaton for M can be written as

$$Q = \{0, 1, 2 \cdots, q-1\},$$

where the initial state is 0, and the only accepting state is k. The state transition function $\delta : Q \times \{0,1\} \to Q$ is defined as

$$\delta(a, x) = (2a + x) \bmod q.$$

Therefore the reliability of an n-component system under this model, or equivalently, the probability that the random n-bit binary number is k modulo q, can be computed in $O(qn)$ time.

4.4. *Logic Circuits*

Regular models can also be applied to predict the output of a logic circuit (with or without memory) on processing random binary input strings. For example, the following logic circuit in Figure 3 can be seen to record the parity of 1-inputs and produce an 1-output for every odd-numbered 1-input. A circle with a dot represents an And-gate. A circle with a + represents an OR-gate. A circle with a $\sim$ represents an inverter.

Assume $X = Y = 0$ in the initial state, and there is sufficient time between changes in input values for signals to propagate and for the network to reach a stable configuration.

In order to predict the final output of the logic circuit after processing an n-bit random binary string where pi is the probability that the ith bit is 1, we define a regular model $M=\{$binary strings, when processed by the logic circuit, cause the final output to be $1\}$.

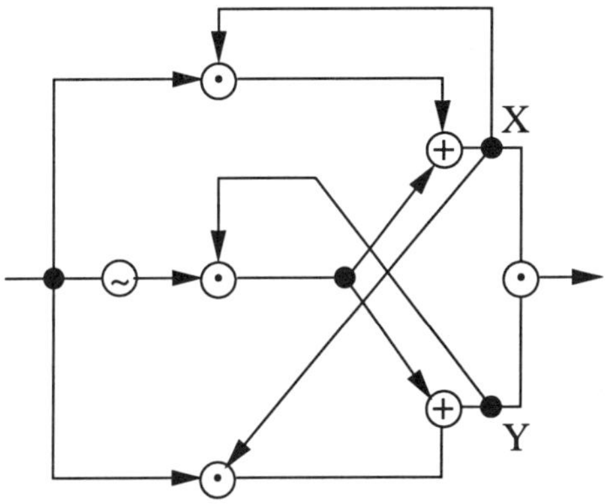

Fig. 3. A logic circuit.

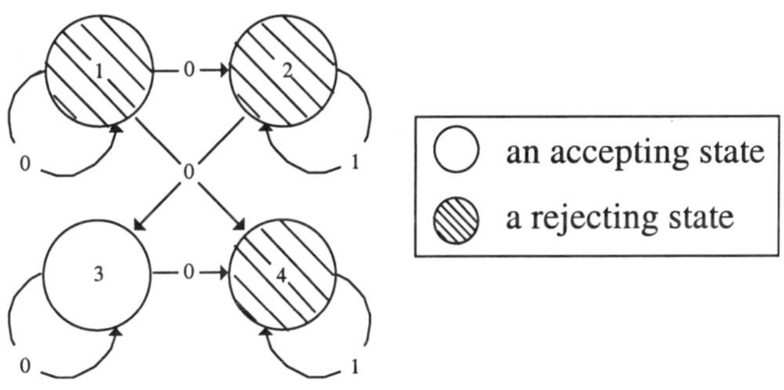

Fig. 4. The state transition diagram of the minimal state automaton for M.

Obviously, M is neither an F nor a G model. But it is regular because we can construct an automaton for it and then transform the automaton to a regular expression. The state transition diagram of the minimal state automaton for M is given in Figure 4.

The number of states in the automaton is 4, a small constant. Therefore, when given $p_1, p_2, \ldots, p_n$, we can compute the probability that the final output is 1 in $O(n)$ time. Such an algorithm achieves the optimal time complexity.

5. Further Research

Based on the regular reliability model, many well-studied systems can be re-analyzed to find the underlying minimal state heterogeneous Markov chains, thus more efficient reliability algorithms are possible. Furthermore, we already know that there exist systems which can not be described and analyzed by the regular reliability model. To extend the application of the regular reliability model is another research topic.

References

1. J.E. Hopcroft and J.D. Ullman, *Introduction to automata theory, languages, and computation*, (Addison Wesley, 1979).
2. Y. Bar-Hillel, M. Perles and E. Shamir, "On formal properties of simples phrase structure grammars", *Z. Phonetic. Sprachwiss. Kommunikations-forsch.*, **14**, 143–172 (1961).
3. J.M. Kontolcon, "Reliability determination of a r-successive-out-of-n systems", *IEEE Trans. Reliability*, **R-29**, 437 (1980).
4. D. Chiang and S.C. Nin, "Reliability of consecutive-k-out-of-n systems", *IEEE trans. Reliability*, **R-30**, 87–89 (1981).
5. W. Kuo, W. Zhang and M. Zuo, "A consecutive-k-out-of-n: G system: The mirror image of a consecutive-k-out-of-n: F system", *IEEE Trans. Reliability*, **39**, 244–253 (1990).
6. W.S. Griffith, "On the consecutive-k-out-of-n failure systems and their generalizations", *Reliability and Quality Control*, Ed: A.P. Basu, Elsevier, Amsterdam, 157–166 (1986).
7. H. Sun and J. Liao, "The reliability of (n, f, k) system", *J. Electron.*, **12**, 436–439 (1980).
8. G.J. Chang, L. Cui and F.K. Hwang, "Reliabilities for (n, f, k) systems", *Stat. and Reliab. Lett.*, **43**, 237–242 (1999).
9. G.J. Chang, L. Cui and F.K. Hwang, *Reliabilities of Consecutive-k Systems*, (Kluwer, Boston, 2000).
10. J.C. Chang, R.J. Chen and F.K. Hwang, "Faster algorithms to compute the reliabilities of (n, f, k) systems", submitted to *Computer Modeling in Engineering and Science*.
11. J.C. Chang, R.J. Chen and F.K. Hwang, "A minimal-automaton-based algorithm for the reliability of $Con(d, k, n)$ systems", to appear in *Methodology and Computing in Applied Probability*.
12. R.E. Barlow and F. Proschan, *Mathematical Theory of Reliability*, 204–211 (SIAM, 1996).

CHAPTER 2

BOUNDING SYSTEM RELIABILITY

Jane Nichols Hagstrom

Department of Information and Decision Sciences
University of Illinois, 601 S. Morgan, Chicago, IL 60607-7124, U.S.A.
E-mail: hagstrom@uic.edu

Sheldon M. Ross

Department of Industrial Engineering and Operations Research
University of California, Berkeley, CA 94720-1777, U.S.A.
E-mail: smross@newton.berkeley.edu

We present upper and lower bound pairs which may be used when computing the reliability of a binary system is difficult. The first bound pair is derived from the conditional expectation inequality for the sum of random binary variables. For some cases, the bounds improve on bounds found in the literature. These conditional expectation bounds can be used even when not all min-cuts and min-paths are known. The second bound pair is applicable for some cases where components are associated and only marginal component reliabilities are known. The bounds are derived by "bounding" the joint distribution of the component states. This second bound pair may fail unless the interdependent components form a structural complement or substitute set. A complement (substitute) set is a set of components no two of which occur together in a min-cut (min-path).

1. Introduction

In this paper, we consider the problem of bounding system reliability for systems of two-state components. For large systems of independent components, it may be difficult to compute the exact reliability. For this case, we introduce a new pair of upper and lower bounds on system reliability. In many systems, it is known that component failures are correlated, although there is inadequate data to establish the exact level of correlation. Under certain circumstances, if we know the marginal reliabilities of the components, we can define upper and lower bounds on the system reliability.

In Section 2, we introduce bounds for system reliability when component states are independent and discuss the behavior of these bounds. In Section 3, we establish results concerning the ability to construct bounds when component states are positively correlated, but the exact nature of the correlation is unknown. In the rest of this section, we review common terminology for binary systems. General background in system reliability is found in Barlow and Proschan[1].

Let the components of a system be indexed $i = 1 \ldots n$. Define $x_i = 1$ if component i is working and $x_i = 0$ otherwise. The *system structure function* is a binary function Φ such that $\Phi(\mathbf{x}) = 1$ if the component state vector $\mathbf{x}$ allows the system to operate correctly and $\Phi(\mathbf{x}) = 0$ otherwise. Given disjoint subsets U and V of the components, we will use the notation $\Phi(\mathbf{1}_U, \mathbf{0}_V, \mathbf{x})$ to denote the function Φ evaluated with the ith component state equal to 1 if $i \in U$, 0 if $i \in V$, and x_i otherwise. When a set $U = \{i\}$, we will often drop the braces.

A system structure function is *monotone* if it is nondecreasing in each argument. For a monotone structure function, the following definitions hold. A *pathset* of Φ is a set P such that $\Phi(\mathbf{1}_P, \mathbf{x}) = 1$ for all $\mathbf{x}$. A *cutset* of Φ is a set K such that $\Phi(\mathbf{0}_K, \mathbf{x}) = 0$ for all $\mathbf{x}$. A *min-path* is a pathset which contains no other pathset. A *min-cut* is a cutset which contains no other cutset.

A monotone structure function Φ has a dual structure function Ψ defined by $\Psi(\mathbf{X}) = 1 - \Phi(\mathbf{1} - \mathbf{X})$. The min-paths for Ψ are the min-cuts for Φ and the min-cuts for Ψ are the min-paths for Φ.

2. Using the Conditional Expectation Inequality to Bound the Reliability Function

For the case in which component states are independent, Barlow and Proschan[1] discuss options for computing bounds when the exact system reliability is difficult to compute. These bounds are defined in terms of the min-cuts and min-paths of the system. We present a new pair of upper and lower bounds which for some cases improve on these previous bounds. This new bound pair can be relaxed, in the sense that partial lists of min-cuts and min-paths still lead to bounds on the system reliability. This is an important property, since it is often extremely difficult to enumerate all min-paths or all min-cuts.

Let

$$N = \sum_{k=1}^{m} X_k$$

where X_k is a Bernoulli random variable, $k = 1, \ldots, m$. The following inequality, called the *Conditional Expectation Inequality*, is proven (Corollary 10.3.2) in Ross, 1996.[5]

The Conditional Expectation Inequality

$$\Pr\{N > 0\} \geq \sum_{k=1}^{m} \frac{E[X_k]}{E[N|X_k = 1]} \tag{1}$$

The conditional expectation inequality, although apparently not well known, is not only stronger than the second moment inequality (for a proof of this statement see Ross, 2001[6]), but is often easier to calculate. We will now show how to use it to bound the reliability function.

Let S_k, $k = 1, \ldots, m$ be the min-paths, and let C_k, $k = 1, \ldots, r$ be the min-cuts of an n component system. The reliability function $r(p_1, \ldots, p_n)$ of this system is the probability that it will function if the components' states are independent, with component i functioning with probability p_i. To obtain a lower bound, let Y_k be the indicator variable for the event that all of the components in path S_k function, and let $N = \sum_{k=1}^{m} Y_k$. The conditional expectation inequality yields

$$\begin{aligned} r(\mathbf{p}) &= \Pr\{N > 0\} \\ &\geq \sum_{k=1}^{m} \frac{E[Y_k]}{E[N|Y_k = 1]} \\ &= \sum_{k=1}^{m} \frac{\prod_{i \in S_k} p_i}{1 + \sum_{j \neq k} \prod_{i \in S_j - S_k} p_i} \end{aligned} \tag{2}$$

To obtain an upper bound, let Z_k be the indicator of the event that all of the components of C_k are failed, and let $M = \sum_{k=1}^{r} Z_k$. With $q_i = 1 - p_i$, the conditional expectation inequality yields that

$$1 - r(\mathbf{p}) = \Pr\{M > 0\} \geq \sum_{k=1}^{r} \frac{\prod_{i \in C_k} q_i}{1 + \sum_{j \neq k} \prod_{i \in C_j - C_k} q_i} \tag{3}$$

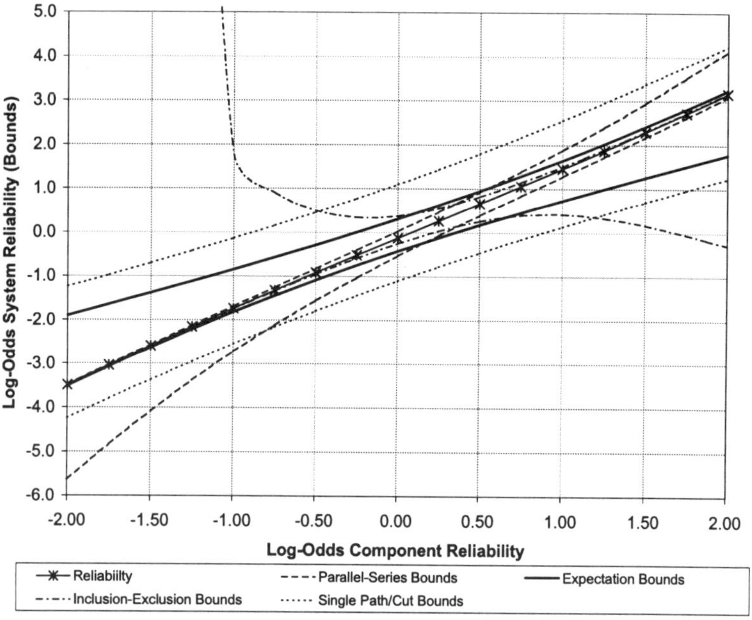

Fig. 1. **Comparison of Bounds for Bridge System**

Example 1: In Figure 1, we compare bounds presented by Barlow and Proschan[1] to these conditional expectation bounds. The bounds have been computed for the bridge network system represented by Figure 2. The system works when there is a path of working links from s to t. We assume that all components have the same reliability. We plot log-odds of the system reliability (or reliability bound) against log-odds of the component reliability. For a component/system with reliability (or reliability bound) p, the log-odds reliability (or reliability bound) of the component/system is $\ln(p/(1-p))$. This transformation maps the interval $(0, 1)$ monotonically to $(-\infty, \infty)$. The transformation clarifies the asymptotic behavior of the bounds at the extremes of the component reliability range. The distance

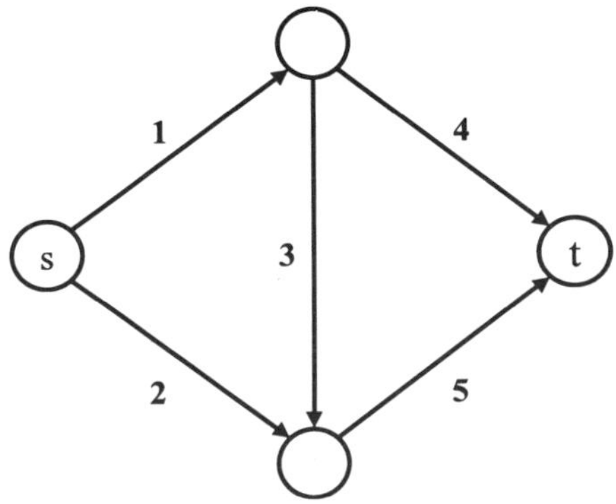

Fig. 2. **Two-Terminal Bridge Network**

between a point on a bound curve and the corresponding point on the reliability curve reflects the ratio of the odds computed from the bound to the odds computed from the reliability.

The bounds computed are (i) the parallel-series bounds, based on considering the system to be either composed of min-cuts in series or min-paths in parallel, (ii) the inclusion-exclusion bounds, computed by truncating an inclusion-exclusion expression for the system reliability in terms of the events that min-cuts fail or min-paths work, (iii) the single path/cut bounds, where the system is approximated by its most reliable min-path or its least reliable min-cut, and (iv) the expectation bounds we introduce here.

We note that all bounds seem to behave well at one end of the range and poorly at the other. This is consistent with the conjecture of Barlow and Proschan that bounds based on cuts may be appropriate when components have high reliability and bounds based on paths may be appropriate when components have low reliability.

Based on this plot, we make the following observations.

(1) The conditional expectation probability bound remains between 0 and 1, unlike the inclusion-exclusion bound, for which we included terms corresponding to basic events and pairs of basic events. It is a straightforward consequence of the derivation[5] that the conditional expectation bounds always stay between 0 and 1.

(2) The conditional expectation bound seems to be very close to the inclusion-exclusion bound when these bounds are both "good." Both bounds require the computation of the same number of terms. The conditional expectation bound may be less prone to numerical errors.
(3) For this example, the conditional expectation bounds seem to dominate the bounds based on the reliability of a single min-cut (min-path), although examples using nonidentical component reliabilities can be constructed for which neither bound dominates the other.
(4) The conditional expectation bounds seem to complement the parallel-series bounds, since the conditional expectation lower (upper) bound is good where the parallel-series upper (lower) bound is good.

If not all min-paths and all min-cuts are known, the parallel-series bounds fail. In contrast, the conditional expectation bounds are usable as long as some min-paths and some min-cuts are known. To compute the lower bound given some of the min-paths of the system, we note that the reliability R^* of a system defined by the known min-paths is smaller than the reliability R of the complete system. Since we can compute the lower bound for R^*, we have a lower bound on R. By considering a system defined by the known min-cuts, we can in a similar fashion obtain an upper bound.

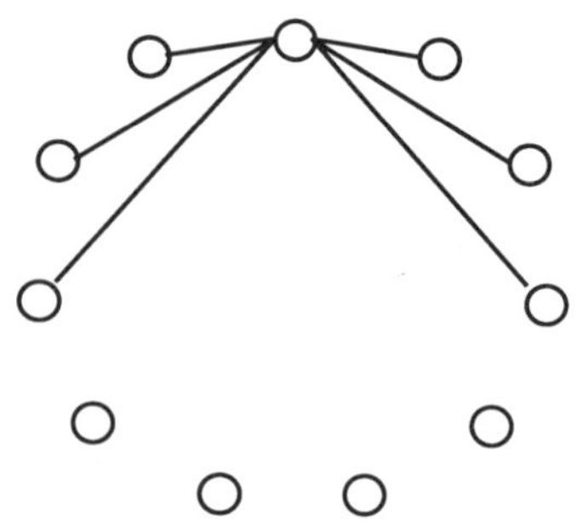

Fig. 3. **Circulant Graph with 11 Nodes of Degree 6**

Example 2: Figure 3 is a partial diagram of a system for which identifying all min-cuts and all min-paths may be too difficult. The figure illustrates a circulant graph with 11 nodes. Each node is connected with its 6 nearest neighbors, as is illustrated for a single node in the diagram. The links are subject to failure. The system works if every node is connected to every other node by a path of working links. Figure 4 illustrates bounds computed

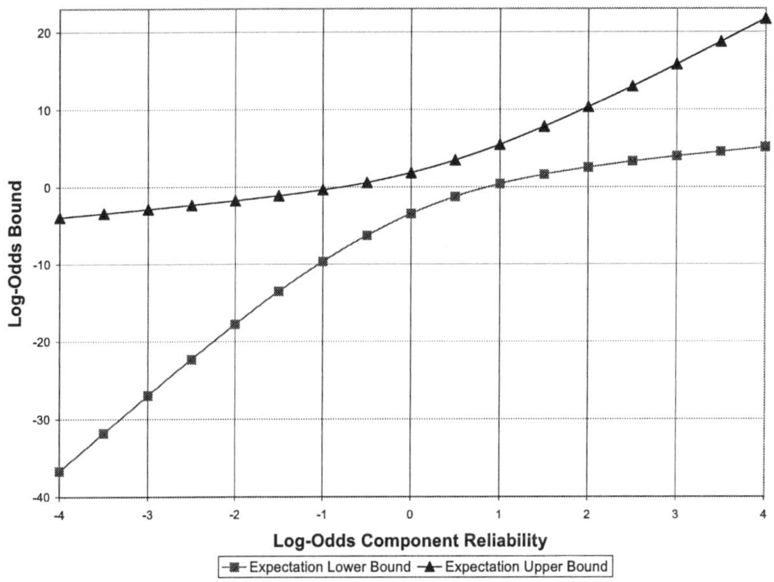

Fig. 4. **Conditional Expectation Bounds for Circulant Graph System**

using only the most easily identified min-paths and min-cuts. As before, all links are assumed to have the same reliability. We can be sure that the system reliability lies between these two bounds.

In general, system structures are not defined in terms of min-paths or min-cuts, but rather in terms of other structures such as fault trees and networks. Given a fault tree or a network, the problem of enumerating all min-paths or all min-cuts is NP-hard[7], and thus we expect the enumeration to require time exponential in the number of components. The ability to obtain bounds even when not all min-cuts and min-paths are known leads to a powerful approach to approximating system reliability.

Therefore we propose an approach where lower and upper bounds are computed based on partial enumeration of min-paths and min-cuts; if the distance between the bounds is too great, enumeration is continued until the bounds approach within an acceptable degree. If m min-cuts and m min-paths are used, a simple version of this process requires $O(mC(m)n)$ time, where $C(m)$ is the time required to enumerate m min-cuts or min-paths, and n is the maximum size of any min-cut or min-path.

We note that the conditional expectation bounds may not improve monotonically as more paths or cuts are included.

Table 1. **Conditional Expectation Lower Bound for Series-Parallel System**

		Component Reliability 0.5	0.9
Number of Paths in Bound	1	0.2500	0.8100
	2	0.4000	0.8950
	3	0.4107	0.8871
	4	0.4444	0.8975
System Reliability		0.5625	0.9801

Example 3: In Table 1, we compute bounds based on 1, 2, 3, and all paths for the series-parallel system shown in Figure 5, where the system works if there is a path of working links from s to t. We assumed all components have

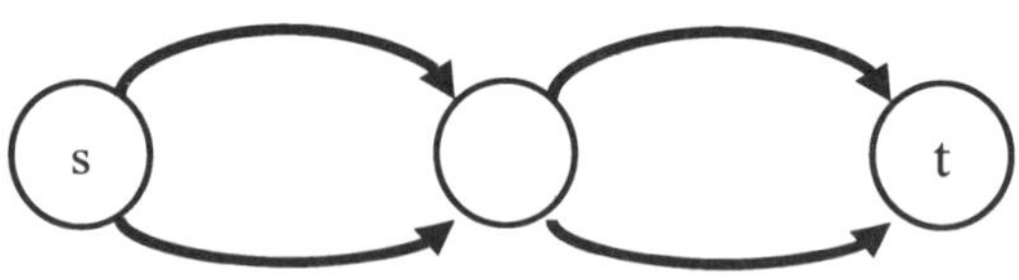

Fig. 5. **Series-Parallel System**

the same reliability. By making our second path disjoint from the first, we obtained the values shown. If all components have reliability 0.5, the lower bound improves monotonically as we increase the number of paths used; however, at component reliability 0.9, using 2 paths is better than using 3 paths.

In this section, we have introduced new bounds based on min-cuts and min-paths which in some cases perform better than previously known bounds for general monotone binary systems. The bounds can be used when only some min-cuts and some min-paths are known, in contrast to the well-known parallel-series bound. Although even-order inclusion-exclusion bounds can be used in the same way, the behavior of the bounds presented here is better, since they remain within the interval [0,1].

3. Bounds When Component States Are Dependent

The problem of dealing with dependence among component states is of great concern to reliability analysts. The analyst may have good enough data to provide marginal component reliabilities, but only vague information about the full joint distribution of the component states. In many cases, the analyst may be quite sure the components exhibit a positive dependence, e.g., they are associated. The question for the analyst is how to say something about the system reliability when the correct joint distributions are unknown. We show that under certain circumstances the marginal reliabilities of the components can be used to bound the system reliability.

3.1. *Effect of Unknown Dependence Relationships on Reliability Estimation*

Association is a type of positive dependence that is commonly used in reliability analysis. A set of random variables $\{X_1, X_2, ..., X_n\}$ is said to be *associated* if for any two nondecreasing binary functions Φ, Ψ on n variables, $\text{Cov}(\Phi(\mathbf{X}), \Psi(\mathbf{X})) \geq 0$. The concept of association was introduced by Esary, Proschan, and Walkup[2]. The properties of association used in Examples 4 and 5 below are straightforward consequences of the definition. They will be used in the proof in Section 3.2. Further consequences are given in the Esary, Proschan, and Walkup paper.

We first consider two simple examples where we can obtain upper and lower bounds on the system reliability using just the marginal component reliabilities.

Example 4: Let the components of a series system be indexed $i = 1 \ldots n$. Let the random variable $X_i = 1$ if component i is working; $X_i = 0$ otherwise. The system is working if $\prod_{i=1}^n X_i = 1$. The reliability of the system is $\mathrm{E}(\prod_{i=1}^n X_i)$. If the random variables X_i are associated,

$$\mathrm{E}(\prod_{i=1}^{n} X_i) \geq \prod_{i=1}^{n}(\mathrm{E}X_i).$$

Also, we have

$$\mathrm{E}(\prod_{i=1}^{n} X_i) \leq \min_{i=1}^{n}(\mathrm{E}X_i).$$

Example 5: Let the components of a parallel system be indexed $i = 1 \ldots n$. Then the system is failed if $\prod_{i=1}^n (1 - X_i) = 1$. The reliability of the system is $1 - \mathrm{E}(\prod_{i=1}^n (1 - X_i))$. Since $\{X_i\}_{i=1}^n$ is a set of associated random variables, so is $\{1 - X_i\}_{i=1}^n$. Then

$$\mathrm{E}(\prod_{i=1}^{n}(1 - X_i)) \geq \prod_{i=1}^{n}(1 - \mathrm{E}X_i).$$

Also

$$\mathrm{E}(\prod_{i=1}^{n}(1 - X_i)) \leq 1 - \max_{i=1}^{n} \mathrm{E}X_i.$$

Thus we have lower and upper bounds on the reliability of this system.

We give another case in which association allows us to develop bounds on system reliability.

Example 6: In Figure 2, let us consider the case where the states of components 1 and 4 are associated, while each of the other components has a state independent of all the rest. The reliability of the system can be expressed as

$$p_2 p_5 + p_1 p_5 (1 - p_2) p_3 + p_{14}[1 - p_5(p_2 + p_3 - p_2 p_3)].$$

$p_{14} \leq \min\{p_1, p_4\}$ and by association $p_{14} \geq p_1 p_4$. Since the coefficient of p_{14} is always nonnegative, we can use these inequalities to bound the system reliability.

In the next section, we will define complement and substitute sets and establish that we can obtain these types of bounds whenever the set of interdependent components are associated and form such a set. The first bound is obtained just by treating all components as independent. The second bound is obtained from generating a highly correlated distribution from the marginal component reliabilities. We formalize the definitions of both these distributions.

Given a distribution for the random vector $\mathbf{X}$, we define the *independent distribution derived from* $\mathbf{X}$ for the random vector $\tilde{\mathbf{X}}$ by $\mathrm{E}\tilde{X}_i = \mathrm{E}X_i$ for all i and $\mathrm{E}\prod_{i\in A}\tilde{X}_i = \prod_{i\in A}\mathrm{E}\tilde{X}_i$ for all subsets A of the indices of $\mathbf{X}$. Given a distribution for the random vector $\mathbf{X}$, we define the *highly correlated distribution derived from* $\mathbf{X}$ for the random vector $\hat{\mathbf{X}}$ by $\mathrm{E}\hat{X}_i = \mathrm{E}X_i$ for all i and $\mathrm{E}\prod_{i\in A}\hat{X}_i = \min_{i\in A}\mathrm{E}\hat{X}_i$ for all subsets A of the indices. One can show that $\mathrm{E}\prod_{i\in A}(1-\hat{X}_i) = 1 - \max_{i\in A}\mathrm{E}\hat{X}_i$.

These two distributions can be used to estimate the reliability. In Examples 4-6, they provide bounds. However, as illustrated in the following example, the estimates do not always provide bounds.

Example 7: In Figure 2, let us consider the case where the states of components 1, 3, and 5 are associated, while the states of 2 and 4 are independent of all other components. We let $p_1 = p_3 = p_5 = 0.5$, $p_{13} = p_{15} = p_{35} = 0.45$, $p_{135} = 0.35$. We set $p_2 = p_4 = p$, and consider the reliability and our estimates of the reliability as a function of p.

Figure 6 shows the reliability and our estimates for the reliability. The plot shows that our estimates do not always form a pair of bounds.

3.2. *Complements and Substitutes and Bounding System Reliability*

In this section, we define structural complement and substitute sets. We then consider our estimates based on the independent and highly correlated distributions. We show that these form bounds for all associative distributions of the component reliabilities if and only if the interdependent components form a complement or substitute set.

Hagstrom[4] developed the concepts of structural substitute and complement pairs in the reliability context. Two components are *structural complements* if they never occur in the same min-cut. They are *structural substitutes* if they never occur in the same min-path. We will define a *structural*

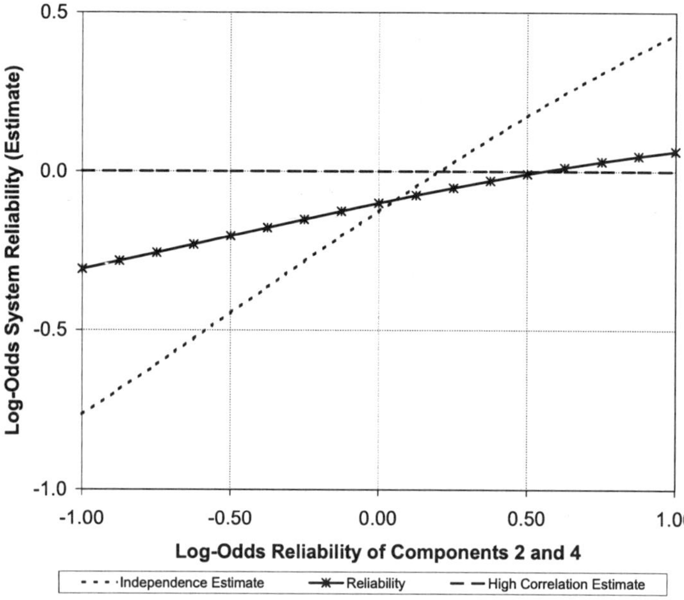

Fig. 6. **Estimates for Reliability Which Do Not Provide Bounds**

complement set to be a set of components such that no two components occur in the same min-cut. A *structural substitute set* is a set of components such that no two components occur in the same min-path. Components that are in series form a structural complement set. Components that are in parallel form a structural substitute set. For the bridge network system in Figure 2, components 1 and 4 form a complement set, while components 2, 3, and 4 form a substitute set.

We now consider a set A of components whose states are associated. We let B contain the remaining components and assume that the random vector of states of the components in A, $\mathbf{X}_A$, is independent of the vector of states of the rest of the components, $\mathbf{X}_B$. We define the random vectors $\tilde{\mathbf{X}}_A$ with the independence distribution for $\mathbf{X}_A$ and $\hat{\mathbf{X}}_A$ with the highly correlated distribution for $\mathbf{X}_A$ as in Section 3.1.

Theorem 8: The following statements are equivalent.

(1) A is a structural complement set.

(2) $E\Phi(\mathbf{X}) \geq E\Phi(\tilde{\mathbf{X}}_A, \mathbf{X}_B)$ for all distributions for $\mathbf{X}$ such that $\mathbf{X}_A$ is independent of $\mathbf{X}_B$ and $\mathbf{X}_A$ is a vector of associated random variables.
(3) $E\Phi(\mathbf{X}) \leq E\Phi(\hat{\mathbf{X}}_A, \mathbf{X}_B)$ for all distributions for $\mathbf{X}$ such that $\mathbf{X}_A$ is independent of $\mathbf{X}_B$.

Proof: Suppose A is a structural complement set. Define sets $\beta_k, k = 1,2,3$, as follows:

$$\mathbf{x}_B \in \begin{cases} \beta_1 \text{ if } \Phi(\mathbf{x}_A, \mathbf{x}_B) = 0 \text{ for all } \mathbf{x}_A \\ \beta_2 \text{ if } \Phi(\mathbf{x}_A, \mathbf{x}_B) = 1 \text{ for all } \mathbf{x}_A \\ \beta_3 \text{ otherwise.} \end{cases}$$

Consider a fixed $\mathbf{x}_B \in \beta_3$. Let $A(\mathbf{x}_B) \subseteq A$ be defined by $i \in A(\mathbf{x}_B)$ if there exists $\mathbf{x}_A$ such that $\Phi(1_i, \mathbf{x}_A, \mathbf{x}_B) > \Phi(0_i, \mathbf{x}_A, \mathbf{x}_B)$. (All $i \notin A(\mathbf{x}_B)$ must be *irrelevant* to the system defined by fixing the states $\mathbf{x}_B$.) Since A is a structural complement set of Φ, no two elements of A occur together in a min-cut. Fixing $\mathbf{x}_B$ does not change this relationship. Then for fixed $\mathbf{x}_B$, it must be the case that

$$\Phi(\mathbf{x}_A, \mathbf{x}_B) = \prod_{i \in A(\mathbf{x}_B)} x_i.$$

Applying Example 4, we have that for all $\mathbf{x}_B \in \beta_3$,

$$E\Phi(\tilde{\mathbf{X}}_A, \mathbf{x}_B) \leq E\Phi(\mathbf{X}_A, \mathbf{x}_B) \leq E\Phi(\hat{\mathbf{X}}_A, \mathbf{x}_B). \tag{4}$$

Conditioning on X_B, we can write

$$\begin{aligned} E\Phi(\mathbf{X}) &= E(\Phi(\mathbf{X})|\mathbf{X}_B \in \beta_1)\Pr\{\mathbf{X}_B \in \beta_1\} + \\ &\quad E(\Phi(\mathbf{X})|\mathbf{X}_B \in \beta_2)\Pr\{\mathbf{X}_B \in \beta_2\} + E(\Phi(\mathbf{X})|\mathbf{X}_B \in \beta_3)\Pr\{\mathbf{X}_B \in \beta_3\} \\ &= \Pr\{\mathbf{X}_B \in \beta_2\} + \sum_{\mathbf{x}_B \in \beta_3} E\Phi(\mathbf{X}_A, \mathbf{x}_B)\Pr\{\mathbf{X}_B = \mathbf{x}_B\} \end{aligned}$$

Applying (4) to this last expression, we have

$$\begin{aligned} &\Pr\{\mathbf{X}_B \in \beta_2\} + \sum_{\mathbf{x}_B \in \beta_3} E\Phi(\tilde{\mathbf{X}}_A, \mathbf{x}_B)\Pr\{\mathbf{X}_B = \mathbf{x}_B\} \\ &\leq E\Phi(\mathbf{X}) \leq \Pr\{\mathbf{X}_B \in \beta_2\} + \sum_{\mathbf{x}_B \in \beta_3} E\Phi(\hat{\mathbf{X}}_A, \mathbf{x}_B)\Pr\{\mathbf{X}_B = \mathbf{x}_B\} \end{aligned}$$

or

$$E\Phi(\tilde{\mathbf{X}}_A, \mathbf{X}_B) \leq E\Phi(\mathbf{X}) \leq E\Phi(\hat{\mathbf{X}}_A, \mathbf{X}_B).$$

Suppose A is not a complement set. Then in particular, there exists a pair $i, j \in A$ which belong to some min-cut K. We now construct a distribution for $\mathbf{X}$ such that the elements in A are associated, and the elements in B are independent of each other and of A. We then show that $\mathrm{E}\Phi(\tilde{\mathbf{X}}_A, \mathbf{X}_B) > \mathrm{E}\Phi(\mathbf{X}) > \mathrm{E}\Phi(\hat{\mathbf{X}}_A, \mathbf{X}_B)$.

(i) Set $\Pr\{X_\ell = 0\} = 1$ for all $\ell \in K$, $\ell \neq i, j$. Since all components in the min-cut except for i, j always fail, we can regard i and j as forming a two-component cut of this system special case.

(ii) Set $\Pr\{X_\ell = 1\} = 1$ for all $\ell \notin K$. Since all components not in the min-cut are perfect, and always work, we can regard the system as consisting of just i and j. Since they form a min-cut for this special-case system, they are in parallel.

(iii) Set $\Pr\{X_i = 1, X_j = 1\} = 0.3$, $\Pr\{X_i = 1\} = 0.5$, $\Pr\{X_j = 1\} = 0.5$. This means $\Pr\{X_i = 0, X_j = 0\} = 0.3$. X_i and X_j are associated.

Now we have

$$\begin{aligned} \mathrm{E}\Phi(\tilde{\mathbf{X}}_A, \mathbf{X}_B) &= 1 - \mathrm{E}(1 - \tilde{X}_i)(1 - \tilde{X}_j) = 1 - 0.25 = 0.75 \\ \mathrm{E}\Phi(\mathbf{X}) &= 1 - \mathrm{E}(1 - X_i)(1 - X_j) = 1 - 0.3 = 0.7 \\ \mathrm{E}\Phi(\hat{\mathbf{X}}_A, \mathbf{X}_B) &= 1 - \mathrm{E}(1 - \hat{X}_i)(1 - \hat{X}_j) = \max\{\mathrm{E}X_i, \mathrm{E}X_j\} = 0.5 \end{aligned}$$ □

Corollary 9: *The following statements are equivalent.*

(1) A is a structural substitute set.

(2) $\mathrm{E}\Phi(\mathbf{X}) \leq \mathrm{E}\Phi(\tilde{\mathbf{X}}_A, \mathbf{X}_B)$ for all distributions for $\mathbf{X}$ such that $\mathbf{X}_A$ is independent of $\mathbf{X}_B$ and $\mathbf{X}_A$ is a vector of associated random variables.

(3) $\mathrm{E}\Phi(\mathbf{X}) \geq \mathrm{E}\Phi(\hat{\mathbf{X}}_A, \mathbf{X}_B)$ for all distributions for $\mathbf{X}$ such that $\mathbf{X}_A$ is independent of $\mathbf{X}_B$.

Proof: Letting Ψ be the dual structure function to Φ, we can write $\Psi(\mathbf{1} - \mathbf{X}) = 1 - \Phi(\mathbf{X})$. Applying the theorem to Ψ as a function of $\mathbf{1} - \mathbf{X}$, we have the equivalence of the following:

(1) A is a structural complement set for Ψ.

(2) $\mathrm{E}[1 - \Phi(\mathbf{X})] \geq \mathrm{E}[1 - \Phi(\tilde{\mathbf{X}}_A, \mathbf{X}_B)]$ for all distributions for $\mathbf{X}$ such that $\mathbf{1}_A - \mathbf{X}_A$ is independent of $\mathbf{1}_B - \mathbf{X}_B$ and $\mathbf{1}_A - \mathbf{X}_A$ is a vector of associated random variables.

(3) $\mathrm{E}[1 - \Phi(\mathbf{X})] \leq \mathrm{E}[1 - \Phi(\tilde{\mathbf{X}}_A, \mathbf{X}_B)]$ for all distributions for $\mathbf{X}$ such that $\mathbf{1}_A - \mathbf{X}_A$ is independent of $\mathbf{1}_B - \mathbf{X}_B$.

A is a structural substitute set with respect to Φ if and only if A is a structural complement set with respect to Ψ. Since $(\mathbf{1} - \mathbf{X})_A$ is associated whenever $\mathbf{X}_A$ is, and independence relationships are not changed by considering $\mathbf{1} - \mathbf{X}$ instead of $\mathbf{X}$, the above equivalence translates to the desired result. □

3.3. *Applying the Dependence Results*

We have established that we can bound reliability when component states are associated if the interdependent components form a complement or substitute set. Instances were illustrated in Section 3.1, in Examples 4, 5, and 6.

In the following example, we show how we can combine these bounds with the conditional expectation bounds.

Example 10: We again use the bridge network system of Figure 2. In this example we let the complement set $\{1, 4\}$ have associated states, while the rest of the components' states are independent. We set all marginal component reliabilities equal. Using the independence and highly correlated distributions, we obtain the bounds shown in Figure 7. If we cannot compute these bounds, we compute looser bounds using the conditional expectation bounds. The lower expectation bound is computed using Inequality (2). We must adapt the upper expectation bound (3) so that it can be used with the highly correlated distribution. A single term in the upper expectation bound has the form

$$\frac{\mathrm{E}Z_k}{\mathrm{E}[M|Z_k = 1]} = \frac{\mathrm{E}Z_k}{\sum_{j=1}^{r} \mathrm{E}[Z_j|Z_k = 1]}$$

Since C_k is a min-cut, the complement set A has at most one component in common with C_k. Therefore the components in C_k are independent and the numerator of the expression above remains the same as in (3). Most terms in the denominator remain the same. We must adjust the term $\mathrm{E}[Z_j|Z_k = 1]$ if $A \cap C_j = \{i_1\} \neq \{i_2\} = A \cap C_k$. In this case,

$$\mathrm{E}[Z_j|Z_k = 1] = \min\left\{1, \frac{q_{i_1}}{q_{i_2}}\right\} \prod_{i \in C_j - C_k - \{i_1\}} q_i.$$

The plot shows the bounds so obtained for the bridge system.

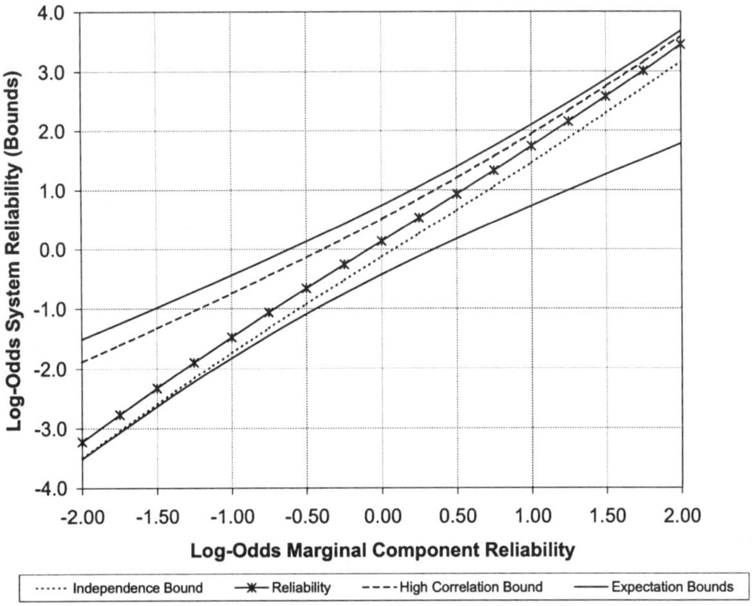

Fig. 7. **Combining Bounds**

The theorem and its corollary lead to an interesting case in which treating components as independent when they are associated contributes no error at all. For this example, approximating the distribution using the highly correlated distribution also induces no error.

Example 11: Consider the network shown in Figure 8. Arcs 3 and 4 never occur together in a min-path, and they never occur together in a min-cut. Therefore $A = \{3, 4\}$ is both a complement and a substitute set. If the states of 3, 4 are associated and independent of the rest of the components, the theorem and its corollary imply that $\mathrm{E}\Phi(\mathbf{X}) = \mathrm{E}\Phi(\tilde{\mathbf{X}}_{\{3,4\}}, \mathbf{X}_{\{1,2,5,6\}}) = \mathrm{E}\Phi(\hat{\mathbf{X}}_{\{3,4\}}, \mathbf{X}_{\{1,2,5,6\}})$. The ability to use an independent distribution is important in applying triconnected decomposition to computing reliability of directed networks. [3]

Example 7 illustrates the necessity of the structural condition of complement or substitute set in order to be able to bound the reliability. We will provide one last example to show that a restriction on the dependence

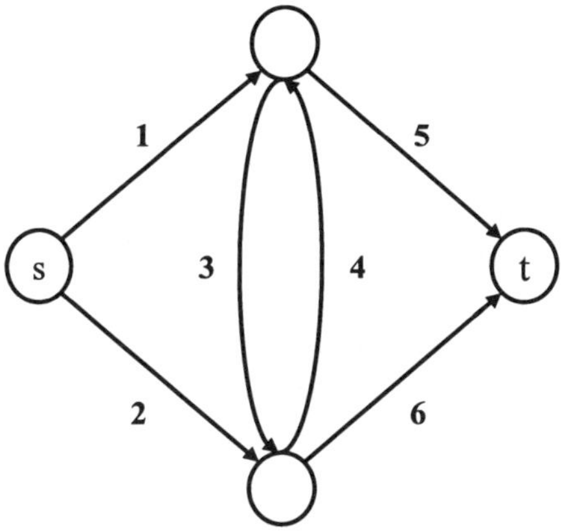

Fig. 8. **Network System Containing a Directed Cycle**

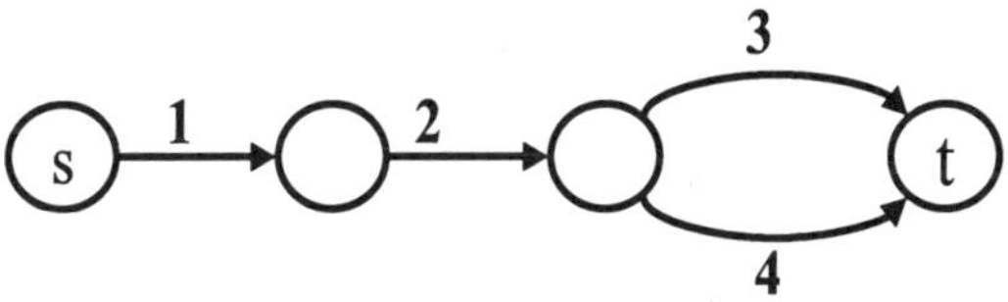

Fig. 9. **Network System Illustrating Necessity of Restriction on Dependence**

is needed to be sure that the independence estimate provides a bound.

Example 12: Consider the system illustrated in Figure 9. The state of component 4 is independent of the others. The set of components $\{1, 2, 3\}$ form a complement set. By directly conditioning on the state of component 4, the reliability of the system is

$$p_4 p_{12} + (1 - p_4) p_{123}.$$

If we assumed independence, and made this calculation, the error would be

$$p_4(p_1 p_2 - p_{12}) + (1 - p_4)(p_1 p_2 p_3 - p_{123}). \qquad (5)$$

Let the states of the components $\{1, 2, 3\}$ have the following distribution:

$$p_1 = p_2 = p_3 = \tfrac{1}{2}$$
$$p_{12} = p_{13} = p_{23} = \tfrac{7}{32}$$

$$p_{123} = \tfrac{5}{32}$$

Expression (5) reduces to $\dfrac{2p_4 - 1}{32}$. The states of the components $\{1,2,3\}$ are not associated and the error induced by assuming independence among $\{1,2,3\}$ ranges between $-\frac{1}{32}$ and $\frac{1}{32}$ depending on the reliability of the truly independent component 4.

We note that the bounds theorem and its corollary could have been stated with somewhat looser conditions than that of association on the dependence. However the dependence condition for a substitute set would then be different from that for a complement set. Association is a well-studied property which captures the kind of dependence analysts expect to find under operating conditions. Stating the dependence condition in terms of this property seems thus to be a useful framing for these results.

4. Summary

We have provided two pairs of bounds on system reliability. The first pair is useful when the complexity of the system prevents the exact computation of the system reliability. The second pair can be used to bound system reliability when component states are associated but the exact dependence relationships are unknown.

We provide a new pair of bounds, based on the conditional expectation inequality (1), when exact reliability is difficult to compute. This pair of bounds has good behavior and in some cases improves on known bounds. The bounds are usable even when not all min-cuts or min-paths of the system are known. They also may be used when component states are dependent and the dependence can be quantified.

In some situations, components exhibit associative dependence, but the dependence cannot be quantified. We present a pair of bounds which may be applied when the set of interdependent components form a complement or substitute set. These bounds require only marginal component reliabilities.

Acknowledgments

The research of Sheldon M. Ross was supported by the National Science Foundation Grant DMI-9901053 with the University of California.

References

1. R. E. Barlow and F. Proschan. *Statistical Theory of Reliability and Life Testing*. Holt, Rinehart and Winston, New York, 1975.
2. J. D. Esary, F. Proschan, and D. W. Walkup. Association of random variables, with applications. *Annals of Mathematical Statistics*, 38:1466–1474, 1967.
3. Jane N. Hagstrom. Using the decomposition tree for directed network reliability computation. *IEEE Trans. on Reliability*, R-33:390–395, 1984.
4. Jane N. Hagstrom. Redundancy, substitutes and complements in system reliability. Technical report, College of Business Administration, University of Illinois, 601 S. Morgan St., Chicago, IL 60607, 1990.
5. Sheldon M. Ross. *Stochastic Processes*. Wiley, 2nd edition, 1996.
6. Sheldon M. Ross. *Probability Models for Computer Science*. Academic Press, 2002.
7. Leslie G. Valiant. The complexity of enumeration and reliability problems. *SIAM Journal on Computing*, 8:410–421, 1979.

CHAPTER 3

LARGE EXCESSES FOR FINITE-STATE MARKOV CHAINS

David Blackwell

Department of Statistics, University of California
Berkeley, CA 94720, U.S.A.
E-mail: davidbl@stat.berkeley.edu

For my colleague, sometime office-mate, and fellow Bayesian Dick Barlow, for his 70th birthday.

1. Summary

Let $X(n)$ be a stationary finite-state indecomposable Markov chain, let f be any real-valued function on the states, and put $Y(n) = f(X(n))$, $S(n) = Y(1) + \ldots + Y(n)$. For any a with $E(Y(1)) < a < \max(f)$, we find the exponential rate at which $P(S(n) > an + d$ for some $n)$ goes to 0 as d becomes infinite.

2. The Rate

Let T be an indecomposable $m \times m$ Markov matrix and let $X(0), X(1), \ldots$ be a stationary Markov chain with transition matrix T. Let f be a real-valued function on the states and put $Y(n) = f(X(n))$, $S(n) = Y(1) + \ldots + Y(n)$, $n \geq 1$. Fix a number a with $E(Y(1)) < a < \max(f)$. We want to estimate the numbers

$$p(i, d) = P(S(n) \geq an + d \text{ for some } n \geq 1 \mid X(0) = i).$$

If for a sequence $s = (x(1), \ldots, x(n))$ of X-values we define the excess of s as $(f(x(1)) + \ldots + f(x(n)) - an)$, then

$$p(i, d) = P(\text{excess of } (X(1), \ldots, X(n)) \geq d \text{ for some } n \mid X(0) = i).$$

For $d > \max(f) - a$, p satisfies the equations

$$p(i,d) = \sum_j T(i,j)p(j, d + a - f(j)), \; i = 1, \ldots, m. \tag{1}$$

We shall approximate p by a function $v(i,d)$ that satisfies the equations for all d. Our v will have the form

$$v(i,d) = c(i)\exp(-zd). \tag{2}$$

For v of this form, the equations (1) become

$$c(i) = \sum_j T(i,j)c(j)\exp(zg(j)), \; i = 1, \ldots, m, \tag{3}$$

where $g(j) = f(j) - a$.

Theorem 1: Suppose there is a state $i0$, starting from which arbitrarily large excesses are possible, i.e., such that $p(i0, d) > 0$ for every d. Then

(a) the system (3) has a unique solution $c(1), \ldots, c(m), z$ with $z > 0$ and $\min(c(1), \ldots, c(m)) = 1$.
(b) for this solution, we have, for $i = 1, \ldots, m$ and $d > 0$,

$$wc(i)\exp(-zd) \le p(i,d) \le c(i)\exp(-zd),$$

where $w = \exp(-z(\max(f) - a))/\max(c)$.

Thus, $p(i,d)$ goes to 0 like $\exp(-zd)$ as d becomes infinite. Large deviations for Markov chains have been extensively studied; see Dembo and Zeitouni [1993]; but our bounds on the chance of large excesses seem to be new.

Proof: (a) asserts that there is a unique $z > 0$ for which the matrix $M(z)$ has largest eigenvalue 1, where

$$M(z)(i,j) = T(i,j)\exp(zg(j)).$$

Denote by $h(z)$ the largest eigenvalue of $M(z)$. Then, (1) $h(0) = 1$, (2) $h'(0) < 0$, (3) h is convex, and (4) $h(z) > 1$ for large z. These four properties imply (a). Property (1) is clear, since $M(0) = T$. For property (2), Steve Evans kindly called my attention to the formula

$$h'(0) = yM'(0)x/yx,$$

where y, x are left and right eigenvectors of $M(0)$ (see Horn and Johnson [1985], p. 372). It gives, with y the stationary distribution for T and $x = (1, 1, \ldots, 1)$,

$$h'(0) = \sum_j y(j)g(j) = E(Y(1)) - a < 0.$$

Property(3), convexity of h, is proved in Miller [1961]. Here is a proof that h has the stronger property log-convexity: $ln(h)$ is convex. The nonzero elements of M are clearly log-convex. The class of log-convex functions is closed under multiplication and positive linear combination, so that the nonzero elements of M^n, the nth power of M, are log-convex. Also the nth root of a log-convex function is log-convex, so that $t(n) = n$th root of any nonzero element of M^n is log-convex. Since $t(n)$ converges to h as n becomes infinite, h is log-convex. For property (4), note that

$$M^n(i, j) = \sum_s Pr(s \mid i) \exp(z(\text{excess}(s)),$$

where the sum is over all sequences s of length n that end in j, and $Pr(s \mid i) = P((X(1), \ldots, X(n)) = s \mid X(0) = i)$. Our hypothesis about $i0$ and indecomposability of T imply that there is a sequence s that ends in $i0$ and has $Pr(s \mid i0) > 0$ and has positive excess. Thus if $i0$ has length $n0$, the $(i0, i0)$ element of M^{n_0} is larger than 1 for large z, which implies that the largest eigenvalue of M^{n_0} exceeds 1 for large z, which implies the same for M.

Now for the proof of (b). It uses the gambling ideas of Dubins and Savage [1965]. Given i and d, here is a stopping problem. Set $X(0) = i$ and watch the process $X(0), X(1), \ldots$ as long as you please, stopping eventually. When you stop, say after n periods with $S(n) = s$ and $X(n) + j$, you receive $v(j, d + an - s)$. Since, for all u,

$$v(i, u) = \sum_j T(i, j)v(j, u + a - f(j)),$$

in any position, your income from stopping equals your expected income if you observe one more X and then stop. It follows that all bounded stopping rules have the same expected income, say H.

Stopping with $X(0)$ gives $H = v(i, d)$. For any positive integer N, here is another stopping rule: stop as soon as $S(n) \geq an + d$ or when $n = N$.

Estimating its expected income from below and above, with $E(Y(1)) < b < a$, gives:

$$\begin{aligned} A(N) &\le H \le A(N)Q + B(N)\max(c) + C(N)R(N), \text{ where} \qquad (4)\\ A(N) &= Pr(S(n) \ge an + d \text{ for some } n \le N),\\ Q &= \text{largest possible income} = \max(c)\exp(z(\max(f) - a)),\\ B(N) &= Pr(S(n) < an + d \text{ for all } n \le N \text{ and } S(N) \ge bN),\\ C(N) &= Pr(S(n) < an + d \text{ for all } n \le N \text{ and } S(N) < bN),\\ R(N) &= \max \text{ over } j \text{ of } v(j, d + (a - b)N)). \end{aligned}$$

As N becomes infinite, $A(N) \to p(i,d)$ and $B(N)$, $R(N) \to 0$, so (4) gives

$$p(i,d) \le v(i,d) \le p(i,d)Q,$$

which is equivalent to (3), completing the proof. □

Amir Dembo has pointed out that the variables $v(X(n), d + an - S(n))$ form a martingale, and that the proof of (b) follows easily from this.

3. Example

Example 2: We take for T the 2×2 matrix

$$\begin{vmatrix} 1-x & x \\ x & 1-x \end{vmatrix}$$

for f the function $f(1) = -1$, $f(2) = 1$, and take $a = .1$. Here are the values of $z, c(1), c(2), r = \exp(-z)$ and w for $x = .2, .5, .8$.

x	z	$c(1)$	$c(2)$	r	w
.2	.0504944	1	1.06209	.95079	.899709
.5	.201346	1	1	.81763	.834259
.8	.769729	1.74941	1	.463139	.285922

The rate z for the independence case $x = .5$ is faster than that for $x = .2$, where states tend to persist, but slower than that for $x = .8$, where states change more often than under independence.

As another example, we look at MCMC sampling. We have a joint distribution of two variables X and Y, say

$x\backslash y$	1	2
1	.1	.2
2	.3	.4

and we want to sample from this distribution to estimate the mean value of f, where $f(1,1) = f(2,2) = 0$ and $f(1,2) = f(2,1) = 1$. We do MCMC sampling. Start with any values x_0, y_0. Then choose y_1 according to $p(y \mid x_0)$. Then choose x_1 according to $p(x \mid y_1)$. Then choose y_2 according to $p(y \mid x_1), \ldots$. We estimate the mean of f as the average of $f(x_0, y_0), f(x_1, y_0), f(x_1, y_1), \ldots, f(x_n, y_n)$.

Our Markov chain has eight states (x, y, ph), where each variable has the two values $1, 2$ and $ph = 1$ means that x is sampled next. The function f does not depend on ph. We shall not exhibit the 8×8 matrix, but give two values. For $i = (1,2,1)$, $j_1 = (1,2,2)$, $j_2 = (2,2,2)$, we have $T(i, j_1) = 1/3$, $T(i, j_2) = 2/3$ and $M(i, j) = 0$ for other j.

For $a = .6$ we find the rate $zmc = .704245$, $\exp(-zmc) = .494482$. For independent sampling from the distribution p, we find $zis = .822171$, $\exp(-zis) = .439477$. The chance of excess d with MCMC sampling is about that of excess cd with independent sampling, where $c = zmc/zis = .852568$. Different a or f would give different values of c. For instance, for $a = .7$, same f, we get $zmc = 1.55439$, $zis = 1.80108$, $c = zmc/zis = .863032$.

References

1. Dembo, A., and Zeitouni, O., *Large Deviation Techniques and Applications*, (Jones & Bartlett, Boston, 1993).
2. Dubins, L., and Savage, L., *How to Gamble if You Must: Inequalities for Stochastic Processes*, (McGraw-Hill, New York, 1965).
3. Horn, R., and Johnson, C., *Matrix Analysis*, (Cambridge University Press, Cambridge, 1985).
4. Miller, H. D., "A convexity property in the theory of random variables defined on a finite Markov chain", *Ann. Math. Statist.* **32**, 1260–1270 (1961).

CHAPTER 4

HARDWARE-SOFTWARE RELIABILITY PERSPECTIVES

Hoang Pham

Department of Industrial Engineering, Rutgers University
96 Frelinghuysen Road, Piscataway, New Jersey 08854, U.S.A.
Email: hopham@rci.rutgers.edu

Nowadays the size and complexity of modern systems, such as nuclear power plants, medical monitoring control, real-time military and air traffic control, are extremely huge and it has been shown that there exist remarkable interactions between hardware and software. Ultra-reliability is a crucial need of complex critical systems that can inflict or prevent death. This chapter discusses (1) recent studies in software reliability that include nonhomogeneous Poisson process and Bayesian models; (2) the interactions between hardware and software failures; and (3) future research directions and challenge issues in reliability engineering.

1. Introduction

New computer and communication technologies are obviously transforming our daily life. They are the basis of many of the changes in our telecommunications systems and also a new wave of automation on the farm, in manufacturing, hospital, transportation, and in the office. Furthermore, computers are being used in diverse areas for various applications, for example, air traffic control, nuclear reactors, aircraft, real-time military, industrial process control, automotive mechanical and safety control, and hospital patient monitoring systems. Computer information products and services have become a major and still rapidly growing component of our global economy.

As the functionality of computer operations becomes more essential and complicated in our modern society, and critical software applications increase in size and complexity, the reliability of computer software not only becomes more important, but faults in software design become more subtle [Pham,1991]. A computer system comprises two major components – hardware and software. Although extensive research has been done in the area of hardware reliability, the growing importance of software dictates the focus shifts to software reliability and/or the interactions between the two, hardware and software issues. Software reliability is different from hardware reliability in the sense that software does not wear out or burn out. The software itself does not fail; rather, flaws within the software can possibly cause a failure in its dependent system.

In recent years, the costs of developing software and the penalty costs of software failures are the major expenses in a system. A research study has shown that professional programmers average 6 software defects for every 1000 lines of code (LOC) written. At this rate, a typical commercial software application of 350,000 lines of code can easily contain over 2,000 programming errors including memory-related errors, memory leaks, language specific errors, errors calling third-party libraries, extra compilation errors, standard library errors, etc. As software projects get larger, the rate of software defects indeed increases geometrically. Finding software faults is not only extremely difficult, but also very expensive [Pham & Zhang,1999a].

Software errors have caused spectacular failures and led to serious consequences in our daily life. For example, an inquiry has revealed that a software design error and insufficient software testing caused an explosion that ended the maiden flight of the European Space Agency's Ariane 5 rocked less than 40 seconds after liftoff on June 4, 1996. The problems occurred in the Ariane 5's flight control system, and were caused by a few lines of Ada code containing three unprotected variables. One of these variable pertained to the rocked launcher's horizontal velocity. The overflow occurred because the horizontal velocity of the Ariane 5 trajectory was much greater than that of the Ariance 4 trajectory [Pham,2000a].

The organization of this chapter is divided into five sections. Section 1 presents the basic concepts of software reliability engineering. Section 2 presents software development process includes software life cycle, software versus hardware reliability, and data analysis. Section 3 presents some existing NHPP software reliability models and its application to illustrate the results. Section 4 presents a recent study on software reliability model considering environmental factors. Section 5 discusses the interactions between the hardware and software failures, the N-version fault tolerant systems and also present several recent Bayesian software reliability models. Finally, Section 6 presents a recent study on a generalized software cost model. Future research directions in software reliability engineering and challenge issues are also discussed in each section.

1.1. *Software Reliability Engineering Concepts*

Research activities in software reliability engineering have been conducted over the past 30 years, and many statistical models have been developed for the estimation of software reliability [Wood,1996; Pham,2000a]. Software reliability is a measure of how closely user requirements are met by a software system in actual operation. Most existing models [Jelinski,1972; Goel & Okumoto,1979; Pham,1993,1996; Pham&Zhang,1997; Pham etc.,1999 (b-c); Ohba,1984; Yamada, etc. 1983; Yamada & Osaki,1985; Yamada etc.,1992; Pham,2000(a-b)] for quantifying software reliability are purely based upon observation of failures during the system test of the software product. Some companies are indeed putting these models to use in their software product development. For examples, Hewlett-Packard company has been used one existing reliability model to estimate the failure intensity expected for firmware, software embedded in the hardware, in two terminals, known as the HP2393A and HP2394A and determine when to release the firmware. The results of the reliability modeling enabled it to test the modules more efficiently and so contributed to the terminals' success by reducing development cycle cost while maintaining high reliability. AT&T software developers also used a software reliability model to predict the quality of software system T. The model consistently predicted and the results were within 10 percent of predictions. AT&T Bell Lab's also predicted requested rate for field maintenance to its 5ESS telephone switching system differed from the company's actual experience by only 5 to 13 percent. Such accuracies could help to make warranties of software performance practical [Ehrlich,1993].

Most existing models however require considerable numbers of failure data in order to obtain an accurate reliability prediction. Information concerning the development of the software product, the method of failure detection, environmental factors, etc., however, are ignored to almost all the existing models. In order to develop a useful software reliability model and to make sound judgments when using the model, one need to have an in-depth understanding of how software is produced; how errors are introduced, how software is tested, how errors occur, types of errors, and environmental factors can help us in justifying the reasonableness of the assumptions, the usefulness of the model, and the applicability of the model under a given user environment [Pham,1995]. In other words, these models would be valuable to software developers, users, and practitioners if they are capable of using information about the software development process, incorporating the environmental factors, and are able to give greater confidence in estimates based on small numbers of failure data.

The following acronyms and notation will be used throughout the paper.

Acronyms

AIC	the Akaike information criterion [Akaike,1974]
EF	environmental factors
LSE	least squared estimate
MLE	maximum likelihood estimate
MVF	mean value function
NHPP	non-homogeneous Poisson process
SRGM	software reliability growth model
SSE	sum of squared errors

Notation

$a(t)$	time dependent fault content function: total number of faults in the software, including the initial and introduced faults
$b(t)$	time dependent fault detection-rate function per fault per unit time
$\lambda(t)$	failure intensity function: faults per unit time
$m(t)$	expected number of error detected by time t ("mean value function")
$N(t)$	random variable representing the cumulative number of errors detected by time t
S_j	actual time at which the jth error is detected
$R(x/t)$	software reliability function, i.e., the conditional probability of no failure occurring during $(t, t+x)$ given that the last failure occurred at time t
$\hat{}$	estimates using maximum likelihood estimation method
y_k:	the number of actual failures observed at time t_k
$\hat{m}(t_k)$:	estimated cumulative number of failures at time t_k obtained from the fitted mean value functions, $k = 1, 2, \ldots, n$.

2. Software Development Process

As software becomes an increasingly important part of many different types of systems that perform complex and critical functions in many applications, such as military defense, nuclear reactors, etc., the risk and impacts of software-caused failures have increased dramatically. There is now general agreement on the need to increase software reliability by eliminating errors made during software development. Software is a collection of instructions or statements in a computer language. It is also called a computer program, or simply a program. A software program is designed to perform specified functions. Upon execution of a program, an input state is translated into an output state. An input state can be defined as a combination of input variables or a typical transaction to the program. When the actual output deviates from the expected output, a failure occurs. The definition of failure, however, differs from application to application and should be clearly defined in specifications. For instance, a response time of 30 seconds could be a serious failure for an air traffic control system, but acceptable for an airline reservation system.

In hardware reliability, the mechanism of failure occurrence is often treated as a black box. The emphasis is on the analysis of failure data. In software reliability, one is interested in the failure mechanism. The emphasis is on the model's assumptions and the interpretation of parameters. Software reliability strives to systematically reduce or eliminate system failures that will adversely affect performance of a software program. Software systems do not degrade over time unless they are modified. Although many of the reliability and testing concepts and techniques of hardware are applicable to software, there are many differences; therefore, further analysis and comparisons of software and hardware reliability would be useful in the development of software reliability modeling.

2.1. *Software Life Cycle*

A software life cycle consists of five successive phases: Analysis, Design, Coding, Testing, and Operation [Pham,2000a]. The *Analysis phase* is, the most important phase, the first step in the software development process and the foundation of building a successful software product. The purpose of the analysis phase is to define the requirements and provide specifications for the subsequence phases and activities. The *Design phase* is concerned with how to build the system to behave as described. There are two parts of designs: system architecture design and detailed design. The system architecture design includes, for example, system structure and the system architecture document. System structure design is the process of partitioning a software system into smaller parts. Before decomposing the system, we need to do further specification analysis, which is to examine the details of performance requirements, security requirements, assumptions and constraints, and the needs for hardware and software.

Coding phase involves translating the design into code in a programming language. Coding can be decomposed into the following activities: identify reusable modules, code editing, code inspection, and final test plan. The final test plan should provide details on what needs to be tested, testing strategies and methods, testing schedules and all necessary resources and be ready at the coding phase. *Testing phase* is the verification and validation activity for the software product. Verification and validation (V&V) are the two ways to check if the design satisfies the user requirements. In other words, verification checks if the product, which is under construction, meets the requirements definition and validation checks if the product's functions are what the customer wants. The objectives of testing phase are to (1) affirm the quality of the product by finding and eliminating faults in the program, (2) demonstrate the presence of all specified functionality in the product, and (3) estimate the operational reliability of the software. During this phase, system integration of the software components and system acceptance tests are performed against the requirements. *Operation phase* is the final phase in the software life cycle. The operation phase usually contains the activities such as installation, training, support, and maintenance. It involves the transfer of responsibility for the maintenance of the software from the developer to the user by installing the software product and becomes the user's responsibility to establish a program to control and manage the software.

2.2. *Data Analysis*

There are two commonly types of failure data: time-domain data and interval-domain data. Some existing software reliability models can handle both types of data. The time- domain data is characterized by recording the individual times at which the failure occurred. For example the real-time control system data [Lyu,1996], there are totally 136 faults reported and the time-between failures (TBF) in second are listed in Table 1. The interval-domain data is characterized by counting the number of failures occurring during a fixed period. The time-domain data always provides better accuracy in the parameter estimates with current existing software reliability models, but involves more data collection efforts than the interval domain approach.

Table 1. The Real-time Control System Data for Time Domain Approach

Fault	TBF	Cum. TBF	Fault	TBF	Cum. TBF	Fault	TBF	Cum. TBF	Fault	TBF	Cum. TBF
1	3	3	35	227	5324	69	529	15806	103	108	42296
2	30	33	36	65	5389	70	379	16185	104	0	42296
3	113	146	37	176	5565	71	44	16229	105	3110	45406
4	81	227	38	58	5623	72	129	16358	106	1247	46653
5	115	342	39	457	6080	73	810	17168	107	943	47596
6	9	351	40	300	6380	74	290	17458	108	700	48296
7	2	353	41	97	6477	75	300	17758	109	875	49171
8	91	444	42	263	6740	76	529	18287	110	245	49416
9	112	556	43	452	7192	77	281	18568	111	729	50145
10	15	571	44	255	7447	78	160	18728	112	1897	52042
11	138	709	45	197	7644	79	828	19556	113	447	52489
12	50	759	46	193	7837	80	1011	20567	114	386	52875
13	77	836	47	6	7843	81	445	21012	115	446	53321
14	24	860	48	79	7922	82	296	21308	116	122	53443
15	108	968	49	816	8738	83	1755	23063	117	990	54433
16	88	1056	50	1351	10089	84	1064	24127	118	948	55381
17	670	1726	51	148	10237	85	1783	25910	119	1082	56463
18	120	1846	52	21	10258	86	860	26770	120	22	56485
19	26	1872	53	233	10491	87	983	27753	121	75	56560
20	114	1986	54	134	10625	88	707	28460	122	482	57042
21	325	2311	55	357	10982	89	33	28493	123	5509	62551
22	55	2366	56	193	11175	90	868	29361	124	100	62651
23	242	2608	57	236	11411	91	724	30085	125	10	62661
24	68	2676	58	31	11442	92	2323	32408	126	1071	63732
25	422	3098	59	369	11811	93	2930	35338	127	371	64103
26	180	3278	60	748	12559	94	1461	36799	128	790	64893
27	10	3288	61	0	12559	95	843	37642	129	6150	71043
28	1146	4434	62	232	12791	96	12	37654	130	3321	74364
29	600	5034	63	330	13121	97	261	37915	131	1045	75409
30	15	5049	64	365	13486	98	1800	39715	132	648	76057
31	36	5085	65	1222	14708	99	865	40580	133	5485	81542
32	4	5089	66	543	15251	100	1435	42015	134	1160	82702
33	0	5089	67	10	15261	101	30	42045	135	1864	84566
34	8	5097	68	16	15277	102	143	42188	136	4116	88682

3. Software Reliability Modeling

Research activities in software reliability engineering have been conducted and a number of NHPP software reliability growth models [Goel & Okumoto,1979; Pham,1996; Pham&Zhang,1997; Pham etc.,1999(b-c); Ohba,1984; Hossain etc. 1993; Yamada, etc. 1983; Yamada & Osaki,1985; Yamada etc.,1992; Pham,2000(a-b)] have been developed to assess the reliability of software. Software reliability models based on the NHPP have been quite successful tools in practical software reliability engineering [Pham,2000a]. In this section, we only discuss software reliability models based on NHPP. These models consider the debugging process as a counting process characterized by its mean value function. Software reliability can be estimated once the mean value function is determined. Model parameters are usually estimated using either the MLE or LSE.

3.1. *A Generalized NHPP Model*

Many existing NHPP models assume that failure intensity is proportional to the residual fault content. A general class of NHPP SRGMs can be obtained by solving the following differential equation [Pham,1997]:

$$\frac{dm(t)}{dt} = b(t)[a(t) - m(t)] \tag{1}$$

The general solution of the above differential equation is given by

$$m(t) = e^{-B(t)}\left[m_0 + \int_{t_0}^{t} a(\tau)b(\tau)e^{B(\tau)}d\tau\right] \tag{2}$$

where $B(t) = \int_{t_0}^{t} b(\tau)d\tau$ and $m(t_0) = m_0$ is the marginal condition of Eq.1 with t_0 representing the starting time of the debugging process. The reliability function based on the NHPP is given by:

$$R(x/t) = e^{-[m(t+x)-m(t)]} \tag{3}$$

Many existing NHPP models can be considered as a special case of the above general model. An increasing function $a(t)$ implies an increasing total number of faults (note that this includes those already detected and removed and those inserted during the debugging process) and reflects imperfect debugging. An increasing $b(t)$ implies an increasing fault detection rate, which could be either attributed to a learning curve phenomenon, or to software process fluctuations, or a combination of both. Different $a(t)$

and $b(t)$ functions also reflect different assumptions of the software testing processes. A summary of the most NHPP existing models is presented in Table 2.

3.2. *Application 1: The Real-Time Control System*

We perform the model analysis using the real-time control system data given in Table 1. The first 122 data points are used for the goodness of fit test and the remaining data are used for the predictive power test. The results for fit and prediction are listed in Table 3.

Although software reliability models based on the NHPP have been quite successful tools in practical software reliability engineering [Pham,2000a], there is a need to fully validate their validity with respect to other applications such as in communications, manufacturing, medical monitoring and defense systems.

4. Generalized Models with Environmental Factors

Notation

$\tilde{z}$	Vector of environmental factors
$\tilde{\beta}$	Coefficient vector of environmental factors
$\Phi(\tilde{\beta}\tilde{z})$	Function of environmental factors
$\lambda_0(t)$	Failure intensity rate function without environmental factors
$\lambda(t,\tilde{z})$	Failure intensity rate function with environmental factors
$m_0(t)$	Mean value function without environmental factors
$m(t,\tilde{z})$	Mean value function with environmental factors
$R(x/t,\tilde{z})$	Reliability function with environmental factors

The proportional hazard model (PHM), which was first proposed by Cox (1972), has been successfully utilized to incorporate environmental factors in survival data analysis in medical field and in hardware system reliability area. The basic assumption for the PHM is that the hazard rates of any two items associated with the settings of environmental factors, say z_1 and z_2 respectively, will be proportional to each other. The environmental factors are also known as covariates in PHM. When the PHM applied to the non-homogeneous Poisson process, it becomes the proportional intensity model (PIM). A general fault intensity rate function incorporating the

Table 2. Summary of the Mean Value Functions [Pham etc.,1999b]

Model Name	Model Type	MVF (m(t))	Comments
Goel-Okumoto (G-O)	Concave	$m(t) = a(1 - e^{bt})$ $a(t) = a$ $b(t) = b$	Also called exponential model.
Delayed S-shaped	S-shaped	$m(t) = a(1 - (a + bt)e^{bt})$	Modification of G-O model to make it S-shaped
Inflection S-shaped SRGM	S-shaped	$m(t) = \frac{a(1-e^{bt})}{1+\beta e^{bt}}$ $a(t) = a$ $b(t) = \frac{b}{1+\beta e^{bt}}$	Solves a technical condition with the G-O model. Becomes the same as G-O if $\beta = 0$
Yamada Exponential	S-shaped	$m(t) = a(1 - e^{-r\alpha(1-e^{(-\beta t)})})$ $a(t) = a$ $b(t) = r\alpha\beta e^{-\beta t}$	Attempt to account for testing-effort
Yamada Rayleigh	S-shaped	$m(t) = a(1 - e^{-r\alpha(1-e^{(-\beta t^2/2)})})$ $a(t) = a$ $b(t) = r\alpha\beta t e^{-\beta t^2/2}$	Attempt to account for testing-effort
Yamada Exponential Imperfect debugging model (Y-ExpI)	Concave	$m(t) = \frac{ab}{\alpha+b}(e^{\alpha t} - e^{-bt})$ $a(t) = ae^{\alpha t}$ $b(t) = b$	Assume exponential fault content function and constant fault detection rate
Yamada Linear Imperfect debugging model (Y- LinI)	Concave	$m(t) = a[1 - e^{-bt}][1 - \frac{\alpha}{b}] + \alpha at$ $a(t) = a(1 + \alpha t)$ $b(t) = b$	Assume constant introduction rate α and the fault detection rate
Pham-Nordmann-Zhang (P-N-Z) Model	S-shaped and concave	$m(t) = \frac{a[1-e^{-bt}][1-\frac{\alpha}{b}]+\alpha at}{1+\beta e^{-bt}}$ $a(t) = a(1 + \alpha t)$ $b(t) = \frac{b}{1+\beta e^{-bt}}$	Assume introduction rate is a linear function of testing time, and the fault detection rate function is non-decreasing with an inflexion S-shaped model.
Pham-Zhang (P-Z) Model	S-shaped and concave	$m(t) = \frac{1}{(1+\beta e^{-bt})}[(c + a)(1 - e^{-bt}) - \frac{a}{b-\alpha}(e^{-\alpha t} - e^{-bt})]$ $a(t) = c + a(1 - e^{-\alpha t})$ $b(t) = \frac{b}{1+\beta e^{-bt}}$	Assume introduction rate is exponential function of the testing time, and the fault detection rate is non-decreasing with an inflexion S-shaped model.

Table 3. Parameter Estimation and Model Comparison

Model Name	SSE (fit)	SSE(Predict)	AIC	MLEs
G-O Model	7615.1	704.82	426.05	$\hat{a} = 125$ $\hat{b} = 0.00006$
Delayed S-shaped	51729.23	257.67	546	$\hat{a} = 140$ $\hat{b} = 0.00007$
Inflexion S-shaped	15878.6	203.23	436.8	$\hat{a} = 135.5$ $\hat{b} = 0.00007$ $\hat{\beta} = 1.2$
Yamada Exponential	6571.55	332.99	421.18	$\hat{a} = 130$ $\hat{\alpha} = 10.5$ $\hat{\beta} = 5.4 \times 10^{-6}$
Yamada Rayleigh	51759.23	258.45	548	$\hat{a} = 130$ $\hat{\alpha} = 5 \times 10^{-10}$ $\hat{\beta} = 6.035$
Y-ExpI model	5719.2	327.99	450	$\hat{a} = 120$ $\hat{b} = 0.00006$ $\hat{\alpha} = 1 \times 10^{-5}$
Y-LinI model	6819.83	482.7	416	$\hat{a} = 120.3$ $\hat{b} = 0.00005$ $\hat{\alpha} = 3 \times 10^{-5}$
P-N-Z Model	5755.93	106.81	415	$\hat{a} = 121$ $\hat{b} = 0.00005$ $\hat{\alpha} = 2.5 \times 10^{-6}$ $\hat{\beta} = 0.002$
P-Z model	14233.88	85.36	416	$\hat{a} = 20$ $\hat{b} = 0.00007$ $\hat{\alpha} = 1.0 \times 10^{-5}$ $\hat{\beta} = 1.922$ $\hat{c} = 125$

environmental factors based on PIM can be constructed using the following assumptions:

(a) The new fault intensity rate function consists of two components: the fault intensity rate functions without environmental factors, $\lambda_0(t)$, and the environmental factor function, $\Phi(\tilde{\beta}\tilde{z}$.

(b) The fault intensity rate function $\lambda_0(t)$ and the function of the environmental factors are independent. The function $\lambda_0(t)$ is also called the baseline intensity function.

Assume that the fault intensity function $\lambda(t, \tilde{z})$ is given in the form as follows:

$$\lambda(t, \tilde{z}) = \lambda_0(t) \cdot \Phi(\tilde{\beta}\tilde{z})$$

Typically, $\Phi(\tilde{\beta}\tilde{z})$ takes an exponential form such as:

$$\Phi(\tilde{\beta}\tilde{z}) = \exp(\beta_0 + \beta_1 z_1 + \beta_2 z_2 + \cdots)$$

The mean value function with environmental factors then can be easily obtained:

$$m(t, \tilde{z}(= \int_0^t \lambda_0(s)\Phi(\tilde{\beta}\tilde{z})ds = \Phi(\tilde{\beta}\tilde{z}) \int_0^t \lambda_0(s)ds = \Phi(\tilde{\beta}\tilde{z})m_0(t)$$

Therefore, the reliability function with environmental factors is given by [Pham,1999c]:

$$\begin{aligned} R(x)/t, \tilde{z} &= e^{-[m(t+x,\tilde{z}]-m(t,\tilde{z})]} = e^{-[\Phi(\tilde{\beta}\tilde{z})m_0(t+x,\tilde{z})-\Phi(\tilde{\beta}\tilde{z})m_0(t,\tilde{z})]} \\ &= \{\exp(-[m_0(t+x) - m_0(t)]\}^{\Phi(\tilde{\beta}\tilde{z})} \\ &= [R_0(x/t)]^{\Phi(\tilde{\beta}\tilde{z})} \end{aligned}$$

4.1. *Parameters Estimation*

The MLE method is a widely used method to estimate the unknowns in the reliability models and will be used to estimate the model parameters presented in Section 4. Since the environmental factors are considered here, the parameters need to be estimated include not only the ones in baseline intensity rate function $\lambda_0(t)$, but also the coefficients β_i's in the link function of the introduced environmental factors. For example, we have m unknown parameters in function $\lambda_0(t)$ and we introduced k environmental factors into the model, $\beta_1, \beta_2 \ldots, \beta_k$, then we have $(m+k)$ unknown parameters to estimate. The maximum likelihood function for this model can be expressed as follows:

$$\begin{aligned} L(\tilde{\theta}, \tilde{\beta}, t, z) = P\{\prod_{j=1}^{n}[m(0, z_j) = 0, m(t_{1,j}, z_j) = y_{1,j}, m(t_{2,j}, z_j) \\ = y_{2,j}, \ldots, m(t_{n,j}, z_j) = y_{n,j}]\} \end{aligned}$$

$$= \prod_{j=1}^{n} \prod_{i=1}^{k_j} \frac{[m(t_{i,j}, z_j) - m(t_{i-1,j}, z_j)]^{(y_{i,j} - y_{i-1,j})}}{(y_{i,j} - y_{i-1,j})!}$$
$$\times\, e^{-[m(t_{i,j}, z_j) - m(t_{i-1,j}, z_j)]}$$

where

n	the number of total failure data groups
k_j	the number of faults in group $j, j = 1, 2, \ldots, n$
z_j	the vector variable of the environmental factors in data group j
$m(t_{i,j}, z_j$	the mean value function incorporating the environmental factors.

The logarithm likelihood function is given by

$$\ln[L(\tilde{\theta}, \tilde{\beta}, t, \tilde{z})] = \sum_{j=1}^{n} \sum_{i=1}^{k_j} \{(y_{i,j} - y_{i-1,j} \ln[m(t_{i,j}, z_j) - m(t_{i-1,j}, z_j)]$$
$$- \ln[y_{i,j} - y_{i-1,j})!] - [m(t_{i,j}, z_j) - m(t_{i-1,j}, z_j)]\}$$

A series of differential equations can be constructed by taking the derivatives of the log likelihood function with respect to each parameter, and set them equal to zero. The estimates of unknown parameters can be obtained by solving these differential equations.

A widely used method, which is known as partial likelihood estimate method, can be used to facilitate the parameter estimate process. The partial likelihood method estimates the coefficients of covariates, the β_1's, separately from the parameters in the baseline intensity function. The likelihood function of partial likelihood method is given by [Cox, 1972]:

$$L(\beta) = \prod_i \frac{\exp(\beta z_i)}{[\sum_{m \in R} \exp(\beta z_m)]^{d_i}}$$

where d_i represents the tie failure times.

4.2. *Application 2: The Real-Time Monitor Systems*

In this section, we illustrate the software reliability model with environmental factors based on PIM method using the software failure data collected from the real-time monitor systems [Tohma,1991]. The software consists of about 200 modules and each module has, on average, 1000 lines of a high-level language like FORTRAN. Total 481 software faults were detected during 111-day testing period. Both the information of testing team size and the software failure data are recorded.

The only environmental factor available in this application is the testing team size. Team size is one of the most useful measures in the software development process. It has close relationship with the testing effort, testing efficiency and the development management issues. From the correlation analysis of the thirty-two environmental factors [Zhang and Pham,2000], team size is the only factor correlated to the program complexity, which is the number one significant factor according to our environmental factor study. Intuitively, the more complex the software, the larger team is required. Since the testing team size ranges from 1 to 8, we first categorize the factor of team size into two levels. Let z_1 denote the factor of team size as follows:

$$z_1 = \begin{cases} 0, & \textit{team size ranges from } 1-4 \\ 1, & \textit{team size ranges from } 5-8 \end{cases}$$

After carefully examining the failure data, we find that after day 61, the software turns stable and the failures occur with a much slower frequency. Therefore, we use the first 61 data points for testing the goodness-of-fit and estimating the parameters, and use the remaining 50 data points (from day 62 to day 111) as real data for examining the predictive power of software reliability models.

In this application, we use the P-Z model listed in Table 2 as the baseline mean value function, that is,

$$m_0(t) = \frac{1}{(1+\beta e^{-bt})}[(c+a)(1-e^{-bt}) - \frac{ab}{b-\alpha}(e^{-\alpha t} - e^{-bt})]$$

and the corresponding baseline intensity function is:

$$\lambda_0(t) = \frac{1}{(1+\beta e^{-bt}}[(c+a)(1-e^{-bt}) - \frac{ab}{b-\alpha}(e^{-\alpha t} - e^{-bt})] \\ + \frac{-\beta b e^{-bt}}{(1+\beta e^{-bt})^2}[(c+a)(1-e^{-bt}) - \frac{ab(e^{-\alpha t} - e^{-bt})}{b-\alpha}]$$

Therefore, the intensity function with environmental factor is given by [Zhang,1999]:

$$\lambda(t) = \lambda_0(t) \cdot e^{\beta_1 z_1}$$

$$= \left\{ \frac{1}{(1+\beta e^{-bt})} [(c+a)(1-e^{-bt}) - \frac{ab}{b-\alpha}(e^{-\alpha t} - e^{bt})] \right.$$
$$\left. + \frac{-\beta b e^{-bt}}{(1+\beta e^{-bt})^2} [(c+a)(1-e^{-bt} - \frac{ab(e^{-\alpha t} - e^{-bt}}{b-\alpha}] \right\} e^{\beta_1 z_1}$$

The estimate of β_1 using partial likelihood estimate method is $\hat{\beta}_1 = 0.0246$ which indicates that this factor is significant to consider. The estimates of parameters in the baseline intensity function are given as:

$$\hat{a} = 40.0, \hat{b} = 0.09, \hat{\beta} = 8.0, \hat{\alpha} = 0.015, \hat{c} = 450.$$

The results of several existing NHPP models are also listed in Table 4.

The results show that incorporating the factor of team size into the P-Z model that explains the fault detection better and thus enhances the predictive power of this model. Further research is needed to incorporate application complexity, test effectiveness, test suite diversity, test coverage, code reused, and real application operational environments into the NHPP software reliability models and into the software reliability model in general. Future software reliability models must account for these important situations.

5. Hardware & Software Systems

A lot of research has been done in the area of hardware reliability and software reliability, but most of it has been limited to either the hardware sub-system alone or the software sub-system alone. Welke et al (1995) tried to establish a reliability model for the whole system - including both hardware and software, but they assume that hardware and software subsystems are totally independent of each other, i.e., there is no interaction between the two.

Nowadays the size and complexity of modern systems are extremely huge, and it has been shown that there exist remarkable interactions between hardware and software. Hardware and software have been heavily involved with each other, and it is becoming more and more difficult to distinguish between hardware failures and software failures in many systems where the hardware-software independence assumption does not hold, and consequently, the system reliability cannot be considered as the multiplication of these two.

The interactions between hardware and software have two aspects: one is positive and another is negative. Iyer (1985) mentioned that the mechanical problems of the Hubble telescope were mitigated by software changes. This

Table 4. Model Evaluation

Model Name	MVF (m(t))	SSE (Prediction)	AIC
G-O model	$m(t) = a(1 - e^{-bt})$ $a(t) = a$ $b(t) = b$	1052528	978.14
Delayed S-shaped	$m(t) = a(1 - (1 + bt)e^{-bt})$ $a(t) = a$ $b(t) = \frac{b^2 t}{1+bt}$	83929.3	983.90
Inflexion S-shaped	$m(t) = \frac{a(1-e^{-bt})}{1+\beta e^{-bt}}$ $a(t) = a$ $b(t) = \frac{b}{1+\beta e^{-bt}}$	1051714.7	980.14
Yamada Exponential	$m(t) = a(1 - e^{-r\alpha(1-e^{(-\beta t)})})$ $a(t) = a$ $b(t) = r\alpha\beta e^{-\beta t}$	1085650.8	979.88
Yamada Rayleigh	$m(t) = a(1 - e^{-r\alpha(1-e^{(-\beta t^2/2)})})$ $a(t) = a$ $b(t) = r\alpha\beta t e^{-\beta t^2/2}$	86472.3	967.92
Y-ExpI model	$m(t) = \frac{ab}{\alpha+b}(e^{\alpha t} - e^{-bt})$ $a(t) = ae^{\alpha t}$ $b(t) = b$	791941	981.44
Y-LinI model	$m(t) = a[1 - e^{-bt}][1 - \frac{\alpha}{b}] + \alpha a t$ $a(t) = a(1 + \alpha t)$ $b(t) = b$	238324	984.62
P-N-Z model	$m(t) = \frac{a}{1+\beta e^{bt}}[(1 - e^{-bt})(1 - \frac{\alpha}{b} + \alpha t]$ $a(t) = a(1 + \alpha t)$ $b(t) = \frac{b}{1+\beta e^{-bt}}$	94112.2	965.37
P-Z model	$m(t) = \frac{1}{(1+\beta e^{-bt})}[(c + a)(1 - e^{-bt})$ $- \frac{ab}{b-\alpha}(e^{-\alpha t} - e^{-bt})]$ $a(t) = c + a(1 - e^{-\alpha t})$ $b(t) = \frac{b}{1+\beta e^{-bt}}$	86180.8	960.68
Environmental Factor model	$m(t) = \frac{1}{(1+\beta e^{-bt})}[(c + a)(1 - e^{-bt})$ $- \frac{ab}{b-\alpha}(e^{-\alpha t} - e^{-bt})]e^{\beta_1 z_1}$ $a(t) = c + a(1 - e^{-\alpha t})$ $c(t) = 1 - \frac{(1+\beta)}{(e^{bt}+\beta)}$	560.82	890.68

means that software methods can be used to fix hardware problems, which is a positive aspect of the interactions between hardware and software. Software remedies for hardware problems build fault-tolerance into the system, which improves the system reliability. On the other hand, unanticipated hardware failure modes can cause the software to be executed according to an atypical operational profile for which the software may not have been rigorously tested against. As a result, the system may be less reliable when operating in the masked failure state.

5.1. *Hardware Failures*

Hardware can fail due to a number of causes such as wear-out or fatigue, temperature or electrical stress, or design susceptibilities. Good commercial systems are usually designed around these problems by using such practices as applying component derating criteria, the use of preferred parts, and concurrent engineering techniques, etc. A scheme for classification of failures is proposed by Bodsberg (1993) to pinpoint main differences between hardware and software reliability models. For example, according to Table 5, failure of a system to deliver proper service can be categorized into (1) natural aging failure, (2) stress failure, (3) intervention failure, or (4) input failure. Natural aging and stress failures are physical failures, whereas intervention and input failure are functional failures.

Keene and Lane (1992) also investigate the failures of the same type of circuit cards. The interesting observation was that these failures did not fit the random failure models. These failures were deterministic and resulted from a defect in the design. Thus, the design defect is also an important cause of hardware failures. In summary, we have following causes of hardware failures:

(1) Natural Aging (Wear out) - Degradation
(2) High Stress
(3) Intervention and disturbance (From outside)
(4) Design defect
(5) Wrong System Input

Natural aging is considered as the main cause of hardware failures in many hardware reliability models, and many reliability models have been developed under this category. High stress is also another cause of hardware failures. Many hardware components work well when the stress level is low, but fail when the work-load is high. In a computer system, many hardware components are potentially liable to fail by high stress. Interventions and disturbances from outside environment may also prevent the hardware component from functioning properly. For example, outside electromagnetic disturbance may alter the data inside some communication units, and consequently cause the whole system to fail. In this case, the hardware components may not be considered (permanently) failed since those hardware components can perform tasks correctly if those interventions and disturbances are eliminated from the context.

The architecture of hardware system has become more and more complicated, and now researchers have begun to pay more attention to hardware design defects. For example, there is a design bug in the first generation of Pentium chips, and this could lead to a failure. The wrong input can also cause hardware components to fail, even destroy the hardware components. It can be avoided by adding a validation unit, which checks the input before it is sent to hardware components. Hardware designers should anticipate such situations and it can be considered as a design defect of the hardware system when the designers fail to do so.

According to the duration of failures, hardware failures can be divided

Table 5. Distinctions Between Failure Categories for Automatic Systems

<table>
<tr><td rowspan="3">Dimension</td><td colspan="4">Failure Category</td></tr>
<tr><td colspan="2">Physical failures</td><td colspan="2">Functional failures</td></tr>
<tr><td>Natural aging failure</td><td>Stress Failure</td><td>Intervention failure</td><td>Input failure</td></tr>
<tr><td rowspan="3">Failure definition / Triggering event characteristics</td><td>Internal</td><td colspan="3">External</td></tr>
<tr><td colspan="2">Physical degradation</td><td colspan="2">No Physical degradation</td></tr>
<tr><td colspan="3">System state change</td><td>No Change</td></tr>
<tr><td rowspan="2">Failure Description</td><td rowspan="2">Wear out (Within design envelope)</td><td colspan="2">External “shocks” (Outside design envelope)</td><td>Unacceptable input</td></tr>
<tr><td>Nature/human</td><td>Human</td><td></td></tr>
<tr><td>Variable affecting reliability</td><td>Inherent degradation characteristics</td><td colspan="2">Inherent characteristics and ‘shock’ rate/profile</td><td>Inherent characteristics and execution rate/profile</td></tr>
<tr><td rowspan="2">Terms used in the literature</td><td>Random failures/ independent failures</td><td colspan="3">Systematic failures/common cause failures/dependent failures</td></tr>
<tr><td colspan="3">Operational failure</td><td>Design failure</td></tr>
</table>

into following categories:

(1) Transient Hardware Failures
(2) Permanent Hardware Failures
    (a) Total Hardware Failures
    (b) Partial Hardware Failures

Transient hardware failures are caused by disturbances to the operation environment. For example, the central processing unit (CPU) stops functioning if the environment temperature is higher than some specific value. The data in communication systems may be altered by strong electromagnetic disturbances. Consequently, the system may stop functioning or give an incorrect computation result. Permanent hardware failures are hardware failures or degradations that can not be recovered by simply redoing the tasks. We can divide permanent hardware failures into two categories: total hardware failures and partial hardware failures.

Total hardware failures are those failures for which the entire hardware component cannot function any more, no matter how many times we redo the task on the component. For example, if the CPU burns out, then the whole system fails. Partial hardware failures are those failures for which only part of the hardware component can not function any more. For example, the memory in a computer actually consists of many blocks, and those blocks may fail independently. If some memory blocks fail while other memory blocks still work, this is a partial hardware failure. In this case, the computer may still work if those failed memory blocks are not used again by the software, but the system may be degraded - it responds very slowly.

5.2. *Software Faults and Failures*

Software failures have begun to dominate over hardware failures in many modern systems. Following are possible reasons for this phenomenon:

(1) Software development is very labor intensive.
(2) There are more unverified interfaces in software than in hardware.
(3) Software has less restricted inputs between its functional component areas.
(4) Human element is more interactive with software than hardware.

Because software systems have become more and more complicated and the software development is quite labor intensive, it is almost impossible to deliver a software application without any faults. To improve the software reliability, it is very important to find and remove software faults from the software. The basic premise of software reliability growth models (SRGM) is that once a software fault is found and removed from the software, failures related to the fault will not occur any more and the reliability of software system improves (grows).

There are many different ways to classify software faults. In order to analyze the interactions between hardware and software, we divide all software faults into the following two categories:

(1) Hardware-related faults.
(2) Hardware-unrelated faults.

Hardware-related faults are faults in those software modules that in a large extent depend on the hardware components. For example, many software applications cannot work properly on old hardware platforms. This can be viewed as a mismatch between hardware and software. We consider this is a software problem, because in general software is easily modified to match hardware. It has been shown that there exist remarkable interactions between hardware and software. Hardware-related faults are manifestations of the interactions between hardware and software. Hardware-unrelated faults are faults in those software modules that are unrelated to the hardware components. These faults are 'pure' software faults. For example, some numerical algorithm modules can be executed on almost all hardware platforms. If there are some faults in those modules, then they are considered hardware-unrelated.

Iyer (1985) analyzed hardware-related software (HW/SW) errors on an MVS/SP operating system at Stanford University. He examined the operating system's handling of HW/SW errors and also the effectiveness of recovery management. Nearly 35 percent of all observed software failures were found to be hardware-related. The hardware-related errors are actually caused by the interactions between the hardware and software components. This section we also use the term "hardware-related software failure" to describe the negative interactions between the hardware and software.

5.3. *Hardware/Software Interactions*

All hardware-related software failures can be put into two categories: Transient and permanent hardware-related software failures. Temporary hardware-

Table 6. Failure types of Hardware System Components

Failure Type	Hardware Failure			Software Failures		
Duration	Permanent		Temporary	Hardware-involved		Non-hardware involved
Degree	Total	Partial		Hardware does not work properly*	Hardware works Properly**	e.g. Algorithm Bugs

* : *Hardware-related software failures*
**: Software design does not match initial hardware configuration

related software failures are mostly caused by transient hardware problems, such as the data inconsistency in the presence of hardware transients. In this case, hardware is not considered (permanently) failed, and the software also works well, but the computation result is not correct. The reason is that the hardware system is not robust enough to resist all disturbances from outside environment. Since this kind of failure is transient, the problem can be solved simply by redoing the tasks (roll-back) after the hardware transients are over.

Permanent hardware-related software failures are mostly caused by the degradation of hardware and the design defects of software. In this case, hardware components are partially failed, but the system can still perform normal operations (in a degradation manner) if work-load is not very high. However, if the work-load (stress) is high for the system and the hardware degradation is undetected by the software, then software may try to perform operations on the failed hardware component and the system will fail. This kind of system failures is permanent, and cannot be solved by a roll-back scheme. A good software design should avoid permanent hardware-related software failures. Because of the difficulties in distinguishing between hardware failures and software failures, this classification is not absolute. Table 6 gives the general idea of how we distinguish between failure types.

Hardware failures can be put into two categories by their duration: permanent failure and temporary (transient) failures. Once a hardware component no longer functions properly, then we say that the component is failed permanently. Inside the hardware subsystem, many hardware components may fail in this way. The failure of some components, such as CPU, (assume only one CPU in the system) may lead to failure of the whole system. In such case, we say this is a total hardware failure. However, there are also

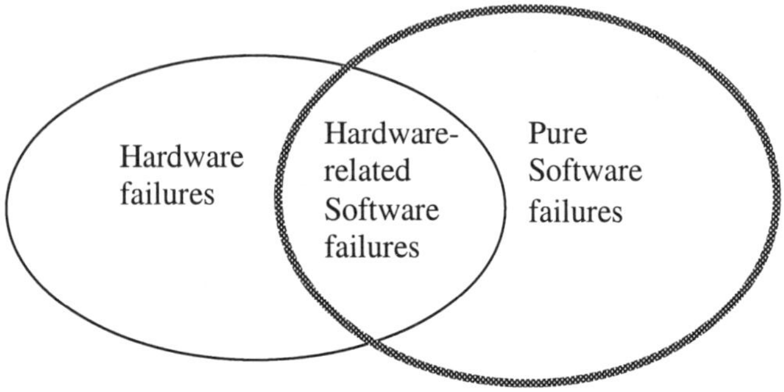

Fig. 1. System failure categories

some other components, such as memory and disks, which consist of many sub-components. The failure of some sub-components does not necessarily lead to the failure of the entire system since the system can still function properly in a degraded manner under some circumstances. In this case, we say that the hardware component is degraded, or only partially failed.

Software failures can be categorized as hardware-involved and non-hardware-involved. Generally speaking, all software is involved with hardware in some way. Here, "non- hardware-involved" software failure means that the software does not depend on specific hardware configurations. They are "pure" software failures.

Software is designed to work with hardware. For the original hardware system, if the software does not work well, then the problems are considered as pure software problems. However, if originally hardware and software match each other very well, and after some time, the hardware component goes into a degradation state, then the software may fail due to the changes of hardware configuration. This kind of software failures is considered as hardware-related software failure.

Although some research has been done to combine hardware and software reliability modeling, most of the proposed models assume no interactions between hardware and software. Figure 1 shows the system failure categories, and there is an overlap region between hardware failures and software failures, which we call hardware-related software failures (HW/SW).

Several reliability models have been studied for evaluating the reliability of hardware and software systems with the assumption that hardware and software are independent of each other. But very little work has been done in modeling the interactions between hardware and software. We are currently developing an integrated system reliability model with considerations of hardware, software and the interactions between them.

5.4. *N* - *Version Fault Tolerant Software*

Fault-tolerant software approach can be applied to achieve the entire system reliability goal by using protective redundancy at the software level [Pham,1992]. For examples, the Boeing 777 flight-control system consists of about 120 distinct functional microprocessors and at least 475 when they account for redundancy. The Elektra Austrian railway signaling system uses design diversity N-version fault tolerant software. Several other fault tolerant software examples include French train system, Airbus fly-by-wire aircraft and NASA space shuttles.

Numerous papers [Lyu, 1993; Nicola, 1990, Littlewood, 1989; Scott, 1984; Tai, 1993; Avizienis, 1995; Voas, 2001; Pham, 1992; 2001] have been devoted to the modeling, analysis and evaluation of software reliability of N-version programming systems over the last three decades. Some reliability models focus on software diversity modeling, and they are based on the detailed analysis of the dependencies in diversified software. Some other models focus on modeling fault-tolerant software system behavior, aim primarily to evaluate some dependability measures for specific types of software systems.

However, few of them consider software reliability growth during testing and debugging due to continuous removal of faults from the components of the NVP systems. The modifications to software versions cause their reliability to grow, which in turn causes the reliability of N-version programming software to grow. Kanoun [1993] proposed the first reliability growth model for NVP systems by using a hyper-exponential model. The hyper-exponential model parameters characterizing reliability growth due to the removal of the faults. This model, however, assumes that each time a failure occurs, the errors which cause it is immediately removed and no new errors are introduced (perfect debugging). Other SRGM models, but not NVP-SRGM, recently consider imperfect debugging [Pham, 1996, 1999b], i.e., faults are not always successfully removed and new ones can be introduced into the software during the debugging process.

Different types of faults in an NVP system have different roles to the system. If they are s-independent of each other among different versions, then they can be successfully masked or tolerated by NVP scheme. But if those faults are common to multiple versions, they may be activated by the same input, then several versions will fail at the same time, thus the NVP system will fail. The role of faults in NVP systems may change due to imperfect debugging, some potential common faults may reduce to low-level common faults or independent faults.

Common Faults and Independent Faults

Although diverse software versions are developed by using different specifications, designs, programming teams, programming languages, etc, many researchers have revealed that those independently developed software versions do not necessarily fail s- independently [Nicola, 1990; Scott, 1984, 1987; Vouk, 1985]. Some experiments show that the use of different languages and design philosophies has little effect on the reliability in N-Version Programming because people tend to make similar logical mistakes in difficult-to-program parts of the software. Laprie [1990] classifies faults according to their independence into either related or independent. Related faults manifest themselves as similar errors and lead to common-mode failures, whereas independent faults usually cause distinct errors and separate failures.

Due to the difficulty of distinguishing the related faults from the unrelated faults, in recent paper [Pham, 2001] they simplify this fault classification as follows: If two or more versions give identical but all wrong results, then the failures are caused by the related faults between versions; if two or more versions give dissimilar but wrong results, then the faults are caused by independent software faults. Figure 2 illustrates the common faults and the independent faults in an NVP with N=2 (2VP).

Common Faults are those that located in the functionally equivalent modules among two or more software versions because their programmers are prone to making the same or similar mistakes although they develop different versions independently. Those faults will be activated by the same input to cause those versions to fail simultaneously, and these failures by common faults are called *Common Failures. Independent Faults* are usually located in different or functionally unequivalent modules between or among different software versions. Since they are independent of each other and are considered harmless to the fault-tolerant systems because their resulting failures are typically distinguishable to the decision mechanism. However,

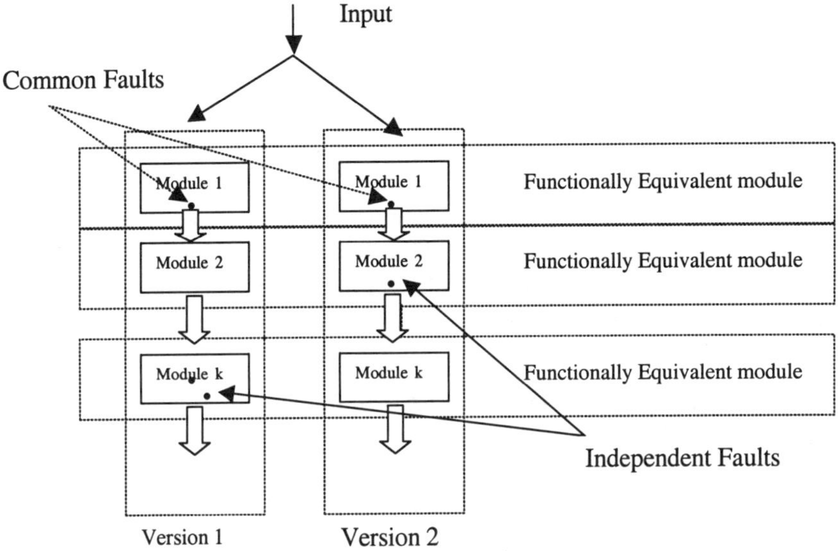

Fig. 2. Common Faults and Independent Faults

there is still a probability, though very small compared with that of *Common Failures*, that an unforeseeable input activates two independent faults in different software versions that will lead those versions to fail at the same time. These failures by independent faults are called *Concurrent Independent Failures.* Table 7 shows the differences between the common failures and the concurrent independent failures.

Pham (2001) develops a software reliability growth model for N-Version Programming (NVP) Systems (NVP-SRGM) based on the NHPP. This model is the first reliability growth model for NVP Systems with consid-

Table 7. Common Failures and Concurrent Independent Failures

	Common Failures	**Concurrent Independent Failures**
Fault Type	Common faults	Independent Faults
Output	Usually the same	Usually different
Fault Location (Logically)	Same	Different
Voting Result (Majority Voting)	Choose wrong solution	Unable to choose correct solution

erations of the error introduction rate and the error removal efficiency. During testing and debugging, when a software fault is found, a debugging effort will be spent to remove this fault. Due to the high complexity of the software, this fault may not be successfully removed, and new faults can also be introduced into the software. A simplify software control logic for a water reservoir control system is also presented to illustrate how to apply their proposed software reliability model. They also provide the confidence bounds on the system reliability estimation. More application is needed to validate fully their NVP-SRGM for quantifying the reliability of fault-tolerant software systems in a general industrial setting.

5.5. *Bayesian Software Reliability Models with Pseudo-Failures*

Within the non-Bayes framework, the parameters of the assumed distribution are thought to be unknown but fixed and are estimated from past inter-failure times utilizing the maximum likelihood estimation technique. A time interval for the next failure is then obtained from the fitted model. However, the times between failures models based on this technique tend to give results that are grossly optimistic [Littlewood, 1989] due to the use of MLE. To overcome this problem, various Bayes predictive models [Mazzuchi, 1988; Csenki, 1990] have been proposed since if the prior accurately reflects the actual failure process, model prediction on software reliability can be improved while there is a reduction in total testing time or sample size requirements. Pham [2000b] presents Bayesian software reliability models with stochastically decreasing hazard rate. Within any given failure time interval, the hazard rate is a function of both total testing time as well as number of encountered failures. To improve the predictive performance of previous models in [Pham, 2000b], Pham [2001] recently presents a modified Bayesian model introducing pseudo-failures whenever there is a period when the failure-free execution estimate equals interval-percentile of a predictive distribution.

6. Cost Modeling

The quality of the software system usually depends on how much time testing takes and what testing methodologies are used. On the one hand, the longer time people spend in testing, the more errors can be removed, which leads to the more reliable software; however, the testing cost of the software will also increase. On the other hand, if the testing time is too short, the cost of the software could be reduced, but the customers may take a higher risk of buying unreliable software [Pham,1999a; Zhang & Pham,1998,1999b]. This will also increase the cost during the operational phase, since it is much more expensive to fix an error during operational phase than testing phase. Therefore, it is important to determine when to stop testing and release the software. In this section, we present a recent generalized cost model and also other existing cost models as well.

6.1. *Generalized Cost Models*

In order to improve the reliability of software products, testing serves as the main tool to remove faults in software products. However, efforts to increase reliability will require an exponential increase in cost, especially after reaching a certain level of software refinement. Therefore, it is important to determine when to stop testing based on the reliability and cost assessment. Several software cost models and optimal release policies have been studied in the past two decades. Okumoto and Goel (1980) discussed a simple cost model addressing linear developing cost during testing and operational period. Ohtera and Yamada (1990) also discussed the optimum software-release time problem with a fault-detection phenomenon during operation. They introduced two evaluation criteria for the problem: software reliability and mean time between failures. Leung (1992) discussed optimal software release time with consideration of a given cost budget. Dalal and McIntosh (1994) studied the stop-testing problem for large software systems with changing code using graphical methods. They reported the details of a real time trial of a large software system that had a substantial amount of code added during testing. Yang and Chao (1995) proposed two criteria of making decisions on when to stop testing. According to them, software products are released to market when (1) the reliability has reached a given threshold, and (2) the gain in reliability cannot justify the testing cost.

Pham (1996) developed a cost model with an imperfect debugging and random life cycle as well as a penalty cost to determine the optimal release policies for a software system. Hou et al. (1997) discussed optimal release times for software systems with scheduled delivery time based on the hypergeometric software reliability growth model. The cost model included the penalty cost incurred by the manufacturer for the delay in software release. Recently, Pham and Zhang (1999d) developed the expected total net gain in reliability of the software development process, as the economical net gain in software reliability that exceeds the expected total cost of the software development, that uses to determine the optimal software release time which maximizes the expected total gain of the software system.

Pham and Zhang (1999a) also recently developed a generalized cost model addressing the fault removal cost, warranty cost and software risk cost due to software failures for the first time. The following cost model calculates the expected total cost:

$$E(T) = C_0 + C_1 T^{\alpha} + C_2 m(T)\mu_y + C_3 \mu_w [m(T+T_w) - m(T)] + C_R[1 - R(X/T)]$$

where

C_0	set-up cost for software testing
C_1	software test cost per unit time
C_2	cost of removing each fault per unit time during testing
C_3	cost to remove an fault detected during warranty period
C_R	loss due to software failure
$E(T)$	expected total cost of a software system at time T
μ_y	expected time to remove a fault during the testing period
μ_w	expected time to remove a fault during the warranty period.

The details how to obtain the optimal software release policies that minimize the expected total cost can be obtained in [Pham & Zhang,1999a]. The benefits of using the above cost models are that they provide:

(1) assurance that the software has achieved safety goals, and
(2) a means of rationalizing when to stop testing the software.

In addition, with this type of information, a software manager can determine whether more testing is warranted or whether the software is sufficiently tested to allow its release or unrestricted use [Pham,1999a]. Further research is interested and needed in determining the following:

(1) How should resources be allocated from the beginning of the software lifecycle to ensure that the on-time and efficient delivery of a software product?
(2) What information during the system test does a software engineer need to determine when to release the software?
(3) What is the risk cost due to the software failures after release? and
(4) How should marketing efforts – including advertising, public relations, trade- show participation, direct mail, and related initiatives – be allocated to support the release of a new software product effectively?

7. Further Reading

There are several survey papers and books on software reliability research, software cost, and fault tolerant systems, published in the **last five years**, that can be read at an introductory/intermediate stage. Interested readers are referred to the following articles by Pham (1999c) and Whittaker and Voas (2000), Voas (IEEE Software, July/August 2001) and the handbooks, *Handbook of Software Reliability Engineering* by M. Lyu (ed.), (McGraw-Hill and IEEE CS Press, 1996); Handbook of Reliability Engineering by H. Pham (ed.) (Springer, Summer 2002).

The books *Software Reliability* by H. Pham, (Springer-Verlag, 2000); *Software- Reliability-Engineered Testing Practice* by J. Musa, (John Wiley & Sons, 1998), are recently new and, excellent textbooks and references for students, researchers, and practitioners.

This list is by no means exhaustive, but I believe it will help readers get started learning about the subjects.

Acknowledgements

This research was supported in part by the FAA William J. Hughes Technical Center and the U.S. National Science Foundation.

References

1. H. Akaike, "A new look at statistical model identification", *IEEE Trans. on Automatic Control*, 19, 1974, pp 716–723
2. A. Avizienis, "The Methodology of N-Version Programming," *Software Fault-Tolerance*, M.R. Lyu, ed. John Wiley & Sons, New York, 1995, pp 23–46
3. L. Bordsberg, "Comparative Study of Quantitative Models for Hardware, Software and Human Reliability Assessment", Quality and Reliability Engineering International, Vol.9, 1993, pp 501–518
4. D.R. Cox, "Regression models and life-tables (with discussion)," *J. R. Statist. Soc. B34*, 1972, pp 187-220
5. A. Csenki, "Bayes Predictive Analysis of a Fundamental Software Reliabiltiy Model", IEEE Trans. on Reliability, Vol.R-39, June 1990, pp 177–183.
6. S.R. Dalal, and A.A. McIntosh, "When to Stop Testing For Large Software Systems With Changing Code," IEEE Trans. on Software Engineering, vol 20, no. 4, 1994, pp 318–323
7. W. Ehrlich, B. Prasanna, J. Stampfel, and J. Wu, "Determining the Cost of a Stop-Testing Decision," IEEE Software, vol 10, 1993 March, pp 33–42

8. A. L. Goel, and K. Okumoto, "Time-Dependent Error-Detection Rate Model For Software and Other Performance Measures", *IEEE Trans. on Reliability*, 28, 1979, pp 206–211
9. S. A. Hossain and C. D. Ram, "Estimating the Parameters of a Non-Homogeneous Poisson Process Model For Software Reliability", *IEEE Trans. on Reliability*, vol. 42, num 4,1993, pp 604–612
10. R. H. Hou, S.Y. Kuo and Y.P. Chang, "Optimal Release Times for Software Systems With Scheduled Delivery Time Based on the HGDM", *IEEE Trans. on Computers*, vol. 46, no. 2, 1997, pp 216–221
11. R.K. Iyer, "Hardware-Related Software Errors: Measurement and Analysis", IEEE Trans. on Software Engineering, Vol. SE-11, No.2, February 1985, pp 223–230
12. Z. Jelinski and P. B. Moranda. "Software Reliability Research," in Statistical Computer Performance Evaluation, W. Freiberger, Ed. New York: Academic, 1972
13. S. Keene and C. Lane, "Combined Hardware and Software Aspects of Reliability", Quality and Reliability Engineering International, Vol.9, 1992, pp 419–426
14. K. Kanoun, M. Kaaniche, C. Beounes, J-C. Laprie, and J. Arlat, "Reliability Growth of Fault-Tolerant Software," *IEEE Trans. on Reliability*, vol. 42 no. 2, pp 205–219, Jun 1993
15. J.-C. Laprie, J. Arlat, C. Beounes, and K. Kanoun, "Definition and Analysis of Hardware- and Software-Fault-Tolerant Architectures", IEEE Computer, June 1990, pp 39-51
16. Y.W. Leung, "Optimal Software Release Time With a Given Cost Budget", *Journal of Systems and Software*, vol. 17, pp 233–242, 1992
17. B. Littlewood and A. Softer, "A Bayesian Modification to the Jelinski-Moranda Software Reliability Growth Model," *Software Engineering Journal*, pp 23-37, March 1989
18. M. R. Lyu, "Improving the N-version Programming Process Through the Evolution of a Design Paradigm," *IEEE Trans. on Reliability*, vol. 42, no. 2, pp 179–189, Jun 1993
19. M. Lyu (Ed.), *Software Reliability Engineering Handbook*, McGraw-Hill, 1996
20. T. A. Mazzuchi and R. Soyer, "A Bayes Empirical Bayes Model for Software Reliability", IEEE Trans. on Reliability, Vol.R-37, June 1988, pp 248–254
21. V. F. Nicola and A. Goyal, "Modeling of Correlated Failures and Community Error Recovery in Multiversion Software," *IEEE Trans. on Software Engineering*, Vol. 16, no. 3, pp 350–359, Mar 1990
22. M. Ohba, "Software Reliability Analysis Models", *IBM Journal of Research Development*, 28, 1984, pp 428–443.
23. H. Ohtera and S. Yamada. "Optimal Allocation and Control Problems for Software- Testing Resources," *IEEE Trans. on Reliability*, Vol. 39, 1990 June, pp 171–176.

24. K. Okumoto and A. L. Goel, "Optimum Release Time for Software Systems Based on Reliability and Cost Criteria," *Journal of Systems and Software*, vol. 1, pp 22-31, 1980
25. H. Pham, "Software Reliability Assessment: Imperfect Debugging and Multiple Failure Types in Software Development", *EG&G-RAAM-10737; Idaho National Engineering Laboratory*, 1993
26. H. Pham, Software Reliability and Testing, IEEE Computer Society Press, 1995
27. H. Pham, "A Software Cost Model With Imperfect Debugging, Random Life Cycle and Penalty Cost", *International Journal of Systems Science*, vol. 27, Number 5, 1996, pp 455-463
28. H. Pham and X. Zhang, "An NHPP Software Reliability Models and Its Comparison", *International Journal of Reliability, Quality and Safety Engineering*, vol.4, no. 3, 1997, pp 269–282.
29. H. Pham (ed.), *Fault-Tolerant Software Systems: Techniques and Applications*, IEEE Computer Society Press, 1992
30. H. Pham and M. Pham, *Software Reliability Models For Critical Applications*, EGG- 2663, 1991, Idaho Falls, Idaho
31. H. Pham and X. Zhang, "A Software Cost Model With Warranty and Risk Costs", *IEEE Trans. on Computers*, vol. 48, no. 1, 1999a, pp 71–75
32. H. Pham, L. Nordmann, and X. Zhang, "A General Imperfect Software Debugging Model With S-Shaped Fault Detection Rate", *IEEE Trans. on Reliability*, vol. 48, no. 2, 1999b, pp 169–175
33. Pham, H., Software Reliability, *Wiley Encyclopedia of Electrical and Electronics Engineering*, John G. Webster (ed.), vol. 19, Wiley & Sons, 1999c, pp 565–578
34. Pham, H., and Zhang, X., "Software Release Policies With Gain in Reliability Justifying the Costs", *Annals of Software Engineering*, vol.8, 1999d, pp 147-166
35. H. Pham, *Software Reliability*, Springer-Verlag, 2000a
36. Pham, L. and Pham, H., "Software Reliability Models With Time-Dependent Hazard Function Based on Bayesian Approach", *IEEE Transactions on Systems, Man, and Cybernetics* - Part A: Systems and Humans, vol 30, no. 1, January 2000b
37. H. Pham and X. Teng, "A Software Reliability Growth Model for N-Version Programming Systems," IE Working Report, Rutgers University, 2001a
38. L. Pham and H. Pham, "A Bayesian Predictive Software Reliability Model with Pseudo- Failures", IEEE Trans. on Systems, Man & Cybernetics – Part A, Vol.31, No.3, May 2001b, pp 233–238
39. R.K. Scott, J.W. Gault, D.F. McAllister and J. Wiggs, "Experimental Validation of Six Fault-Tolerant Software Reliability Models," *Proc. of IEEE 14th Fault-Tolerant Computing Symposium*, pp 102–107, 1984
40. R.K. Scott, J.W. Gault and D.F. McAllister, "Fault-Tolerant Reliability Modeling," *IEEE Trans. on Software Engineering*, vol. SE-13, no. 5, pp

582–592, 1987
41. A. T. Tai, J. F. Meyer, and A. Aviziems, "Performability Enhancement of Fault-Tolerant Software," *IEEE Trans. on Reliability*, vol. 42 no. 2, pp 227–237, Jun 1993
42. Y. Tohma, H. Yamano, M. Ohba and R. Jacoby, "The Estimation of Parameters of the Hypergeometric Distribution and Its Application to the Software Reliability Growth Model", *IEEE Trans. on Software Engineering*, vol. 17, no. 5, 1991, pp 483–489
43. J. Voas, "Fault Tolerance," *IEEE Software*, July/August 2001, pp 54–57
44. A.A. Vouk, D.D. McAllister and Tai, K.C., "Identification of Correlated Failures of Fault-Tolerant Software Systems," *Proc. COMPSAC 85*, pp 437–444, 1985
45. S. Yamada, M. Ohba, and S. Osaki, "S-Shaped Reliability Growth Modeling For Software Error Detection", *IEEE Trans. on Reliability*, 12,1983, pp 475–484.
46. S. Yamada, and S. Osaki, "Software Reliability Growth Modeling: Models and Applications", *IEEE Trans. on Software Engineering*, 11, 1985, pp 1431–1437.
47. S. Yamada, K. Tokuno, and S. Osaki, "Imperfect Debugging Models With Fault Introduction Rate For Software Reliability Assessment", *International Journal of Systems Science*, vol. 23, no. 12, pp 56-69, 1992
48. M.C. Yang and A. Chao, "Reliability-Estimation and Stopping Rules For Software Testing Based on Repeated Appearances of Bugs", *IEEE Trans. on Reliability*, 22, no.2,1995, pp 315–326
49. S. R. Welke, B.W. Johnson, and J. H. Aylor, "Reliability Modeling of Hardware/Software Systems", IEEE Tran. on Reliability, Vol. 44, No. 3, 1995 September, pp 413–418
50. J. A. Whittaker and J. Voas, "Toward a More Reliable Theory of Software Reliability", *IEEE Computer*, December 2000, pp 36–42
51. A. Wood, "Predicting Software Reliability", *IEEE Computer*, 11, 1996, pp 69–77
52. X. Zhang and H. Pham, "A Software Cost Model With Error Removal Times and Risk Costs," *International Journal of Systems Science*, vol. 29, no. 4, 1998, pp 435–442
53. X. Zhang, Software Reliability and Cost Models With Environmental Factors, Ph.D. Dissertation, 1999a, Rutgers University, unpublished.
54. X. Zhang, and H. Pham, "A Software Cost Model With Warranty and Risk Costs," *IIE Trans. on Quality and Reliability Engineering*, vol. 30, pp 1135-1142, 1999b
55. X. Zhang and H. Pham, "An Analysis of Factors Affecting Software Reliability," *Journal of Systems and Software*, vol. 50, pp 43-56, 2000

CHAPTER 5

INSPECTION-AGE-REPLACEMENT POLICY AND SYSTEM AVAILABILITY

Jie Mi* and Hassan Zahedi†

Department of Statistics, Florida International University
University Park, Miami, FL 33199, U.S.A.
*E-mail: *mi@fiu.edu,*
E-mail: †zahedih@fiu.edu

Availability is an important measure of system performance. In practice, sometimes it is not feasible to monitor a system continuously due to technical difficulty or lack of resources. Hence, in this case the failure of the system can be detected only by inspection. The periodic inspection is considered in this article. One preventive maintenance policy is to replace the system when it reaches a predetermined age. Combining inspection policy with age replacement policy, we study the average long-run system availability. The lower/upper bound of age policy is obtained for the system that has a bathtub shaped failure rate function. The long-run average system availabilities associated with different lifetime distribution are compared via failure rate. The long-run average gain function is derived, and the associated optimal age policy is studied for a cost structure. At the end, the estimation problem of optimal age policy is discussed.

1. Introduction

Let $\mathcal{S}$ be a system that can be in one of two states, namely 'up' or 'down'. By 'up' we mean that $\mathcal{S}$ is still working and by 'down' we mean that $\mathcal{S}$ is not working, i.e. failed. Assume that a new system $\mathcal{S}$ starts working at time $t = 0$. In the literature, it is common to denote the elapsed time from zero until the instant at which $\mathcal{S}$ failed by U_1, and the elapsed time from then until $\mathcal{S}$ is up again by D_1. The random variable D_1 can be interpreted as the time needed for replacement or complete repair. Here, replacement means replacing the failed system by a statistically independent and identical one, and complete repair means that the failed system will be repaired as good as new, that is the lifetime of the repaired system is iid as the original one. The interval $[0, U_1 + D_1)$ is called cycle 1, and at that point, cycle 2 starts and the sequence is repeated so on and so forth. In general, the up time and down time in the jth cycle are denoted as U_j and D_j, respectively, for $j \geq 1$.

For the system described in the above, many practical problems require finding the probability, which is defined as the (instant) availability of $\mathcal{S}$ at t, that $\mathcal{S}$ is up at time t. More details can be found in Barlow and Proschan[3]. Another way to measure how often the given system is available is to use the ratio of the total up time of $\mathcal{S}$ in $[0, t]$ to the length of the time interval t. Obviously, the greater this ratio is, the better the system's performance is as far as its availability is concerned. Because of the importance of system availability, Bergman[5] mentioned that recent buyers require the vendors to guarantee the availability performance.

In the aforementioned models it is assumed that the failure of $\mathcal{S}$ is self-announcing. However, in some situations the failure of a system can be detected only by inspection. Such is the case in various protective systems such as fire detector, pressure detector, circuit breakers, protective relays, and safety valves. In these cases it is also assumed that the system found to be failed at an inspection is replaced immediately by an iid system (or completely repaired), and the process repeats. The study of inspection-replacement policy can be dated back to Barlow, Hunter and Proschan[1], Barlow and Proschan[2], and Barlow and Proschan[3]. Valdez-Flores and Feldman[10] provides an excellent references for research on preventive maintenance policy between 1975 and 1987. Recent research on this subject includes Wortman and Klutke[11], Wortman, Klutke, and Ayhan[12], Yeh[14], and Klutke, Wortman and Ayhan[4].

In the present article we assume the following:

(a) The failure of the system can be detected only by inspection;
(b) The inspection is scheduled at time $k\tau, k \geq 1$, where $\tau > 0$ is a constant, so-called inspection policy;
(c) The system found to be failed at an inspection will be replaced by an i.i.d. one immediately;
(d) The times for implementing replacement and inspection are negligible

when compared with the lifetime of the system;

(e) The system will operate without any intervention if it is found to be working at an inspection;

(f) There is an age policy characterized by η, where $\eta > 0$ is a fixed integer, or $\eta = \infty$, such that at time $\eta\tau$ the system is replaced regardlees of the status of the system at that time.

Combining inspection-replacement policy with age policy, we will study the long-run average availability of the system.
Sarkar and Sarkar[9] studied this model and, in their paper, it is referred to Model B. Yang and Klutke[13] also considered the same model. However, only the case $\eta = \infty$ is studied in their paper. Assuming $\eta = \infty$, the limiting average availability of the system was derived in both of the above two papers.

In Section 2, the expression of the long-run average system availability is derived. Age policy is studied in Section 3 in the case when the lifetime of system has a bathtub shaped failure rate function. The dependence of this availability on the lifetime distribution function of the system is discussed in Section 4. Intuitively, it is easy to see that when the inspection policy $\tau \to 0$ then the long-run average system availability will increases to 1. To avoid infinitesimal inspection policy, which is impractical, a cost structure is proposed in Section 5, and then the optimal inspection policy is studied there. Finally, in Section 6, we consider the problem of estimating the optimal inspection policy.

2. Model and Preliminary Results

Let a system have lifetime X with distribution function $F(t)$. Suppose that the mean life of the system is $\mu = E(X) < \infty$.

Denote the elapsed times between the successive replacement by C, C_k, $k \geq 1$, and call the time interval between two consecutive replacements as an replacement cycle. Let the number of inspections in the kth replacement cycle by N_k, $k \geq 1$, and $T \equiv \eta\tau$. From the assumptions (a)-(f) made in the previous section we see that the up time of the system in each replacement cycle, which is the time interval between two consecutive replacements is given by $U = \min\{X, T\}$ and $U_k = \min\{X_k, T\}$, $k \geq 1$. The average system availability in the interval $[0, t]$ is defined as

$$A_{av}(t) = \frac{\text{The total up time of system in [0,t]}}{t}$$

The replacement cycles form a renewal process, hence the Renewal Reward Theorem yields (see, e.g. Ross (1996))

$$\lim_{t \to \infty} A_{av}(t) = \frac{E(U)}{E(C)}$$

with probability one, and

$$A \equiv \lim_{t \to \infty} E(A_{av}(t)) = \frac{E(U)}{E(C)} \tag{1}$$

We call $A \equiv E(U)/E(C)$ as the long-run average availability of the system.

Since $E(U),\ E(C)$ depend on $F,\ \tau$ and η, so the long-run average system availability A obviously depends on $F,\ \tau$ and η too. In the following, $A(\tau)$ or $A(F)$, or $A(\eta)$ may be used when we want to emphasize the dependence of A on τ, or F, or η, respectively.

Theorem 1: The long-run average system availability A defined in (1) is given by

$$A = \frac{\int_0^{\eta\tau} \bar{F}(t)dt}{\tau \sum_{k=0}^{\eta-1} \bar{F}(k\tau)}, \tag{2}$$

where $\bar{F}(t) \equiv 1 - F(t)$ is the survival function of $F(t)$.

Proof See Mi[7].

In the case when the underlying lifetime distribution function $F(t)$ has a bathtub shaped failure rate function, lower/upper bound of optimal age policy can be obtained. Let's first introduce two definitions.

Definition 2: A failure rate function $r(t)$ defined on $(0, \infty)$ is said to have a bathtub shape if there exist $0 \le t_1 \le t_2 \le \infty$ such that

$$r(t) = \begin{cases} \text{strictly decreases in } 0 < t < t_1 \\ \text{is a constant for } t_1 \le t < t_2 \\ \text{strictly increases in } t \ge t_2. \end{cases}$$

In this case, t_1 and t_2 are called the first and second change point of $r(t)$, respectively.

Definition 3: Let $A(\eta)$ be the long-run average availability of a system. Age policy η^* is said to be optimal if $A(\eta^*) = \max_{\eta\geq 1} A(\eta)$.

The following result gives lower/upper bound of optimal age policy η^*. The proof of it can be found in Mi[7].

Theorem 4: Suppose that the lifetime distribution function $F(t)$ of a system has a bathtub shaped failure rate function $r(t)$ with change points $t_1 \leq t_2$. Then the optimal age policy η^* must satisfy $\eta^* \geq [t_2/\tau]$, where $[c]$ is the largest integer part of real number c. If further $r(0) < r(\infty)$, then $[t_2/t\tau] \leq \eta^* \leq [t_3/\tau] + 1$, where t_3 is uniquely determined by $r(t_3) = r(0)$.

At the end of this section we will compare the long-run average system availability associated with different lifetime distribution functions. To this end, an auxiliary result is needed.

Lemma 5: *Suppose that $a_i \geq 0$ and $b_i > 0$ for all $i \geq 1$. Then*

$$\inf_{k\geq 1} \frac{a_k}{b_k} \leq \frac{\sum\limits_{k=1}^{\infty} a_k}{\sum\limits_{k=1}^{\infty} b_k} \leq \sup_{k\geq 1} \frac{a_k}{b_k},$$

where the equality holds if and only if all the ratios a_k/b_k, $k \geq 1$ are equal.

Proof See Mi[6].

Theorem 6: Let lifetime distribution functions $F(t)$ and $G(t)$ have failure rate functions $r_F(t)$ and $r_G(t)$, respectively. If $r_1 \leq r_2$, where $r_1 \equiv \sup_{s\geq 0} r_G(s)$, and $r_2 \equiv \inf_{s\geq 0} r_F(s)$, then $A(F) \leq A(G)$.

Proof For any $k \geq 0$ we denote $a_k = \int_{k\tau}^{(k+1)\tau} \bar{F}(t)dt$, and $b_k = \bar{F}(k\tau)$. Then

$$\begin{aligned} \frac{a_k}{b_k} &= \int\limits_{k\tau}^{(k+1)\tau} e^{-\int_{k\tau}^{t} r_F(u)du} dt \\ &= \int_0^{\tau} e^{-\int_0^x r_F(k\tau+s)ds} dx \end{aligned}$$

$$\leq \int_0^\tau e^{-\int_0^x r_2 ds} dx = \int_0^\tau e^{-r_2 x} dx, \ \forall k \geq 0. \tag{3}$$

This inequality and Lemma 5 imply

$$A(F) = \frac{\int_0^{\eta\tau} \bar{F}(t)dt}{\tau \sum\limits_{k=0}^{\eta-1} \bar{F}(k\tau)}$$

$$= \frac{1}{\tau} \cdot \frac{\sum\limits_{k=0}^{\eta-1} \int\limits_{k\tau}^{(k+1)\tau} \bar{F}(t)dt}{\sum\limits_{k=0}^{\eta-1} \bar{F}(k\tau)}$$

$$\leq \frac{1}{\tau} \cdot \int\limits_0^\tau e^{-r_2 x} dx. \tag{4}$$

Similar to the inequality (3), it can be shown that

$$A(G) = \frac{\int\limits_0^{\eta\tau} \bar{G}(t)dt}{\sum\limits_{k=0}^{\eta-1} \bar{G}(k\tau)}$$

$$\geq \frac{1}{\tau} \cdot \int\limits_0^\tau e^{-r_1 x} dx. \tag{5}$$

From (4) and (5) we see that $A(F) \leq A(G)$ since $r_1 \leq r_2$.

3. Optimal Inspection Policy

Some simple properties of the long-run average system availability defined in (1) are given in the following theorem.

Theorem 7: Let $A(\tau)$ be the long-run average system availability defined as in (1). Suppose that $F(t)$ has failure rate function $r(t)$ and $\mu = E(X) < \infty$, then

(i) $\lim\limits_{\tau \to \infty} A(\tau) = 0$
(ii) $\lim\limits_{\tau \to 0+} A(\tau) = 1$

Proof We define

$$B(\tau) \equiv \tau \sum_{k=0}^{\eta-1} \bar{F}(k\tau). \tag{6}$$

It is easy to see that

$$\lim_{\tau\to\infty} B(\tau) \geq \lim_{\tau\to\infty} \tau\bar{F}(0) = \infty \tag{7}$$

and

$$\lim_{\tau\to\infty} \int_0^{\eta\tau} \bar{F}(t)dt = \int_0^{\infty} \bar{F}(t)dt = \mu < \infty.$$

Hence $\lim_{\tau\to\infty} A(\tau) = 0$ holds no matter whether $\eta < \infty$ or $\eta = \infty$, and consequently (i) follows.

In order to show (ii) we need to consider $\eta < \infty$ and $\eta = \infty$ separately.

First, let's consider the case of $\eta = \infty$. In this case

$$A(\tau) = \frac{\int_0^\infty \bar{F}(t)dt}{\tau \sum_{k=1}^{\infty} \bar{F}(k\tau)} = \frac{\mu}{B(\tau)}. \tag{8}$$

Note that

$$\begin{aligned} B(\tau) &= \sum_{k=0}^{\infty} \int_{k\tau}^{(k+1)\tau} \bar{F}(k\tau)dt \\ &\geq \sum_{k=0}^{\infty} \int_{k\tau}^{(k+1)\tau} \bar{F}(t)dt = \mu, \ \forall\tau > 0. \end{aligned} \tag{9}$$

On the other hand,

$$\begin{aligned} B(\tau) &= \sum_{k=0}^{\infty} \tau\bar{F}(k\tau) \\ &= \tau + \sum_{k=1}^{\infty} \int_{(k-1)\tau}^{k\tau} \bar{F}(k\tau)dt \end{aligned}$$

$$< \tau + \sum_{k=1}^{\infty} \int\limits_{(k-1)\tau}^{k\tau} \bar{F}(t)dt = \tau + \mu$$

and thus

$$\limsup_{\tau \to 0+} B(\tau) \leq \mu. \tag{10}$$

From (9) and (10) we obtain $\lim_{\tau \to 0+} B(\tau) = \mu$ and therefore $\lim_{\tau \to 0+} A(\tau) = 1$ by equation (8).

Now, we consider the case of $\eta < \infty$. Let a_k and b_k be the same as defined in the proof of Theorem 6. Clearly, we have

$$A(\tau) = \frac{\int_0^{\eta\tau} \bar{F}(t)dt}{\tau \sum_{k=0}^{\eta-1} \bar{F}(k\tau)} = \frac{1}{\tau} \cdot \frac{\sum_{k=0}^{\eta-1} a_k}{\sum_{k=0}^{\eta-1} b_k}$$

$$\geq \frac{1}{\tau} \inf_{0 \leq k \leq \eta-1} \frac{a_k}{b_k}. \tag{11}$$

The ratio a_k/b_k can be written as

$$\frac{a_k}{b_k} = \int\limits_{k\tau}^{(k+1)\tau} e^{-\int_{k\tau}^{t} r(s)ds} dt.$$

It then follows that

$$\frac{1}{\tau}\frac{a_k}{b_k} \geq \frac{1}{\tau} \int\limits_{k\tau}^{(k+1)\tau} e^{-\int_0^t r(s)ds} dt$$

$$= \frac{1}{\tau} \int\limits_{k\tau}^{(k+1)\tau} \bar{F}(t)dt$$

$$\geq \frac{1}{\tau} \int\limits_{k\tau}^{(k+1)\tau} \bar{F}\left((k+1)\tau\right) dt$$

$$= \bar{F}\left((k+1)\tau\right), \ \forall\, 0 \leq k \leq \eta - 1. \tag{12}$$

The inequality (11) and (12) yield

$$\begin{aligned} A(\tau) &\geq \inf_{0\leq k\leq \eta-1} \frac{1}{\tau}\cdot\frac{a_k}{b_k} \\ &\geq \inf_{0\leq k\leq \eta-1} \bar{F}\left((k+1)\tau\right) \\ &= \bar{F}(\eta\tau). \end{aligned} \tag{13}$$

Taking limit as $\tau \to 0+$, we obtain

$$\liminf_{\tau\to 0+} A(\tau) \geq 1.$$

However, it is obvious that $\limsup_{\tau\to 0+} A(\tau) \leq 1$, and therefore (ii) is proved.

Example 8: Let $\eta = \infty$ and $\bar{F}(t) = e^{-\lambda t}$, i.e., the lifetime of the system follows exponential distribution with mean $\mu = 1/\lambda$. The quantity $B(\tau)$ in (6) is then given by

$$B(\tau) = \tau\sum_{k=0}^{\infty}\bar{F}(k\tau) = \tau\sum_{k=0}^{\infty}e^{-\lambda k\tau} = \frac{\tau}{1-e^{-\lambda\tau}}.$$

Notice that

$$B'(\tau) = \frac{1-(1+\lambda\tau)e^{-\lambda\tau}}{(1-e^{-\lambda\tau})^2} > 0,\ \forall\ \tau > 0$$

since $1+\lambda\tau < e^{\lambda\tau}$, $\forall\tau > 0$. This means that $B(\tau)$ strictly increases in $\tau > 0$. Consequently, $A(\tau) = \mu/B(\tau)$ strictly decreases in $\tau > 0$.

The result of Theorem 7 validates our intuitive thinking that the long-run average system availability can be as close to one as possible by reducing $\tau > 0$. However, too frequent inspection is very costly, so certain cost structure will be necessary under some circumstances to eliminate the choice of infinitesimal inspection policy.

Theorem 9: Suppose that the cost per inspection is c_0, and proportional to up time of the system in any interval $[0,t]$ a profit is obtained with proportional constant c_1, where $c_0 \geq 0$ and $c_1 > 0$. Then the long-run average gain function is

$$G(\tau,\eta,F) = c_1 A(\tau,\eta,F) - \frac{c_0}{\tau}, \tag{14}$$

where $A(\tau,\eta,F)$ is defined in (1).

Proof In time interval $[0, t]$, the gain is given by

$$\begin{aligned} &c_1 \cdot \ \{\text{ The up time of the system in } [0,t]\} \\ -&c_0 \cdot \ \{\text{ The number of inspections in } [0,t]\}. \end{aligned}$$

and so the average gain in $[0, t]$ is

$$G_{av}(t) \equiv c_1 A_{av}(t) - c_0 \frac{\text{The number of inspection in } [0,t]}{t}, \tag{15}$$

where $A_{av}(t)$ is defined at the beginning of Section 2. Applying the Renewal Reward Theorem to $G_{av}(t)$, we obtain

$$\lim_{t \to \infty} G_{av}(t) = c_1 A(\tau, \eta, F) - \frac{c_0 E(N)}{\tau E(N)} = c_1 A - \frac{c_0}{\tau} \tag{16}$$

since the number of inspections in each replacement cycle is N, and the mean length of each replacement cycle is $\tau E(N)$. Therefore, the long-run average gain is given by (14).

To study the behavior of the gain function G defined in (14) we need the following result.

Lemma 10: *Suppose that the lifetime X of a system has distribution function $F(t)$ and failure rate function $r(t)$. If $r^* = \sup_{s>0} r(s) < \infty$, and $E(X^2) < \infty$, then*

$$\frac{d}{d\tau}\left(\sum_{k=1}^{\infty} \bar{F}(k\tau)\right) = \sum_{k=0}^{\infty} \frac{d}{d\tau} \bar{F}(k\tau) = -\sum_{k=0}^{\infty} k f(k\tau).$$

Proof A sufficient condition for exchanging d/dt with $\sum$ is the uniform convergence of $\sum_{k=0}^{\infty} k f(k\tau)$ on any interval $[a, b]$ with $0 < a < b < \infty$.

The uniform convergence of $\sum_{k=0}^{\infty} k f(k\tau)$ is equivalent to the uniform convergence of $\sum_{k=0}^{\infty} k\tau^2 f(k\tau)$ on $[a, b]$.

We have

$$\left| \int_{n\tau}^{\infty} t f(t) dt - \sum_{k=n}^{\infty} k\tau^2 f(k\tau) \right|$$

$$= \left| \sum_{k=n}^{\infty} \int_{k\tau}^{(k+1)\tau} (tf(t) - k\tau f(k\tau))\, dt \right|$$

$$\leq \sum_{k=n}^{\infty} \int_{k\tau}^{(k+1)\tau} |t - k\tau|\, f(t)dt + \sum_{k=n}^{\infty} k\tau \int_{k\tau}^{(k+1)\tau} |f(t) - f(k\tau)|\, dt$$

$$\leq \sum_{k=n}^{\infty} \tau \int_{k\tau}^{(k+1)\tau} f(t)dt + \sum_{k=n}^{\infty} k\tau \int_{k\tau}^{(k+1)\tau} |f(t) - f(k\tau)|\, dt$$

$$= \tau \bar{F}(n\tau) + \sum_{k=n}^{\infty} k\tau \int_{k\tau}^{(k+1)\tau} |f(t) - f(k\tau)|\, dt.$$

Note that

$$\int_{k\tau}^{(k+1)\tau} |f(t) - f(k\tau)|\, dt \leq \int_{k\tau}^{(k+1)\tau} |r(t) - r(k\tau)|\, e^{-\int_0^t r(s)ds} dt$$

$$+ \int_{k\tau}^{(k+1)\tau} r(k\tau) e^{-\int_0^t r(s)ds} \left| 1 - e^{\int_{k\tau}^t r(s)ds} \right| dt$$

$$\leq r^* \int_{k\tau}^{(k+1)\tau} \bar{F}(t)dt + r^* e^{r^*\tau} \int_{k\tau}^{(k+1)\tau} \bar{F}(t)dt$$

$$= r^*(1 + e^{r^*\tau}) \int_{k\tau}^{(k+1)\tau} \bar{F}(t)dt.$$

Hence,

$$\sum_{k=n}^{\infty} k\tau \int_{k\tau}^{(k+1)\tau} |f(t) - f(k\tau)|\, dt \leq r^*(1 + e^{r^*\tau}) \sum_{k=n}^{\infty} k\tau \int_{k\tau}^{(k+1)\tau} \bar{F}(t)dt$$

$$\leq r^*(1 + e^{r^*\tau}) \int_{n\tau}^{\infty} t\bar{F}(t)dt.$$

Therefore,

$$\left|\int_{n\tau}^{\infty} tf(t)dt - \sum_{k=n}^{\infty} k\tau^2 f(k\tau)\right| \leq \tau\bar{F}(n\tau) + r^*(1+e^{r^*\tau}) \int_{n\tau}^{\infty} t\bar{F}(t)dt$$

For any given $\epsilon > 0$, there exists n_0 such that $\bar{F}(n_0 a) < \epsilon/3b$, $\int_{n_0 a}^{\infty} t\bar{F}(t)dt < \epsilon(r^*(1+e^{r^*b}))^{-1}/3$ and $\int_{n_0 a}^{\infty} tf(t)dt < \epsilon/3$ since $\int_0^{\infty} t\bar{F}(t)dt < \infty$ and $\int_0^{\infty} tf(t)dt < \infty$.
Thus, for any $\tau \in [a, b]$, $n \geq n_0$

$$\begin{aligned}\left|\int_{n\tau}^{\infty} tf(t)dt - \sum_{k=n}^{\infty} k\tau^2 f(k\tau)\right| &\leq b \cdot \bar{F}(n_0 a) + r^*(1+e^{r^*b}) \int_{n_0 a}^{\infty} t\bar{F}(t)dt \\ &\leq b \cdot \frac{\epsilon}{3b} + r^*(1+e^{r^*b})\frac{\epsilon}{3}\left[r^*(1+e^{r^*b})\right]^{-1} \\ &= \frac{2\epsilon}{3}, \ \forall n \geq n_0, \ \forall \tau \in [a,b].\end{aligned}$$

This yields

$$\begin{aligned}0 < \sum_{k=n}^{\infty} k\tau^2 f(k\tau) &\leq \int_{n\tau}^{\infty} tf(t)dt + \frac{2\epsilon}{3}, \ \forall\, n \geq n_0, \ \forall\, \tau \in [a,b] \\ &\leq \int_{n_0 a}^{\infty} tf(t)dt + \frac{2\epsilon}{3} < \epsilon, \ \forall\, n \geq n_0, \ \forall\, \tau \in [a,b]\end{aligned}$$

That is, the series $\sum_{k=0}^{\infty} k\tau^2 f(k\tau)$ uniformly converges for $\tau \in [a, b]$.

In the rest of this section and the next section we introduce the notation

$$\text{Arg}(\max_{\tau>0} G(\tau)) \equiv \{\tau^* > 0 : G(\tau^*) = \max_{\tau>0} G(\tau)\}. \tag{17}$$

Theorem 11: Suppose that $\alpha \equiv c_1/c_0$, $\alpha > er^*$, where $r^* = \sup_{s>0} r(s) < \infty$. Then $\text{Arg}(\max_{\tau>0}) \subset (0, \infty)$ and $G(\tau^*) > 0$, $\forall\, \tau^* \in \text{Arg}(\max_{\tau>0} G(\tau))$.

Proof From (14) we have

$$G(\tau) = c_1 \frac{\int_0^{\eta\tau} \bar{F}(t)dt}{\tau \sum_{k=0}^{\eta-1} \bar{F}(k\tau)} - \frac{c_0}{\tau}$$

$$= \frac{c_0 \sum_{k=0}^{\eta-1} \int_{k\tau}^{(k+1)\tau} [\alpha \bar{F}(t) - \bar{F}(k\tau)/\tau] dt}{\tau \sum_{k=0}^{\eta-1} \bar{F}(k\tau)}. \tag{18}$$

Note that for $k\tau \le t \le (k+1)\tau$ we have

$$\begin{aligned}
\alpha \bar{F}(t) - \frac{1}{\tau}\bar{F}(k\tau) &= \alpha e^{-\int_0^t r(s)ds} - \frac{1}{\tau} e^{-\int_0^{k\tau} r(s)ds} \\
&= e^{-\int_0^t r(s)ds} \left[\alpha - \frac{1}{\tau} e^{\int_{k\tau}^t r(s)ds}\right] \\
&> e^{-\int_0^t r(s)ds} \left[\alpha - \frac{1}{\tau} e^{r^*\tau}\right]
\end{aligned} \tag{19}$$

Consider the equation

$$\frac{1}{\tau} e^{r^*\tau} = \alpha. \tag{20}$$

Let $\varphi(\tau) = e^{r^*\tau}/\tau$ and $\psi(\tau) = \ln \varphi(\tau)$. It follows that $\psi(\tau) = r^*\tau - \ln \tau$ and $\psi'(\tau) = r^* - 1/\tau$.

Clearly, $\psi'(\tau) < 0$ for $0 < \tau < 1/r^*$, $\psi'(\tau) > 0$ for $\tau > 1/r^*$. This implies that $\varphi(\tau)$ strictly decreases in $0 < \tau < 1/r^*$, strictly increases in $\tau > 1/r^*$, and $\min_{\tau>0} \varphi(\tau) = \varphi(1/r^*) = er^*$. From the assumption α r^* we see the equation (20) has exactly two solutions τ_1 and τ_2 and they satisfy $0 < \tau_1 < 1/r^* < \tau_2 < \infty$.

For any $\tau_1 \le \tau \le \tau_2$, the function $\varphi(\tau)$ is smaller than α. This and (19) imply

$$\alpha\ \bar{F}(t) - \frac{1}{\tau}\bar{F}(k\tau) > 0,\ \forall\ k \ge 0,\ \forall\ k\tau \le t \le (k+1)\tau. \tag{21}$$

Thus, form (19) and (21) we obtain $G(\tau) > 0,\ \forall\ \tau_1 \le \tau \le \tau_2$.

By Theorem 7 it is easy to see that $G(0+) = -\infty$ and $G(\infty) = 0$. Therefore, $\max_{\tau>0} G(\tau) > 0$ and $\mathrm{Arg}(\max_{\tau>0} G(\tau)) \subset (0, \infty)$.

Theorem 12: Under the same conditions as in Lemma 10: $r^* < \infty$, $E(X^2) < \infty$, the gain function $G(\tau)$ strictly increases in $\tau \in (0, 1/\alpha)$, and $\mathrm{Arg}(\max_{\tau>0} G(\tau)) \subset (1/\alpha, \infty)$.

Proof From Lemma 10, the derivative of $G(\tau)$ with respect to τ can be obtained as

$$G'(\tau) = \frac{d}{d\tau}\left\{ c_1 \frac{\int_0^{\eta\tau} \bar{F}(t)dt}{\tau \sum\limits_{k=0}^{\eta-1} \bar{F}(k\tau)} - \frac{c_0}{\tau} \right\}$$

$$= c_1 \frac{\eta \bar{F}(\eta\tau)\cdot \tau \sum\limits_{k=0}^{\eta-1} \bar{F}(k\tau) - \int\limits_0^{\eta\tau} \bar{F}(t)dt \left[\sum\limits_{k=0}^{\eta-1} \bar{F}(k\tau) - \tau \sum\limits_{k=0}^{\eta-1} k f(k\tau) \right]}{\left(\tau \sum\limits_{k=0}^{\eta-1} \bar{F}(k\tau) \right)^2} + \frac{c_0}{\tau^2}$$

$$> \frac{1}{\tau^2}\left\{ c_0 - c_1 \frac{\int\limits_0^{\eta\tau} \bar{F}(t)dt}{\sum\limits_{k=0}^{\eta-1} \bar{F}(k\tau)} \right\}$$

$$= \frac{c_0}{\tau^2}\left\{ 1 - \alpha\, \tau \frac{\int\limits_0^{\eta\tau} \bar{F}(t)dt}{\tau \sum\limits_{k=0}^{\eta-1} \bar{F}(k\tau)} \right\}$$

$$> \frac{c_0}{\tau^2}\left\{ 1 - \alpha\, \tau \right\}.$$

Obviously, if $0, \tau \leq 1/\alpha$, then $1 - \alpha\, \tau \geq 0$ and hence $G'(\tau) > 0,\ \forall\, 0 < \tau \leq 1/\alpha$. This implies that $G(\tau)$ strictly increases in $\tau \in (0, 1/\alpha)$ and consequently $\mathrm{Arg}(\max\limits_{\tau>0} G(\tau)) \subset (1/\alpha,\ \infty)$.

Corollary 13: *If $r^* < \infty$, $E(X^2) < \infty$, and $\alpha > e\ r^*$, then $\tau_1 > 1/\alpha$, where τ_1 is the smaller root of equation (20).*

Proof Let $\varphi(\tau) = e^{r^*\tau}/\tau$ be the function defined in the proof of Theorem 11. The value of $\varphi(\tau)$ at $\tau = 1/\alpha$ is $\varphi(1/\alpha) = \alpha\ \tau^{r^*/\alpha} > \alpha$. Hence, from the proof of Theorem 11 we see that $1/\alpha$ must satisfy either $\tau_1 > 1/\alpha$ or $\tau_2 < 1/\alpha$. However, in the proof of Theorem 11 we have obtained $\tau_1 < 1/r^* < \tau_2$. Now from the assumption $\alpha > er^*$ it follows that $1/r^* > e/\alpha$. Therefore, $1/\alpha < e/\alpha < 1/r^* < \tau_2$, and consequently it must be true that $1/\alpha < \tau_1$.

As an application, let'us consider exponential distribution once again.

Example 14: Let the underlying distribution is exponential with mean $\mu = 1/\lambda$ and $c_0 > 0$. It this case it can be seen that the long-run average availability is given as

$$A(\tau) = \frac{1 - e^{-\lambda\tau}}{\lambda\tau}.$$

Hence the associate gain function is

$$G(\tau) = \frac{(c_1\mu - c_0) - c_1\mu e^{-\lambda\tau}}{\tau}.$$

It then follows that

$$G'(\tau) = \frac{c_1(\tau + \mu)e^{-\lambda\tau} - (c_1\mu - c_0)}{\tau^2}.$$

From this it is easy to see that if $c_1\mu - c_0 \leq 0$, then $G'(\tau) > 0, \forall\tau > 0$. That is, in this case $\tau^* = \infty$ which can be interpreated as no inspection is needed at all. Thus, let's assume $c_1\mu - c_0 > 0$ below. We define $\psi(\tau) \equiv c_1(\tau + \mu)e^{-\lambda\tau}$. It can be verified that $\psi(0) = c_1\mu > c_1\mu - c_0$, $\psi(\infty) = 0$, and $\psi'(\tau) = -c_1\lambda\tau e^{-\lambda\tau} < 0$. This shows that the function $\psi(\tau)$ strictly decreases in $\tau > 0$ and the equation $G'(\tau) = 0$, or $\psi(\tau) = c_1\mu - c_0$ has unique solution τ^* at which the gain function $G(\tau)$ is maximized.

In particular, suppose that $c_0 = 1, c_1 = 3$ and $\lambda = 1$. Then

$$G(\tau) = \frac{2 - 3e^{-\tau}}{\tau}.$$

Numerical calculation gives $\tau^* \approx 1.19$ and $G(\tau^*) \approx 0.91$.

4. Estimation of Optimal Inspection Policy

In the present section, we assume that there is a sequence of lifetime distributions $\{F_n, n \geq 1\}$ which satisfies the following conditions

(A) $F_n \to F$ in distribution
(B) $\mu_n \equiv \int_0^\infty \bar{F}_n(t)dt \to \mu \equiv \int_0^\infty \bar{F}(t)dt$.

It can easily be verified that for any $0 \leq a < b \leq \infty$, it is true that

$$\lim_{n\to\infty} \int_a^b \bar{F}_n(t)dt = \int_a^b \bar{F}(t)dt. \tag{22}$$

Lemma 15: *Suppose that* $0 < a_n \to a \in (0, \infty)$. *Then* $\sum_{k=0}^{\infty} \bar{F}_n(ka_n) \to \sum_{k=0}^{\infty} \bar{F}(ka)$.

Proof For any fixed $b \in (0, a)$, there exists N_1 such that

$$a_n > b \quad \forall\, n \geq N_1. \tag{23}$$

For any given $\epsilon \in (0, a)$, let positive integer L be large enough such that

$$\int_{(L-1)b}^{\infty} \bar{F}(t)dt < \epsilon. \tag{24}$$

This is possible since $\mu = \int_0^\infty \bar{F}(t)dt < \infty$. It is also easy to see that

$$\lim_{n\to\infty} \int_{(L-1)b}^{\infty} \bar{F}_n(t)dt = \int_{(L-1)b}^{\infty} \bar{F}(t)dt. \tag{25}$$

by the Assumptions (A) and (B). From (22) for the chosen $\epsilon > 0$, there exists N_2 such that

$$\int_{(L-1)b}^{\infty} \bar{F}_n(t)dt \leq \int_{(L-1)b}^{\infty} \bar{F}(t)dt + \epsilon,\ \forall\, n \geq N_2. \tag{26}$$

Let $N = \max\{N_1, N_2\}$. Then we have

$$\begin{aligned} b\sum_{k=L}^{\infty} \bar{F}_n(kb) &= \sum_{k=L}^{\infty} \int_{(k-1)b}^{kb} \bar{F}_n(kb)dt \\ &\leq \sum_{k=L}^{\infty} \int_{(k-1)b}^{kb} \bar{F}_n(t)dt \\ &\leq \int_{(L-1)b}^{\infty} \bar{F}(t)dt + \epsilon \quad \forall\, n \geq N \\ &\leq \epsilon + \epsilon = 2\epsilon \quad \forall\, n \geq N \end{aligned} \tag{27}$$

where the last two inequalities are from (26) and (24), respectively. It is easy to see that

$$\left| b\sum_{k=1}^{\infty} \bar{F}_n(ka_n) - b\sum_{k=1}^{\infty} \bar{F}(ka) \right|$$

$$\leq b\sum_{k=1}^{L-1}\left|\bar{F}_n(ka_n)-\bar{F}(ka)\right|+b\sum_{k=L}^{\infty}\bar{F}_n(ka_n)+b\sum_{k=L}^{\infty}\bar{F}(ka). \tag{28}$$

For any $n \geq N$, we have

$$b\sum_{k=L}^{\infty}\bar{F}_n(ka_n) \leq b\sum_{k=L}^{\infty}\bar{F}_n(kb) \leq 2\epsilon, \tag{29}$$

where the first inequality comes from (23), and the second one comes from (27). On the other hand, it holds that

$$\begin{aligned} b\sum_{k=L}^{\infty}\bar{F}(ka) \leq b\sum_{k=L}^{\infty}\bar{F}(kb) &\leq \sum_{k=L}^{\infty}\int_{(k-1)b}^{kb}\bar{F}(t)dt \\ &= \int_{(L-1)b}^{\infty}\bar{F}(t)dt \leq \epsilon \end{aligned} \tag{30}$$

by the fact that $0 < b < a$ and the inequality (24).

Now, from (28), (29), and (30), we obtain

$$b\left|\sum_{k=1}^{\infty}\bar{F}_n(ka_n)-\sum_{k=1}^{\infty}\bar{F}(ka)\right| \leq b\sum_{k=1}^{L-1}\left|\bar{F}_n(ka_n)-\bar{F}(ka)\right|+3\epsilon,$$

and thus

$$\limsup_{n\to\infty}\left|\sum_{k=1}^{\infty}\bar{F}_n(ka_n)-\sum_{k=1}^{\infty}\bar{F}(ka)\right| \leq \frac{3\epsilon}{b} \tag{31}$$

since $\bar{F}_n(t) \to \bar{F}(t),\ \forall\, t \geq 0$. The arbitrariness of $\epsilon > 0$ in (31) implies

$$\lim_{n\to\infty}\sum_{k=1}^{\infty}\bar{F}_n(ka_n) = \sum_{k=1}^{\infty}\bar{F}(ka).$$

Lemma 16: *Suppose that* $0 < a_n \to \infty$ *an* $n \to \infty$. *Then*

$$\lim_{n\to\infty}\int_0^{\eta a_n}\bar{F}_n(t)dt = \mu = \int_0^{\infty}\bar{F}(t)dt.$$

Proof The case of $\eta = \infty$ is clearly true. In the following let $\eta < \infty$. For any given $\epsilon > 0$, there exists $0 < B < 0$ such that

$$\int_B^\infty \bar{F}(t)dt < \epsilon \tag{32}$$

since $\mu = \int_0^\infty \bar{F}(t)dt < \infty$. For this chosen $0 < B < \infty$, there exists N such that

$$\eta a_n > B, \ \ \forall\, n \geq N$$

which is possible since $\eta a_n \to \infty$ as $n \to \infty$. It is obvious that

$$\begin{aligned}\left|\int_0^{\eta a_n} \bar{F}_n(t)dt - \mu\right| &\leq \left|\int_0^\infty \bar{F}_n(t)dt - \int_0^\infty \bar{F}(t)dt\right| + \int_{\eta a_n}^\infty \bar{F}_n(t)dt \\ &\leq \left|\int_0^\infty \bar{F}_n(t)dt - \int_0^\infty \bar{F}(t)dt\right| + \int_B^\infty \bar{F}_n(t)dt, \ \ \forall\, n \geq N\end{aligned}$$

Therefore,

$$\begin{aligned}\limsup_{n\to\infty}\left|\int_0^{\eta a_n} \bar{F}_n(t)dt - \mu\right| &\leq \lim_{n\to\infty}\left|\int_0^\infty \bar{F}_n(t)dt - \int_0^\infty \bar{F}(t)dt\right| \\ &\quad + \lim_{n\to\infty}\int_B^\infty \bar{F}_n(t)dt = \int_B^\infty \bar{F}(t)dt < \epsilon,\end{aligned}$$

where the last inequality follows from (32). This shows that

$$\lim_{n\to\infty}\int_0^{\eta a_n} \bar{F}_n(t)dt = \int_0^\infty \bar{F}(t)dt = \mu.$$

Theorem 17: Suppose that the conditions (A) and (B) hold, and $\alpha > er^*$. Let μ_n^* be such that $G_n(\tau_n^*) = \max_{\tau>0} G_n(\tau)$. Then, the limit, denoted as τ_0, of any converging subsequence of $\{\tau_n^*, n \geq 1\}$ must satisfy $0 < \tau_0 < \infty$ and $G(\tau_0) = \max_{\tau>0} G(\tau)$.

Proof For the simplicity of notation without loss of generality we may assume $\{\tau_n^*,\ n \geq 1\}$ itself converges with $\lim_{n\to\infty} \tau_n^* = \tau_0$.

We first show that $\tau_0 \neq \infty$. Suppose the contrary, then we would have

$$0 \leq \frac{\int\limits_0^{\eta\tau_n^*} \bar{F}_n(t)dt}{\tau_n^* \sum\limits_{k=0}^{\eta-1} \bar{F}_n(k\tau_n^*)} \leq \frac{\int\limits_0^{\eta\tau_n^*} \bar{F}_n(t)dt}{\tau_n^*} \to 0$$

as $n \to \infty$ since $\lim\limits_{n\to\infty} \int_0^{\eta\tau_n^*} \bar{F}_n(t)dt = \mu < \infty$ and $\lim\limits_{n\to\infty} \tau_n^* = \infty$. It thus implies

$$\lim_{n\to\infty} G_n(\tau_n^*) = \lim_{n\to\infty} \left\{ c_1 \frac{\int_0^{\eta\tau_n^*} \bar{F}_n(t)dt}{\tau_n^* \sum\limits_{k=0}^{\eta-1} \bar{F}_n(k\tau_n^*)} - \frac{c_0}{\tau_n^*} \right\} = 0. \tag{33}$$

On the other hand, for any $\tau > 0$ we see that

$$\begin{aligned} \lim_{n\to\infty} G_n(\tau) &= \lim_{n\to\infty} \left\{ c_1 \frac{\int_0^{\eta\tau} \bar{F}_n(t)dt}{\tau \sum\limits_{k=0}^{\eta-1} \bar{F}_n(k\tau)} - \frac{c_0}{\tau} \right\} \\ &= c_1 \frac{\int_0^{\eta\tau} \bar{F}(t)dt}{\tau \sum\limits_{k=0}^{\eta-1} \bar{F}(k\tau)} - \frac{c_0}{\tau} \\ &= G(\tau) \end{aligned} \tag{34}$$

where the second inequality is true because it is trivial for the case of $\eta < \infty$, and it follows from Lemma 14 for the case $\eta = \infty$.

Now, by the definition of τ_n^* we have

$$G_n(\tau_n^*) \geq G_n(\tau), \quad \forall\, n \geq 1,\ \forall\, \tau > 0. \tag{35}$$

Letting $n \to \infty$ in (35), we obtain

$$0 \geq G(\tau), \quad \forall\, \tau > 0 \tag{36}$$

by (33) and (34). However, the result of Theorem 11 shows that $\max\limits_{\tau>o} G(\tau) > 0$, so (36) is impossible. This contradiction demonstrates that $\tau_0 \neq \infty$.

We also want to show that $\tau_0 \neq 0$. Suppose the contrary. It is easy to see that

$$\frac{\int\limits_0^{\eta\tau_n^*} \bar{F}_n(t)dt}{\tau_n^* \sum\limits_{k=0}^{\eta-1} \bar{F}_n(k\tau_n^*)} \leq 1$$

Thus,

$$c_1 - \frac{c_0}{\tau_n^*} \geq G_n(\tau_n^*) \geq G_n(\tau), \quad \forall\, n \geq 1, \ \forall\, \tau > 0 \tag{37}$$

where the last inequality is true because of the definition of τ_n^*. Letting $n \to \infty$ in (37), we observe that

$$-\infty \geq G(\tau), \quad \forall\, \tau > 0$$

this contradiction shows that $\tau_0 \neq 0$.

In the above we have shown that $\tau_0 \equiv \lim_{n\to\infty} \tau_n^* \neq 0, \infty$. Now, once again by the definition of τ_n^* we have

$$\begin{aligned} G_n(\tau_n^*) &= c_1 \frac{\int_0^{\eta\tau_n^*} \bar{F}_n(t)dt}{\tau_n^* \sum\limits_{k=0}^{\eta-1} \bar{F}_n(k\tau_n^*)} - \frac{c_0}{\tau_n^*} \\ &\geq c_1 \frac{\int_0^{\eta\tau} \bar{F}_n(t)dt}{\tau \sum\limits_{k=0}^{\eta-1} \bar{F}_n(k\tau)} - \frac{c_0}{\tau} \\ &= G_n(\tau), \quad \forall\, \tau > 0. \end{aligned} \tag{38}$$

Taking $n \to \infty$ in (38), we obtain

$$\begin{aligned} \lim_{n\to\infty} G_n(\tau_n^*) &= c_1 \frac{\int_0^{\eta\tau_0} \bar{F}(t)dt}{\tau_0 \sum\limits_{k=0}^{\eta-1} \bar{F}(k\tau_0)} - \frac{c_0}{\tau_0} \\ &\geq c_1 \frac{\int_0^{\eta\tau} \bar{F}(t)dt}{\tau \sum\limits_{k=0}^{\eta-1} \bar{F}(k\tau)} - \frac{c_0}{\tau} \\ &= G(\tau), \quad \forall\, \tau > 0 \end{aligned}$$

That is

$$G(\tau_0) \geq G(\tau), \quad \forall\, \tau > 0.$$

Therefore, the desired result follows.

Let $X_1, X_2, \ldots, X_n$ be a sample from the population of the lifetime X of the system. In particular, we let F_n be the empirical CDF based on $X_1, \ldots X_n$. Then it is well known that $\{F_n, n \geq 1\}$ converges to F with probability one. Moreover, $\int_0^\infty \bar{F}_n(t)dt = \sum_{i=1}^{n} X_i/n$ converges to $\mu = \int_0^\infty \bar{F}(t)dt$ with probability one by the Strong Law of Large Numbers. Thus, the conditions (A) and (B) are satisfied except a set of probability zero. Hence the following result follows from Theorem 16.

Corollary 18: *Let F_n be the empirical CDF based on sample $X_1, X_2, \ldots, X_n$, and τ_n^* be such that $G_n(\tau_n^*) = \max_{\tau>0} G_n(\tau)$. Suppose that $\alpha > er^*$. Then, except a set of probability zero, the limit of any converging subsequence of $\{\tau_n^*, n \geq 1\}$, denoted as τ_0, must satisfy $0 < \tau_0 < \infty$ and $G(\tau_0) = \max_{\tau>0} G(\tau)$.*

5. Possible Model Extension

The following problems could be consider for further studies.

(1) In which way the behavior of the functions $A(\tau)$ and $G(\tau)$ depends on the classification of the underlying lifetime distribuiton function? F

(2) How will $A(\tau)$ and $G(\tau)$ behave when the system is replaced by an iid one at the age $\eta\tau$ and by an independet but not identical one at the time when the inspection finds the system failed before the age limit $\eta\tau$?

(3) How about the results in this paper when minimal repare is applied instead of replacement (or equivalently, complete repare)?

(4) How about general inspection policy $\tau_1 < \tau_2 < \tau_3 \cdots$ instead of equidistance inspection policy $j\tau, j \geq 1$ adopted in the present article?

References

1. R.E. Barlow, L. Hunter and F. Proschan *Journal of the Society for Industrial and Applied Mathematics*, **11**, 1078 (1963).
2. R.E. Barlow and F. Proschan *Mathematical Theory of Reliability*, SIAM series in applied mathematics, (John Wiley & Sons 1965).

3. R.E. Barlow and F. Proschan *Statistical Theory of Reliability and Life Testing*, (Holt, Rinehart & Winston, New York 1975).
4. G.-A. Klutke, M.A. Wortman and H. Ayhan The Availability of Inspected Systems Subject to Random Deterioration. *Probability in the Engineering and Informational Sciences*, **10**, 109 (1996).
5. B. Bergman On Reliability and Its Applications, *Scan. J. Statist*, **12**, 1 (1985).
6. J. Mi Age-Replacement Policy and Optimal Work Size (2001a, submitted).
7. J. Mi On Bounds to Some Optimal Policies in Reliability (2001b, submitted).
8. S. Ross *Stochastic Processes*, (New York: Wiley, 1996).
9. J. Sarkar and S. Sarkar Availability of a Periodically Inspected System under Perfect Repair. *Journal of Statistical Planning and Inference*, **91**, 77-90.
10. C. Valdez-Flores and R. Feldman (1989). Survey of Preventive Maintenance Models for Stochastically Deteriorating Single-Unit Systems. *Naval Research Logistics*, **36**, 419 (2000).
11. M.A. Wortman and G.-A. Klutke On Maintained Systems Operating in a Random Environment. *Journal of Applied Probability*, **31**, 589 (1994).
12. M.A. Wortman, G.-A. Klutke and H. Ayhan A Maintenance Strategy for Systems Subjected to Deterioration Governed by Random Shocks. *IEEE Transactions on Reliability*, **43**, 439 (1994).
13. Y. Yan and G.-A. Klutke Improved Inspection Schemes for Deteriorating Equipment. *Probability in the Engineering and Informational Sciences*, **14**, 445 (2000).
14. L. Yeh An Optimal Inspection-Repair-Replacement Policy for Standby Systems. *Journal of Applied Probability*, **32**, 212 (1995).

CHAPTER 6

BEHAVIOR OF FAILURE RATES OF MIXTURES AND SYSTEMS

Henry W. Block*, Yulin Li† and Thomas H. Savits‡

Department of Statistics, University of Pittsburgh
Pittsburgh, PA 15260, U.S.A.
E-mail: **hwb@stat.pitt.edu,*
E-mail: *†yulst12@pitt.edu,*
E-mail: *‡ths@stat.pitt.edu*

In this paper we review results concerning the behavior of the failure rate of mixtures and systems as a function of failure rates of the components. Our general results on mixtures give that the limiting failure rate of a mixture behaves as the strongest limiting failure rate of the components. For example, under mild conditions, the limiting failure rate of a mixture is the same as the limiting failure rate of its strongest component. Also, if the failure rate of the stromgest component is increasing, so is the failure rate of the mixture. Similar results hold for initial behavior and for systems.

1. Introduction

Mixtures arise from heterogeneous populations. A typical case is where a population consists of two subpopulations (which we sometimes refer to as components of the mixture): the strong subpopulation which has a long lifetime and the weak subpopulation which has a short lifetime. Mixtures are important for burn-in (see Block and Savits, 1997). Our first general result for mixtures, given in Theorem 1, is that if the component failure rates converge uniformly (condition (i)), then the limiting failure rate of the mixture is the limiting failure rate of the strongest component. If some of these component failure rates go to infinity, then a growth condition (condition (ii)) is needed. In Section 2.1, we discuss this result and give various examples to show that the conditions are needed.

Another general result (see Block and Joe, 1997) is that under certain regularity conditions, the ultimate monotonicity of a mixture is the same as the ultimate monotonicity of its strongest component. For example, if one component has failure rate which is ultimately smaller than the failure rates of all other components and is increasing, then the failure rate of the mixture is ultimately increasing. That is, if the strongest component of a mixture eventually wears out, so does the mixture. We give a new version of this result in Theorem 4.

In Section 2.2 we discuss the initial, as opposed to the final, behavior of the failure rate of a mixture. The initial monotonicity is also considered.

For coherent systems (see Barlow and Proschan, 1975) having independent components, similar results to the results for mixtures hold. In Section 3.1, we discuss the asymptotic behavior and monotonicity of the system failure rate. Our major result is that the limiting failure rate of the system is the minimum over all min path sets of the sum of the limiting failure rates for components in each path set. A corresponding result for the direction of the eventual monotonicity of the system failure rate also holds. In Section 3.2, we discuss the initial behavior of the system failure rate.

Unless a specific reference is given, verification of details for the examples and theorems will appear in a forthcoming paper (Block, Li and Savits, 2001).

2. Failure Rate Behavior of Mixtures

In order to discuss mixtures, we use the following notation. Let $f(t,w)$, $\overline{F}(t,w)$ and $r(t,w)$ be the density, survival function and failure (or hazard) rate of a subpopulation indexed by $w \in W$. If P is a probability on W, then the failure rate of the mixture is given by

$$r(t) = \frac{\int_W f(t,w)P(dw)}{\int_W \overline{F}(t,w)P(dw)} .$$

We are interested in the asymptotic and initial behavior of $r(t)$.

2.1. Asymptotic Limit of $r(t)$

In Clarotti and Spizzichino (1990), it is stated that for the exponential case, i.e., $r(t,w) = w$, $r(t) \to \alpha = \inf\{w : w \in W\}$ as $t \to \infty$. This result gives that the failure rate of the mixture approaches that of the strongest subpopulation. This is intuitively explained since the weaker subpopulations die out first leaving only the strongest subpopulation. Their result was extended by Block, Mi and Savits (1993) to the general case. A slightly streamlined version of Theorem 4.1 of that paper is given below.

Theorem 1: Let $\{r(t,w)\}$ be a family of failure rates indexed by $w \in W$.

(i) Assume $r(t,w)$ converges to $a(w) \in [0,\infty]$ uniformly on $S \subset W$ as $t \to \infty$, where $P(S) = 1$.
(ii) Let $I = \{w \in S : a(w) = \infty\}$. If $0 < P(I) < 1$, assume there exists $L, T \geq 0$ such that $r(t,w) \leq e^{Lt}$ for all $w \in I, t \geq T$.

Then $\lim_{t\to\infty} r(t) = \alpha$, where $\alpha = \inf\{0 \leq \lambda < \infty : P(w \in S : a(w) \leq \lambda) > 0\}$.

Remarks:

1. If $P(I) = 0$ or 1, condition (ii) is not needed.
2. The quantity α in the last line of the theorem is called the essential infimum and is essentially $\inf\{a(w) : w \in S\}$ except for inessential $w \in S$.
3. We call the condition (ii) a growth condition.

We now give some examples to show that the result Theorem 1 does not hold without conditions (i) and (ii). In particular, we show that in some sense, the "growth condition" (ii) is the best possible.

Example 1: (Gupta and Gupta, 1996). Let $r_0(t) = \gamma t^{\gamma-1}, \gamma > 1$ be an IFR Weibull failure rate, and set $r(t,w) = wr_0(t)$, for $w > 0$. In the survival analysis literature, this is called a frailty model and $r_0(t)$ is called the baseline failure rate. Consider a gamma mixture, i.e.,

$$P(dw) = \frac{1}{\beta^\lambda \Gamma(\lambda)} w^{\lambda-1} e^{-w/\beta} dw \; .$$

It then is easy to show that $r(t) = \frac{\lambda\beta\gamma t^{\gamma-1}}{1+\beta t^\gamma}$. Consequently, $r(t) \to 0$ even though $r(t,w) \to \infty$ for all $w > 0$. Here both conditions (i) and (ii) fail.

Example 2: As above, we consider the specific Weibull baseline IFR failure rate $r_0(t) = 2t$ and set $r(t,w) = wr_0(t)$. This time, consider a uniform mixing distribution P on $(0,1)$. A straightforward calculation gives that

$$r(t) = \frac{2[1-(t^2+1)e^{-t^2}]}{t[1-e^{-t^2}]}.$$

Again $r(t,w) \to \infty$ for all $w \in (0,1)$, but $r(t) \to 0$. Here condition (ii) is satisfied, but not (i).

Remark: In the above example, if P is any probability on $(0,\infty)$ whose support is bounded away from zero and infinity, then conditions (i) and (ii) are satisfied and so $r(t) \to \infty$.

Example 3: Let $y_0 = 1$ and $y_n = \exp(2ny_{n-1})$ for $n \geq 1$. We introduce the notation $n' = n - (2y_{n-1})^{-1} + (2y_n)^{-1}$. Let $g(x) = y_{n-1}$ for $n-1 \leq x < n$ where $n \geq 1$. We make this function continuous by redefining it, on the interval (n', n) to be

$$\frac{1/2}{n + 1/(2y_n) - x}.$$

It is easy to check that this function, which we call h, increases to infinity but does not satisfy the growth condition. Now consider the distribution with density

$$f_1(t) = h(t) \exp\left(-\int_0^t h(x)dx\right).$$

If we mix this equally (i.e., $P(w=1) = P(w=2) = 0.5$) with the exponential distribution with mean 1, we find that the failure rate of the mixture $r(t)$ satisfies $r(n) \to \infty$ as $n \to \infty$, while $r(n') \to 1$ as $n \to \infty$. Consequently, the limit of the mixture failure rate does not exist. In this case, condition (i) is obviously satisfied since we are dealing with a finite mixture. As noted before, however, the growth condition (ii) fails.

We now give a result which shows that the growth condition is the best possible. Condition (ii) can be rewritten as

$$\limsup_{t\to\infty} \left[\frac{\ln r(t,w)}{t}\right] \leq L$$

uniformly for $w \in I$ for some finite L. Thus, in the case I is finite, if (ii) is not satisfied we must have

$$\limsup_{t\to\infty} \left[\frac{\ln r(t,w)}{t}\right] = \infty$$

for some $w \in I$.

Theorem 2: Let $\phi(t)$ be a positive function increasing to ∞ as $t \to \infty$ and such that $\limsup_{t\to\infty}[\ln \phi(t)/t] = \infty$. Then there exists a failure rate $r_1(t)$ with $r_1(t) \to \infty$ and $r_1(t) \leq \phi(t)$ for all t such that

$$r(t) = \frac{.5f_1(t) + .5e^{-t}}{.5\overline{F}_1(t) + .5e^{-t}}$$

does not converge as $t \to \infty$.

Proof: Given the function $\phi(t)$, let $x_0 = 0$ and inductively choose x_n such that $x_{n+1} - \{2\phi(x_n)\}^{-1} > x_n + 1$ and $\ln \phi(x_{n+1}) > 2x_{n+1}\phi(x_n)$. Let $y_n = \phi(x_n)$ and set $x'_{n+1} = x_{n+1} - (2y_n)^{-1} + (2y_{n+1})^{-1}$ for $n = 0, 1, \ldots$. Define

$$g(x) = \begin{cases} y_n & \text{if } x \in [x_n, x'_{n+1}), \\ \frac{0.5}{x_{n+1}+(2y_{n+1})^{-1}-x} & \text{if } x \in [x'_{n+1}, x_{n+1}). \end{cases}$$

Next define

$$h(x) = \begin{cases} 0 & \text{if } 0 \leq x < (2y_0)^{-1}, \\ g\{x - (2y_0)^{-1}\} & \text{if } x \geq (2y_0)^{-1}. \end{cases}$$

It can be shown that the distribution with density

$$f_1(t) = h(t) \exp\left(\int_0^t h(x)dx\right)$$

satisfies the requirements. □

The next result essentially shows that either the failure rate of a finite mixture converges to the limiting failure rate of the strongest subpopulation, or it does not converge.

Theorem 3: Let $r_1(t) \to \infty, r_2(t) \to \lambda, 0 \leq \lambda < \infty$. Suppose $0 < p < 1$ and

$$r(t) = \frac{pf_1(t) + (1-p)e^{-t}}{p\overline{F}_1(t) + (1-p)e^{-t}} \to \alpha$$

as $t \to \infty$, where $0 \leq \alpha \leq \infty$. Then $\alpha = \lambda$.

2.2. Asymptotic Monotonicity of $r(t)$

Consider the mixture of finitely many subpopulations:

$$r(t) = \frac{\sum_{i=1}^{n} p_i f_i(t)}{\sum_{i=1}^{n} p_i \overline{F}_i(t)}, \quad 0 < p_i < 1,$$

where $r_i(t) = \frac{f_i(t)}{\overline{F}_i(t)} \to a_i < \infty$. From Theorem 1, we know that $r(t) \to \alpha = \min_{1 \le i \le n} a_i$. It is of interest to study how $r(t)$ approaches α. One particular question is under what conditions does the mixture wearout, i.e., $r(t)$ ultimately increases.

Remarks:

1. It is well known (Barlow, Marshall and Proschan, 1963) that if all $r_i(t)$ are decreasing, then $r(t)$ is decreasing and so $r(t) \downarrow \alpha$.
2. Gurland and Sethurman (1995) give many examples of a mixture of two IFR distributions for which $r(t)$ is ultimately decreasing.
3. Marshall and Olkin (2001) exhibit an example of two strictly IFR distributions whose mixture has a failure rate which is strictly decreasing.
4. The failure rate of the mixture of two IFR gamma distributions is ultimately increasing.

Block and Joe (1997) give some sufficient conditions on the individual failure rates for determining the ultimate direction of monotonicity of the failure rate of the mixture using the notion of regularly varying functions. A different approach is given below. For convenience, we first introduce a class of functions. We use the word "ultimately" to mean for t large.

Definition: We say that a function $g(t), t \ge 0$ belongs to the class $\mathcal{L}$ if for every $\lambda > 0, e^{-\lambda t}|g(t)| \to 0$ as $t \to \infty$.

Remark: Most of the standard life distributions have failure rates belonging to the class $\mathcal{L}$. For example, see Block and Joe (1997).

Theorem 4: Assume $r_i(t) \to a_i < \infty$, $1 \le i \le n$ with $a_1 < a_j$ for $j = 2, \ldots, n$, that $r_1(t)$ is ultimately strictly monotone and suppose $\frac{1}{r_1'}, r_2', \ldots, r_n' \in \mathcal{L}$. Then $r(t)$ ultimately approaches $\alpha = a_1$ in the same strict direction as $r_1(t)$.

Remark: What is really required for the proof is that

(i) $\frac{r'_j(t)}{r'_1(t)} = o(e^{\eta_j t})$ where $\eta_j = a_j - a_1, j = 2, \ldots, n$ and
(ii) $\frac{1}{r'_1(t)} = o(e^{\eta t})$ where $\eta = \min_{j=2,\ldots,n}(a_j - a_1)$.

2.3. Initial Behavior of $r(t)$

The initial behavior of the failure rate of a finite mixture is fairly easy to determine. The results are summarized below.

Theorem 5: Let $r_i(t)$, $1 \le i \le n$, be the failure rates of finitely many subpopulations.

(i) Assume $r_i(0+) \equiv f_i(0+)$ exists for $1 \le i \le n$. Then

$$r(0+) = \sum_{i=1}^{n} p_i r_i(0+).$$

(ii) Assume $r_i(0+)$ and $r'_i(0+)$ exist and are finite for $1 \le i \le n$. Then

$$r'(0+) = \sum_{i=1}^{n} p_i r'_i(0+) - \sum_{i \ne j} p_i p_j r_i(0+)[r_i(0+) - r_j(0+)].$$

3. Failure Rate Behavior of Systems

Consider a coherent system ϕ of n independent components having failure rates $r_1(t), \ldots, r_n(t)$ respectively. We can represent ϕ in terms of its min path sets $P_1, \ldots, P_p$ or its min cut sets $K_1, \ldots, K_k$ as

$$\phi(x_1, \ldots, x_n) = \max_{1 \le i \le p} \min_{j \in P_i} x_j = \min_{1 \le i \le k} \max_{j \in K_i} x_j \; .$$

We denote the system failure rate by $r_\phi(t)$.

For systems, the role of the failure rates of the subpopulations in the mixture case are replaced by the failure rates of the min path sets, which we denote by $s_i(t), 1 \le i \le p$. Since we are assuming independence, it follows that

$$s_i(t) = \sum_{j \in P_i} r_j(t).$$

Hence, if we assume that $r_j(t) \to a_j$ for $1 \le j \le n$ as $t \to \infty$, then it follows that $s_i(t) \to \sum_{j \in P_i} a_j$ as $t \to \infty$. Denote this limit by b_i.

3.1. Asymptotic Behavior of $r_\phi(t)$

The following result gives that the limiting failure rate of a system is the limiting failure rate of the strongest path set.

Theorem 6: Let ϕ be a coherent system of n independent components having failure rates $r_1(t), \ldots, r_n(t)$.

(i) Assume $r_i(t) \to a_i$, $1 \le i \le n$, as $t \to \infty$. Then $r_\phi(t) \to \beta = \min_{1 \le i \le p} b_i$, where $b_i = \lim_{t \to \infty} s_i(t) = \sum_{j \in P_i} a_j$.

(ii) In addition, assume $b_1 < b_j < \infty$, $j = 2, \ldots, p$, that $s_1(t)$ is ultimately strictly monotone, and that $\frac{1}{s_1'}, s_2', \ldots, s_p' \in \mathcal{L}$. Then $r_\phi(t)$ approaches $\beta = b_1$ in the same strict direction as $s_1(t)$.

3.2. Initial Behavior of $r_\phi(t)$

Keeping the same notation as above, we state our final result.

Theorem 7:

(i) Assume $r_i(0+)$ exists for $1 \le i \le n$. Then

$$r_\phi(0+) = \sum_{\{i\}} r_i(0+)\ .$$

(ii) Assume $r_i(0+), r_i'(0+)$ exist and are finite, $1 \le i \le n$. Then

$$r_\phi'(0+) = \sum_{\{i\}} r_i'(0+) + 2 \sum_{\{j,k\}} r_j(0+) r_k(0+)\ .$$

Here $\sum_{\{i\}}, \sum_{\{j,k\}}$ means to sum over all min cut sets of size one, two respectively.

Acknowledgements

This work was partially supported by NSF Grant DMS0072207. We would also like to thank a referee who suggested that we consider the asymptotic behavior of mixtures and systems as a function of changing parameters and system structure. This will be the subject of future research.

References

1. R. E. Barlow, A. W. Marshall and F. Proschan, Properties of probability distributions with monotone hazard rate, *Annals of Mathematical Statistics* **34**, 375-389 (1963).
2. R. E. Barlow and F. Proschan, *Statistical Theory of Reliability*, Holt, Rinehart & Winston, New York (1975).
3. H. Block and H. Joe, Tail behavior of the failure rate functions of mixtures, *Lifetime Data Analysis* **3**, 269-288 (1997).
4. H. W. Block, Y. Li and T. H. Savits, On the failure rates of mixtures and systems, Under preparation (2001).
5. H. W. Block, J. Mi and T. H. Savits, Burn–in and mixed populations, *Journal Applied Probability* **30**, 692–702 (1993).
6. H. W. Block and T. H. Savits, Burn-in, *Statistical Science* **12**, 1-19 (1997).
7. C. A. Clarotti and F. Spizzichino, Bayes burn-in decision procedures, *Probability in the Engineering and Information Sciences* **4**, 437-445 (1990).
8. P. I. Gupta and R. C. Gupta, Ageing characteristics of the Weibull mixtures, *Probability in the Engineering and Informational Sciences*, **10**, 591-600 (1996).
9. J. Gurland and J. Sethuraman, How pooling failure data may reverse increasing failure rates, *Journal American Statistical Association* **90**, 1416–1423 (1995).
10. A. W. Marshall and I. Olkin, Personal communication (2001).

CHAPTER 7

A GENERAL MAXIMAL PRECEDENCE TEST

N. Balakrishnan* and H. K. T. Ng†

Department of Mathematics and Statistics, McMaster University, Hamilton, Ontario, Canada L8S 4K1

*E-mail: * bala@mcmail.cis.mcmaster.ca,*

E-mail: † ngh@math.mcmaster.ca

In this paper, we introduce a general maximal precedence test for testing the hypothesis that two distribution functions are equal, which is an extension of the precedence life test first proposed by Nelson[1]. The maximal precedence test involves observing the number of failures in one sample before the r-th failure from the second sample. We derive the null distribution of the general maximal precedence test statistic M. Critical values for some combination of sample sizes for $r = 2(1)6$ are presented. Then, we derive the exact power function under the Lehmann alternative. We also examine the power properties of the general maximal precedence test under a location-shift alternative through Monte Carlo simulations. Finally, the power comparisons are made with the precedence test and Wilcoxon's rank-sum test.

1. Introduction

The maximal precedence test is a distribution-free, two-sample life test based on the order of early failures which is an extension of the precedence test [see Nelson[1,2,3] and Balakrishnan and Frattina[4]]. Suppose there are two failure time distributions F_X and F_Y. We are interested in testing the hypotheses

$$H_0 : F_X = F_Y$$
$$\text{against } H_1 : F_X > F_Y. \tag{1}$$

Note that some specific alternatives such as the location-shift alternative and the Lehmann alternative are subclasses of the general alternative considered here.

In the context of reliability, suppose an experimenter wishes to test whether the lifetimes of units from two different groups are the same. Further, suppose independent samples of units are placed simultaneously on a life-test and the experiment gets terminated after a pre-fixed number of failures occur (namely, a Type-II right censored sample). Then, the problem of interest is to test whether the lifetimes of units from both groups are the same or not. For this testing problem, Nelson[1] proposed a test procedure based on the number of failures that precede the r-th failure and termed it as a *precedence test.* Here, we discuss some alternatives to this nonparametric test procedure.

Precedence test can determine a location difference based on observing only a few failures from the two samples under life testing. Nelson[1] provided tables of critical values, which cover all combinations of sample sizes up to twenty for one-sided (two-sided) significance levels of 0.05 (0.10), 0.025 (0.05), and 0.005 (0.01). Nelson[3] then examined the power of the precedence test when the underlying distributions were normal. Recently, Balakrishnan and Frattina[4] noted that a 'masking effect' affects the performance of the precedence test and, therefore, proposed a maximal precedence test which is based on the maximum of the numbers of failures of the first sample occurring before the first and between the first and second failures of the second sample. They derived the null distribution of the maximal precedence test statistic with $r = 2$ (observing only up to the second failure of the second sample). They also examined the power properties of the maximal precedence test and compared them with those of the precedence test and Wilcoxon's rank-sum test.

This paper is organized as follows. In Section 2, we review some results on the precedence test and point out the masking effect of the precedence test. In Section 3, we propose the general maximal precedence test which is a generalization of the maximal precedence test with $r = 2$ proposed earlier by Balakrishnan and Frattina[4]. In addition, the null distribution of the general maximal precedence test statistic M is derived and some critical values for $r = 2(1)6$ for some choices of sample sizes are presented. Next, we derive in Section 4 the exact power function of the general precedence test under Lehmann alternative. We then examine the power performance of the general maximal precedence test under a location-shift between the two populations through Monte Carlo simulations. Comparison and discussion of the power properties of the maximal precedence test with those of the precedence test and Wilcoxon's rank-sum test are presented in Section 5. Finally, in Section 6 we suggest some possible directions for further research in this area.

2. Review of the Precedence Test

Assume that a random sample of size n_1 is from F_X, another sample of size n_2 is from F_Y, and that all these sample units are placed simultaneously on a life-testing experiment. We use $X_1, X_2, \ldots, X_{n_1}$ to denote the sample from distribution F_X, and $Y_1, Y_2, \ldots, Y_{n_2}$ to denote the sample from distribution F_Y. Then, the precedence test statistic P_r is defined as the number of failures from the first sample that precede the r-th failure from the second sample. It is obvious that large values of P_r lead to the rejection of H_0 and in favor of H_1 in (1).

For a fixed level of significance α, the critical region will be $\{s, s+1, \ldots, n_1\}$, where

$$\alpha = \Pr(P_r \geq s | F_X = F_Y). \tag{2}$$

For specified values of n_1, n_2, s and r, an expression for α in (2) is given by

$$\alpha = \frac{\sum_{j=s}^{n_1} \binom{s+r-1}{j} \binom{n_1+n_2-s-r+1}{n_1-j}}{\binom{n_1+n_2}{n_2}}$$

with the summation terminating as soon as any of the factorial involve negative arguments.

The critical value s and the exact level of significance α as close as possible to 5% for different choices of the sample sizes n_1 and n_2 and $r = 1(1)6$ are given in Table 1.

We can see that there is a masking effect when $r \geq 2$. For example, if we had $n_1 = n_2 = 20$ and we were using the precedence test with $r = 3$ and $s = 8$, then the null hypothesis will be rejected if there were at least 8 failures from the X-sample before the third failure from the Y-sample. If only 7 failures occurred from the X-sample before the third failure from the Y-sample, then we will not reject the null hypothesis by P_3. Nevertheless, if all these 7 failures had occurred before the first failure from the Y-sample

Table 1. Near 5% upper critical values and exact levels of significance for the precedence test statistic P_r.

n_1	n_2	$r=1$	$r=2$	$r=3$	$r=4$	$r=5$	$r=6$
10	10	4(0.04334)	6(0.02864)	7(0.03489)	8(0.03483)	9(0.02709)	9(0.06502)
15	15	4(0.04981)	6(0.04004)	7(0.05432)	9(0.03022)	10(0.03280)	11(0.03280)
20	20	4(0.05301)	6(0.04574)	7(0.06369)	9(0.04118)	10(0.04792)	11(0.05267)
30	30	4(0.05620)	6(0.05139)	8(0.03986)	9(0.05208)	10(0.06264)	12(0.04202)
30	50	3(0.04942)	4(0.06293)	5(0.06494)	6(0.06226)	7(0.05752)	8(0.05190)

(the probability of this happening under H_0 is less than 1%), we would have suspected that there is a location-shift between the two populations. In fact, if we had used P_1 with $s = 4$ or P_2 with $s = 6$, we would have correctly rejected the null hypothesis. Information given by $r = 3$ is thus getting masked in this case.

Maximal precedence test is proposed in order to avoid this problem. It is a testing procedure based on the maximum number of the failures occurring from the X-sample before the first, between the first and the second, ..., between the $(r-1)$-th and the r-th failures from the Y-sample. We will look into the details of this test in the next section.

3. Maximal Precedence Test

Maximal precedence test is a testing procedure for testing the hypotheses (1). We use the same notations in Section 2 and further denote the order statistics from the X-sample and the Y-sample by $X_{1:n_1} \leq X_{2:n_1} \leq \ldots \leq X_{n_1:n_1}$ and $Y_{1:n_2} \leq Y_{2:n_2} \leq \ldots \leq Y_{n_2:n_2}$, respectively. Without loss of generality, we assume that $n_1 \leq n_2$.

Moreover, we let M_1 be the number of the X-failures before $Y_{1:n_2}$ and M_i be the number of the X-failures between $Y_{i-1:n_2}$ and $Y_{i:n_2}$, $i = 2, 3, \ldots, r$. Then, $M = \max(M_1, M_2, \ldots, M_r)$ is called the general maximal precedence test statistic. Large values of M lead to the rejection of H_0 and in favor of H_1 in (1). The null distribution of M for the special case when $r = 2$ was derived by Balakrishnan and Frattina[4]. We will present here the null distribution function of the general maximal precedence test statistic M.

Theorem 1: The null cumulative distribution function of $M = \max(M_1, M_2, \ldots, M_r)$ is given by

$$\Pr(M \leq m \mid F_X = F_Y) = \Pr(M_1 \leq m, M_2 \leq m, \ldots, M_r \leq m \mid F_X = F_Y)$$

$$= \sum_{\substack{m_i (i=1,2,\ldots,r)=0 \\ \sum_{i=1}^{r} m_i \le n_1}}^{m} \frac{\binom{n_1 + n_2 - \sum_{i=1}^{r} m_i - r}{n_2 - r}}{\binom{n_1 + n_2}{n_2}}. \qquad (3)$$

Proof: First, conditional on the Y-failures, we consider the probability that there are m_1 X-failures before $y_{1:n_2}$ and m_i X-failures between $y_{i-1:n_2}$ and $y_{i:n_2}$, $i = 2, 3, \ldots, r$, given by

$$\Pr\{m_1 X's \le y_{1:n_2},\ m_2 X's \in (y_{1:n_1}, y_{2:n_2}], \ldots,\ m_r X's \in (y_{r-1:n_2}, y_{r:n_2}],$$
$$\left(n_1 - \sum_{i=1}^{r} m_i\right) X's > y_{r:n_2} | Y_{1:n_2} = y_{1:n_2}, Y_{2:n_2} = y_{1:n_2}, \ldots, Y_{r:n_2} = y_{r:n_2}\}$$
$$= \frac{n_1!}{m_1! m_2! \cdots m_r! \left(n_1 - \sum_{i=1}^{r} m_i\right)!} [F_X(y_{1:n_2})]^{m_1}$$
$$\times \left\{\prod_{i=2}^{r} [F_X(y_{i:n_2}) - F_X(y_{i-1:n_2})]^{m_i}\right\} [1 - F_X(y_{r:n_2})]^{\left(n_1 - \sum_{i=1}^{r} m_i\right)}.$$

We also know that the joint density of the first r order statistics from the Y-sample is [see David[5] and Arnold, Balakrishnan and Nagaraja[6]]

$$f_{1,2,\ldots,r:n_2}(y_{1:n_2}, y_{2:n_2}, \ldots, y_{r:n_2})$$
$$= \frac{n_2!}{(n_2 - r)!} f_Y(y_{1:n_2}) f_Y(y_{2:n_2}) \cdots f_Y(y_{r:n_2}) [1 - F_Y(y_{r:n_2})]^{n_2 - r}.$$

Then, we obtain the unconditional probability of $\{M_1 = m_1, M_2 = m_2, \ldots, M_r = m_r\}$ as

$$\Pr\{M_1 = m_1, M_2 = m_2, \ldots, M_r = m_r\}$$
$$= \frac{n_1!}{m_1! m_2! \cdots m_r! \left(n_1 - \sum_{i=1}^{r} m_i\right)!}$$
$$\times \int_0^{\infty} \int_0^{y_{r:n_2}} \cdots \int_0^{y_{2:n_2}} [F_X(y_{1:n_2})]^{m_1} \left\{\prod_{i=2}^{r} [F_X(y_{i:n_2}) - F_X(y_{i-1:n_2})]^{m_i}\right\}$$
$$\times [1 - F_X(y_{r:n_2})]^{\left(n_1 - \sum_{i=1}^{r} m_i\right)} f_{1,2,\ldots,r:n_2}(y_{1:n_2}, y_{2:n_2}, \ldots, y_{r:n_2})$$
$$\times dy_{1:n_2} dy_{2:n_2} \ldots dy_{r-1:n_2} dy_{r:n_2}$$

$$= C \int_0^{\infty} \int_0^{y_{r:n_2}} \cdots \int_0^{y_{2:n_2}} [F_X(y_{1:n_2})]^{m_1} \left\{ \prod_{i=2}^{r} [F_X(y_{i:n_2}) - F_X(y_{i-1:n_2})]^{m_i} \right\}$$

$$\times [1 - F_X(y_{r:n_2})]^{\left(n_1 - \sum_{i=1}^{r} m_i\right)} \left\{ \prod_{i=1}^{r} f_Y(y_{i:n_2}) \right\} [1 - F_Y(y_{r:n_2})]^{n_2 - r}$$

$$\times dy_{1:n_2} dy_{2:n_2} \ldots dy_{r-1:n_2} dy_{r:n_2}, \tag{4}$$

where

$$C = \frac{n_1! n_2!}{m_1! m_2! \cdots m_r! \left(n_1 - \sum_{i=1}^{r} m_i\right)! (n_2 - r)!}.$$

Under null distribution, $H_0 : F_X = F_Y$, the expression in (4) becomes

$$\Pr\{M_1 = m_1, M_2 = m_2, \ldots, M_r = m_r \mid F_X = F_Y\}$$

$$= C \int_0^{\infty} \int_0^{y_{r:n_2}} \cdots \int_0^{y_{2:n_2}} [F_X(y_{1:n_2})]^{m_1} \left\{ \prod_{i=2}^{r} [F_X(y_{i:n_2}) - F_X(y_{i-1:n_2})]^{m_i} \right\}$$

$$\times [1 - F_X(y_{r:n_2})]^{\left(n_1 + n_2 - \sum_{i=1}^{r} m_i - r\right)} \left\{ \prod_{i=1}^{r} f_X(y_{i:n_2}) \right\}$$

$$\times dy_{1:n_2} dy_{2:n_2} \ldots dy_{r-1:n_2} dy_{r:n_2}.$$

For notational convenience, let us now set

$$u_i = F_X(y_{i:n_2}) \text{ and } du_i = f_X(y_{i:n_2}) dy_{i:n_2} \quad \text{for } i = 1, 2, \ldots, r.$$

Then, the above expression for the unconditional probability becomes

$$\Pr\{M_1 = m_1, M_2 = m_2, \ldots, M_r = m_r \mid F_X = F_Y\}$$

$$= C \int_0^1 \int_0^{u_r} \cdots \int_0^{u_2} u_1^{m_1} \left\{ \prod_{i=2}^{r} (u_i - u_{i-1})^{m_i} \right\}$$

$$\times (1 - u_r)^{\left(n_1 + n_2 - \sum_{i=1}^{r} m_i - r\right)} du_1 \ldots du_r.$$

Using the transformation $w_1 = u_1/u_2$, we obtain

$$\Pr\{M_1 = m_1, M_2 = m_2, \ldots, M_r = m_r \mid F_X = F_Y\}$$

$$= C \int_0^1 w_1^{m_1} (1 - w_1)^{m_2} dw_1 \int_0^1 \int_0^{u_r} \cdots \int_0^{u_3} u_2^{m_1 + m_2 + 1}$$

$$\times \left\{ \prod_{i=3}^{r} (u_i - u_{i-1})^{m_i} \right\} (1 - u_r)^{\left(n_1 + n_2 - \sum_{i=1}^{r} m_i - r\right)} du_2 \ldots du_r$$

$$= CB(m_1, m_2+1)\int_0^1\int_0^{u_r}\cdots\int_0^{u_3} u_2^{m_1+m_2+1}$$
$$\times\left\{\prod_{i=3}^{r}(u_i-u_{i-1})^{m_i}\right\}(1-u_r)^{\left(n_1+n_2-\sum_{i=1}^{r}m_i-r\right)}\,du_2\ldots du_r,$$

where $B(a,b)=\int_0^1 x^{a-1}(1-x)^{b-1}dx$ denotes the complete beta function.

Similarly, upon performing the transformations $w_l = u_l/u_{l+1}$ for $l = 2, 3, \ldots, r-1$, we obtain

$$\Pr\{M_1=m_1, M_2=m_2,\ldots,M_r=m_r \mid F_X=F_Y\}$$
$$= C\left\{\prod_{j=1}^{r-1}B\left(\sum_{i=1}^{j}m_i+j, m_{j+1}+1\right)\right\}$$
$$\times\int_0^1 u_r^{\left(\sum_{i=1}^{r}m_i+r-1\right)}(1-u_r)^{\left(n_1+n_2-\sum_{i=1}^{r}m_i-r\right)}\,du_r$$
$$= C\left\{\prod_{j=1}^{r-1}B\left(\sum_{i=1}^{j}m_i+j, m_{j+1}+1\right)\right\}$$
$$\times B\left(\sum_{i=1}^{r}m_i+r, n_1+n_2-\sum_{i=1}^{r}m_i-r+1\right)$$

$$= \frac{n_1!n_2!\left(n_1+n_2-\sum_{i=1}^{r}m_i-r\right)!}{\left(n_1-\sum_{i=1}^{r}m_i\right)!(n_2-r)!(n_1+n_2)!}$$
$$= \frac{\binom{n_1+n_2-\sum_{i=1}^{r}m_i-r}{n_2-r}}{\binom{n_1+n_2}{n_2}}. \tag{5}$$

Therefore, the cumulative distribution function of $M = \max(M_1, M_2, \ldots, M_r)$ under the null hypothesis $H_0 : F_X = F_Y$ is given by

$$Pr\{M\le m \mid F_X=F_Y\} = \Pr\{M_1\le m, M_2\le m,\ldots,M_r\le m \mid F_X=F_Y\}$$

$$= \sum_{\substack{m_i(i=1,2,\ldots,r)=0 \\ \sum_{i=1}^{r} m_i \le n_1}}^{m} \frac{\binom{n_1+n_2-\sum_{i=1}^{r} m_i - r}{n_2-r}}{\binom{n_1+n_2}{n_2}}$$

which completes the proof of the theorem. □

For specified value of n_1, n_2, r and the level of significance α, the critical value s (corresponding to a level closest to α) for the maximal precedence test can be found readily from (3) as

$$\begin{aligned}\alpha &= \Pr(M \ge s \mid F_X = F_Y) \\ &= 1 - \Pr(M \le s-1 \mid F_X = F_Y) \\ &= 1 - \sum_{\substack{m_i(i=1,2,\ldots,r)=0 \\ \sum_{i=1}^{r} m_i \le n_1}}^{s-1} \frac{\binom{n_1+n_2-\sum_{i=1}^{r} m_i - r}{n_2-r}}{\binom{n_1+n_2}{n_2}}. \qquad (6)\end{aligned}$$

In Table 2, we have presented the critical value s and the exact level of significance α as close as possible to 5% for some sample sizes n_1 and n_2 and $r = 2(1)6$.

Table 2. Near 5% upper critical values and exact levels of significance for the general maximal precedence test statistic M.

n_1	n_2	$r=2$	$r=3$	$r=4$	$r=5$	$r=6$
10	10	5(0.03250)	5(0.04875)	5(0.06498)	6(0.02709)	6(0.03251)
15	15	5(0.04205)	5(0.06292)	6(0.03368)	6(0.04209)	6(0.05050)
20	20	5(0.04691)	6(0.03023)	6(0.04026)	6(0.05026)	6(0.06025)
30	30	5(0.05179)	6(0.03540)	6(0.04707)	6(0.05868)	7(0.03150)
30	50	4(0.03445)	4(0.05138)	4(0.06810)	5(0.02946)	5(0.03529)

These critical values are used later in the Monte Carlo simulations, presented in Section 5, for determining the rejection rates of the general maximal precedence test.

4. Exact Power under Lehmann Alternative

In this section, we will derive an explicit expression for the power function of the general maximal precedence test under the Lehmann alternative

$H_1 : [F_X]^\gamma = F_Y$ for some γ, which was first proposed by Lehmann[7]. For more details on the Lehmann alternative, see Davies[8], Lehmann[9] and Gibbons and Chakraborti[10].

We can see that $H_1 : [F_X]^\gamma = F_Y$ is a subclass of the alternative $H_1 : F_X > F_Y$ when $\gamma > 1$. The power of a test is the probability of rejecting the null hypothesis when the alternative hypothesis is indeed true. In the following theorem, we derive an explicit expression for the power function.

Theorem 2: Under the Lehmann alternative $H_1 : [F_X]^\gamma = F_Y$, the power function is given by

$$\begin{aligned}
&1 - \Pr\{M \le x-1 \mid [F_X]^\gamma = F_Y\} \\
&= 1 - \sum_{\substack{m_i (i=1,2,\ldots,r)=0 \\ \sum_{i=1}^r m_i \le n_1}}^{x-1} \frac{n_1! n_2! \gamma^r}{m_1!(n_2-r)!} \left\{ \prod_{j=1}^{r-1} \frac{\Gamma\left(\sum_{i=1}^{j} m_i + j\gamma\right)}{\Gamma\left(\sum_{i=1}^{j+1} m_i + j\gamma + 1\right)} \right\} \\
&\quad \times \sum_{k=0}^{n_2-r} \binom{n_2-r}{k} (-1)^k \frac{\Gamma\left(\sum_{i=1}^{r} m_i + (r+k)\gamma\right)}{\Gamma(n_1 + (r+k)\gamma + 1)}. \qquad (7)
\end{aligned}$$

Proof: Under the Lehmann alternative $H_1 : [F_X]^\gamma = F_Y, \gamma > 1$, the expression in (4) can be simplified as follows:

$$\begin{aligned}
&\Pr\{M_1 = m_1, M_2 = m_2, \ldots, M_r = m_r \mid [F_X]^\gamma = F_Y\} \\
&= C \int_0^\infty \int_0^{y_{r:n_2}} \cdots \int_0^{y_{2:n_2}} [F_X(y_{1:n_2})]^{m_1} \left\{ \prod_{i=2}^{r} [F_X(y_{i:n_2}) - F_X(y_{i-1:n_2})]^{m_i} \right\} \\
&\quad \times [1 - F_X(y_{r:n_2})]^{\left(n_1 - \sum_{i=1}^r m_i\right)} \left\{ \prod_{i=1}^{r} \gamma [F_X(y_{i:n_2})]^{\gamma-1} f_X(y_{i:n_2}) \right\} \\
&\quad \times \{1 - [F_X(y_{r:n_2})]^\gamma\}^{n_2-r} \, dy_{1:n_2} dy_{2:n_2} \ldots dy_{r-1:n_2} dy_{r:n_2} \\
&= C\gamma^r \int_0^\infty \int_0^{y_{r:n_2}} \cdots \int_0^{y_{2:n_2}} [F_X(y_{1:n_2})]^{m_1+\gamma-1} \\
&\quad \times \left\{ \prod_{i=2}^{r} [F_X(y_{i:n_2})]^{\gamma-1} [F_X(y_{i:n_2}) - F_X(y_{i-1:n_2})]^{m_i} \right\} \\
&\quad \times [1 - F_X(y_{r:n_2})]^{\left(n_1 - \sum_{i=1}^r m_i\right)} \left\{ \prod_{i=1}^{r} f_X(y_{i:n_2}) \right\}
\end{aligned}$$

$$\times \left\{ \sum_{k=0}^{n_2-r} \binom{n_2-r}{k} (-1)^k [F_X(y_{r:n_2})]^{\gamma k} \right\} dy_{1:n_2} \dots dy_{r:n_2}$$

$$= C\gamma^r \sum_{k=0}^{n_2-r} \binom{n_2-r}{k} (-1)^k \int_0^\infty \int_0^{y_{r:n_2}} \dots \int_0^{y_{2:n_2}} [F_X(y_{1:n_2})]^{m_1+\gamma-1}$$

$$\times \left\{ \prod_{i=1}^{r} f_X(y_{i:n_2}) \right\} \left\{ \prod_{i=2}^{r} [F_X(y_{i:n_2})]^{\gamma-1} [F_X(y_{i:n_2}) - F_X(y_{i-1:n_2})]^{m_i} \right\}$$

$$\times [F_X(y_{r:n_2})]^{\gamma k} [1 - F_X(y_{r:n_2})]^{\left(n_1 - \sum_{i=1}^{r} m_i\right)} dy_{1:n_2} \dots dy_{r:n_2}.$$

For notational convenience, let us now set

$$u_i = F_X(y_{i:n_2}) \text{ and } du_i = f_X(y_{i:n_2}) dy_{i:n_2} \quad \text{for } i = 1, 2, \dots, r.$$

Then, the above expression for the unconditional probability becomes

$$\Pr\{M_1 = m_1, M_2 = m_2, \dots, M_r = m_r \mid [F_X]^\gamma = F_Y\}$$

$$= C\gamma^r \sum_{k=0}^{n_2-r} \binom{n_2-r}{k} (-1)^k \int_0^1 \int_0^{u_r} \dots \int_0^{u_2} u_1^{m_1+\gamma-1}$$

$$\times \left\{ \prod_{i=2}^{r} u_i^{\gamma-1} (u_i - u_{i-1})^{m_i} \right\} u_r^{\gamma k} (1 - u_r)^{\left(n_1 - \sum_{i=1}^{r} m_i\right)} du_1 \dots du_r.$$

Using the transformation $w_1 = u_1/u_2$, we obtain

$$\Pr\{M_1 = m_1, M_2 = m_2, \dots, M_r = m_r \mid [F_X]^\gamma = F_Y\}$$

$$= C\gamma^r \sum_{k=0}^{n_2-r} \binom{n_2-r}{k} (-1)^k \int_0^1 w_1^{m_1+\gamma-1} (1 - w_1)^{m_2} dw_1$$

$$\times \int_0^1 \int_0^{u_r} \dots \int_0^{u_3} u_2^{m_1+m_2+2\gamma-1} \left\{ \prod_{i=3}^{r} u_i^{\gamma-1} (u_i - u_{i-1})^{m_i} \right\}$$

$$\times u_r^{\gamma k} (1 - u_r)^{\left(n_1 - \sum_{i=1}^{r} m_i\right)} du_2 \dots du_r$$

$$= C\gamma^r B(m_1 + \gamma, m_2 + 1) \sum_{k=0}^{n_2-r} \binom{n_2-r}{k} (-1)^k$$

$$\times \int_0^1 \int_0^{u_r} \dots \int_0^{u_3} u_2^{m_1+m_2+2\gamma-1} \left\{ \prod_{i=3}^{r} u_i^{\gamma-1} (u_i - u_{i-1})^{m_i} \right\}$$

$$u_r^{\gamma k} (1 - u_r)^{\left(n_1 - \sum_{i=1}^{r} m_i\right)} du_2 \dots du_r,$$

where $B(a,b)$ denotes the complete beta function as before.

Similarly, upon performing the transformations $w_l = u_l/u_{l+1}$ for $l = 2,3,\ldots,r-1$, we obtain

$$\begin{aligned}
&\Pr\{M_1 = m_1, M_2 = m_2, \ldots, M_r = m_r \mid [F_X]^\gamma = F_Y\} \\
&= C\gamma^r \left\{\prod_{j=1}^{r-1} B\left(\sum_{i=1}^{j} m_i + j\gamma, m_{j+1} + 1\right)\right\} \\
&\quad \times \sum_{k=0}^{n_2-r} \binom{n_2-r}{k} (-1)^k \int_0^1 u_r^{\sum_{i=1}^{r} m_i + (r+k)\gamma - 1} (1-u_r)^{\left(n_1 - \sum_{i=1}^{r} m_i\right)} du_r \\
&= C\gamma^r \left\{\prod_{j=1}^{r-1} B\left(\sum_{i=1}^{j} m_i + j\gamma, m_{j+1} + 1\right)\right\} \\
&\quad \times \sum_{k=0}^{n_2-r} \binom{n_2-r}{k} (-1)^k B\left(\sum_{i=1}^{r} m_i + (r+k)\gamma, n_1 - \sum_{i=1}^{r} m_i + 1\right)
\end{aligned}$$

$$\begin{aligned}
&= \frac{n_1! n_2! \gamma^r}{m_1!(n_2-r)!} \left\{\prod_{j=1}^{r-1} \frac{\Gamma\left(\sum_{i=1}^{j} m_i + j\gamma\right)}{\Gamma\left(\sum_{i=1}^{j+1} m_i + j\gamma + 1\right)}\right\} \\
&\quad \times \sum_{k=0}^{n_2-r} \binom{n_2-r}{k} (-1)^k \frac{\Gamma\left(\sum_{i=1}^{r} m_i + (r+k)\gamma\right)}{\Gamma\left(n_1 + (r+k)\gamma + 1\right)}. \qquad (8)
\end{aligned}$$

Therefore, the cumulative distribution function of $M = \max(M_1, M_2, \ldots, M_r)$ under the Lehmann alternative is given by

$$\begin{aligned}
&\Pr\{M \le m \mid [F_X]^\gamma = F_Y\} \\
&= \Pr\{M_1 \le m, M_2 \le m, \ldots, M_r \le m \mid [F_X]^\gamma = F_Y\} \\
&= \sum_{\substack{m_i (i=1,2,\ldots,r)=0 \\ \sum_{i=1}^{r} m_i \le n_1}}^{m} \Pr\{M_1 = m_1, M_2 = m_2, \ldots, M_r = m_r \mid [F_X]^\gamma = F_Y\}
\end{aligned}$$

$$= \sum_{\substack{m_i(i=1,2,\ldots,r)=0 \\ \sum_{i=1}^{r} m_i \le n_1}}^{m} \frac{n_1! n_2! \gamma^r}{m_1!(n_2-r)!} \left\{ \prod_{j=1}^{r-1} \frac{\Gamma\left(\sum_{i=1}^{j} m_i + j\gamma\right)}{\Gamma\left(\sum_{i=1}^{j+1} m_i + j\gamma + 1\right)} \right\}$$

$$\times \sum_{k=0}^{n_2-r} \binom{n_2-r}{k} (-1)^k \frac{\Gamma\left(\sum_{i=1}^{r} m_i + (r+k)\gamma\right)}{\Gamma\left(n_1 + (r+k)\gamma + 1\right)}$$

which completes the proof of the theorem. □

Here, we demonstrate the use of the expression in (7) as well as the Monte Carlo simulation method for the computation of the power of the maximal precedence test under the Lehmann alternative. In the simulation study of the power under Lehmann alternative, we consider the case that $X_1, X_2, \ldots, X_{n_1}$ is a random sample from a power function distribution with cumulative distribution function

$$F_X(x) = x^\gamma, \qquad 0 \le x \le 1, \gamma > 1$$

and $Y_1, Y_2, \ldots, Y_{n_2}$ is a random sample from a uniform distribution with cumulative distribution function

$$F_Y(x) = x, \qquad 0 \le x \le 1.$$

We generated 100,000 sets of data and found the test statistic M for each set. Then the power values are estimated by the rejection rates of the null hypothesis.

Under the Lehmann alternative and for $n_1 = n_2 = 10, r = 2, 3$ and $\gamma = 2(1)6$, the power values computed from the expression in (7) and those estimated through the Monte Carlo simulation method are presented in Table 3. We observe that the estimated values of the power determined from Monte Carlo simulations are quite close to the exact values. In addition to revealing the correctness of the expression in (7), these results also suggest that the Monte Carlo simulation method provides a simple and accurate way to estimate the power of the test.

5. Monte Carlo Simulation and Power Comparison under Location-Shift Alternative

In order to assess the power properties of the maximal precedence test, we consider the general maximal precedence test under the location-shift

Table 3. Power values under Lehmann alternative for $n_1 = n_2 = 10, r = 2, 3$ and $\gamma = 2(1)6$.

r	γ	Power computed from expression (7)	Simulated Power
r=2	2	0.220950	0.221520
	3	0.453424	0.454620
	4	0.636340	0.645573
	5	0.760590	0.760093
	6	0.841193	0.842160
r=3	2	0.241783	0.241640
	3	0.468023	0.469117
	4	0.645608	0.645573
	5	0.766413	0.766080
	6	0.844902	0.845893

alternative $H_1 : F_X(x) = F_Y(x + \theta)$ for some $\theta > 0$, where θ is a shift in location. The power of the precedence test with $r = 1(1)6$, maximal precedence test with $r = 2(1)6$, and the Wilcoxon's rank-sum test were all estimated through Monte Carlo simulations when $\theta = 0.5$ and $\theta = 1.0$. The following lifetime distributions were used in the Monte Carlo simulations in order to demonstrate the power performance of the maximal precedence test under this location-shift alternative:

1. Standard normal distribution;
2. Standard exponential distribution;
3. Gamma distribution with shape parameter a and standardized by mean a and standard deviation $\sqrt{a}$;
4. Lognormal distribution with shape parameter σ and standardized by mean $e^{\sigma^2/2}$ and standard deviation $\sqrt{e^{\sigma^2}(e^{\sigma^2} - 1)}$.

For a detailed discussion on various properties of these distributions, one may refer to Johnson, Kotz and Balakrishnan[11]. For different choices of sample sizes, we generated 100,000 sets of data, utilizing the IMSL subroutines RNNOR, RNEXP, RNGAM and RNLNL under Microsoft FORTRAN Visual Workbench Version 1.0, in order to obtain the estimated rejection rates.

In Tables 4–6, we have presented the estimated power values of the precedence tests with $r = 1(1)6$, maximal precedence tests with $r = 2(1)6$, and the Wilcoxon's rank-sum test for the underlying standard normal, standard exponential, standardized gamma and standardized lognormal distributions, with location-shift being equal to 0.5 and 1.0. For comparison proposes, the corresponding critical values and the exact levels of significance

Table 4. Power of precedence tests, maximal precedence tests and Wilcoxon's rank-sum test for $n_1 = n_2 = 10$.

Test	r	critical value	exact l.o.s.	location shift	Distribution					
					N(0,1)	Exp(1)	Gamma(2)	Gamma(10)	LN(0.1)	LN(0.5)
P	1	4	0.0433	0.5	0.1736	0.7197	0.4141	0.2249	0.1936	0.3944
				1.0	0.4208	0.9779	0.8806	0.5742	0.4817	0.8709
	2	6	0.0286	0.5	0.1496	0.3916	0.2525	0.1731	0.1603	0.2693
				1.0	0.4243	0.8376	0.6879	0.5060	0.4571	0.7253
	3	7	0.0349	0.5	0.1779	0.3073	0.2343	0.1893	0.1828	0.2571
				1.0	0.4813	0.7239	0.6148	0.5177	0.4964	0.6676
	4	8	0.0348	0.5	0.1771	0.2251	0.1936	0.1755	0.1777	0.2149
				1.0	0.4799	0.5767	0.5114	0.4758	0.4783	0.5656
	5	9	0.0271	0.5	0.1473	0.1453	0.1380	0.1388	0.1444	0.1527
				1.0	0.4245	0.3960	0.3725	0.3833	0.4038	0.4144
	6	9	0.0650	0.5	0.2659	0.2385	0.2344	0.2436	0.2567	0.2531
				1.0	0.5832	0.4991	0.4954	0.5266	0.5545	0.5314
M	2	5	0.0325	0.5	0.1417	0.4923	0.2474	0.1611	0.1479	0.2543
				1.0	0.3768	0.9177	0.7303	0.4569	0.4083	0.7327
	3	5	0.0487	0.5	0.1753	0.4949	0.2579	0.1870	0.1781	0.2678
				1.0	0.4171	0.9178	0.7327	0.4765	0.4395	0.7360
	4	5	0.0650	0.5	0.1993	0.4974	0.2653	0.2040	0.1990	0.2768
				1.0	0.4359	0.9181	0.7344	0.4853	0.4538	0.7376
	5	6	0.0271	0.5	0.0901	0.2684	0.1197	0.0895	0.0905	0.1297
				1.0	0.2482	0.7791	0.5136	0.2833	0.2586	0.5284
	6	6	0.0325	0.5	0.0938	0.2690	0.1208	0.0921	0.0938	0.1309
				1.0	0.2496	0.7791	0.5138	0.2839	0.2596	0.5286
W		128	0.0446	0.5	0.2536	0.4332	0.3344	0.2663	0.2605	0.3508
				1.0	0.6480	0.8163	0.7508	0.6713	0.6554	0.7819

Note: P - Precedence Test; M - Maximal Precedence Test; W - Wilcoxon's Rank-sum Test

are also presented.

From Tables 4–6, we see that the power of all tests increase with increasing sample sizes as well as with increasing location-shift. When we compare the power values of the maximal precedence tests with those of Wilcoxon's rank-sum test, we find that the Wilcoxon's rank-sum test performs better than the precedence tests and maximal precedence tests if the underlying distributions are close to symmetry, such as the normal distribution, gamma distributions with large values of shape parameter a, and lognormal distributions with small values of shape parameter σ. However, under some right-skewed distributions such as the exponential distribution, gamma distribution with shape parameter $a = 2.0$, and lognormal distribution with shape parameter $\sigma = 0.5$, the maximal precedence tests have higher power values than the Wilcoxon's rank-sum test. For example, in

Table 5. Power of precedence tests, maximal precedence tests and Wilcoxon's rank-sum test for $n_1 = n_2 = 20$.

Test	r	critical value	exact l.o.s.	location shift	N(0,1)	Exp(1)	Gamma(2)	Gamma(10)	LN(0.1)	LN(0.5)
P	1	4	0.0530	0.5	0.2386	0.9887	0.7880	0.3552	0.2804	0.6855
				1.0	0.5553	1.0000	0.9982	0.8026	0.6543	0.9943
	2	6	0.0457	0.5	0.2662	0.9355	0.6711	0.3638	0.3038	0.6453
				1.0	0.6529	0.9998	0.9914	0.8371	0.7338	0.9901
	3	7	0.0637	0.5	0.3485	0.9034	0.6767	0.4391	0.3833	0.6838
				1.0	0.7615	0.9993	0.9886	0.8891	0.8210	0.9907
	4	9	0.0412	0.5	0.2947	0.7499	0.5317	0.3583	0.3219	0.5614
				1.0	0.7352	0.9932	0.9615	0.8418	0.7877	0.9716
	5	10	0.0479	0.5	0.3317	0.6904	0.5173	0.3808	0.3525	0.5555
				1.0	0.7787	0.9864	0.9487	0.8532	0.8163	0.9648
	6	11	0.0527	0.5	0.3565	0.6291	0.4966	0.3920	0.3706	0.5369
				1.0	0.8055	0.9733	0.9323	0.8558	0.8295	0.9540
M	2	5	0.0469	0.5	0.2434	0.9634	0.6580	0.3235	0.2741	0.5983
				1.0	0.5994	1.0000	0.9934	0.7893	0.6788	0.9882
	3	6	0.0302	0.5	0.1847	0.9051	0.4878	0.2327	0.2016	0.4460
				1.0	0.5235	0.9997	0.9787	0.6876	0.5896	0.9654
	4	6	0.0403	0.5	0.2138	0.9053	0.4949	0.2546	0.2273	0.4572
				1.0	0.5605	0.9997	0.9790	0.7030	0.6170	0.9662
	5	6	0.0503	0.5	0.2360	0.9058	0.5004	0.2709	0.2472	0.4642
				1.0	0.5833	0.9997	0.9793	0.7115	0.6327	0.9666
	6	6	0.0602	0.5	0.2545	0.9062	0.5044	0.2834	0.2626	0.4688
				1.0	0.5977	0.9997	0.9796	0.7171	0.6423	0.9668
W		472	0.0482	0.5	0.4403	0.7011	0.5699	0.4622	0.4462	0.5941
				1.0	0.9136	0.9802	0.9622	0.9275	0.9179	0.9716

Note: P - Precedence Test; M - Maximal Precedence Test; W - Wilcoxon's Rank-sum Test

Table 5, when $n_1 = n_2 = 20$ and the location-shift equals 0.5, the power of the maximal precedence test with $r = 2$ is 0.6580 (exact level of significance is 0.0469) while the power of Wilcoxon's rank-sum test is 0.5699 (exact level of significance is 0.0482). From Tables 4–6, it is evident that the more right-skewed the underlying distribution is, the more powerful the maximal precedence test is.

Moreover, we can compare the power values of the precedence tests and the maximal precedence tests with the same value of r since they are both test procedures based on failures from the X-sample occurring before the r-th failure from the Y-sample. The power values presented in Tables 4–6 show that the precedence test is more powerful than the maximal precedence test under the normal distribution. However, under the exponential distribution, the maximal precedence test performs better than the prece-

Table 6. Power of precedence tests, maximal precedence tests and Wilcoxon's rank-sum test for $n_1 = n_2 = 30$.

Test	r	critical value	exact l.o.s.	location shift	N(0,1)	Exp(1)	Gamma(2)	Gamma(10)	LN(0.1)	LN(0.5)
P	1	4	0.0562	0.5	0.2690	0.9998	0.9359	0.4354	0.3247	0.8259
				1.0	0.6135	1.0000	1.0000	0.8878	0.7300	0.9998
	2	6	0.0514	0.5	0.3181	0.9974	0.8832	0.4754	0.3754	0.8287
				1.0	0.7355	1.0000	0.9999	0.9302	0.8277	0.9997
	3	8	0.0399	0.5	0.3166	0.9851	0.8073	0.4540	0.3702	0.7846
				1.0	0.7755	1.0000	0.9995	0.9337	0.8525	0.9992
	4	9	0.0521	0.5	0.3898	0.9758	0.8088	0.5175	0.4393	0.8091
				1.0	0.8455	1.0000	0.9992	0.9556	0.9014	0.9993
	5	11	0.0358	0.5	0.3468	0.9233	0.7081	0.4524	0.3867	0.7260
				1.0	0.8375	0.9999	0.9965	0.9402	0.8882	0.9978
	6	12	0.0420	0.5	0.3888	0.8977	0.7007	0.4851	0.4263	0.7293
				1.0	0.8718	0.9998	0.9953	0.9481	0.9104	0.9973
M	2	5	0.0518	0.5	0.2898	0.9990	0.8782	0.4284	0.3421	0.7880
				1.0	0.6909	1.0000	1.0000	0.9019	0.7865	0.9996
	3	6	0.0354	0.5	0.2381	0.9956	0.7735	0.3435	0.2802	0.6766
				1.0	0.6556	1.0000	0.9996	0.8577	0.7430	0.9983
	4	6	0.0471	0.5	0.2795	0.9957	0.7797	0.3761	0.3187	0.6910
				1.0	0.7050	1.0000	0.9997	0.8748	0.7788	0.9984
	5	6	0.0587	0.5	0.3119	0.9957	0.7842	0.4013	0.3490	0.7007
				1.0	0.7370	1.0000	0.9997	0.8849	0.8009	0.9985
	6	7	0.0315	0.5	0.2072	0.9866	0.6400	0.2690	0.2312	0.5460
				1.0	0.6096	1.0000	0.9986	0.7844	0.6772	0.9943
W		1027	0.0498	0.5	0.5868	0.8531	0.7339	0.6146	0.5909	0.7571
				1.0	0.9809	0.9985	0.9951	0.9854	0.9822	0.9969

(The columns N(0,1) through LN(0.5) are grouped under the heading "Distribution".)

Note: P - Precedence Test; M - Maximal Precedence Test; W - Wilcoxon's Rank-sum Test

dence test. From the power values of the precedence tests, we observe that the masking effect in the precedence test becomes more obvious when r becomes larger. The maximal precedence test eliminates the masking effect present in precedence test. The power values reveal that larger the value of r, the more superior the maximal precedence test becomes as compared to the precedence test under an exponential distribution. For example, when $n_1 = n_2 = 20$ and the location-shift equals 0.5, we find the power of the precedence test to be 0.9355 (exact level of significance is 0.0457) while the power of the maximal precedence test is 0.9631 (exact level of significance is 0.0469) for $r = 2$; and the power of the precedence test is 0.6291 (exact level of significance is 0.0527) while the power of the maximal precedence test is 0.9062 (exact level of significance is 0.0602) for $r = 6$.

6. Possible Future Research

Even though we have suggested an alternative to the precedence test in the form of maximal precedence test, it too is based on frequencies of failures preceding the r-th failure. One possible extension may be to construct Wilcoxon-type rank based test that takes into account the magnitude of the failure times. Another possible extension to the work presented here is to consider the testing problem when the observed data is progressively Type-II censored (instead of conventional Type-II censoring); for details on progressive censoring and related developments, one may refer to Balakrishnan and Aggarwala[12]. Work in these directions are currently under progress and we hope to report these findings in a future paper.

References

1. Nelson, L. S., Tables of a Precedence Life Test, *Technometrics*, **5**, 491-499, (1963).
2. Nelson, L. S., Precedence Life Test, In *Encyclopedia of Statistical Sciences*, **7** (Eds., S. Kotz and N. L. Johnson), pp. 134-136, New York: John Wiley & Sons, (1986).
3. Nelson, L. S., Tests on early failures - The precedence life test, *Journal of Quality Technology*, **25**, 140-143, (1993).
4. Balakrishnan, N. and Frattina, R., Precedence Test and Maximal Precedence Test, In *Recent Advances in Reliability Theory: Methodology, Practice, and Inference* (Eds. N. Limnios and M. Nikulin), pp. 355-378, Boston: Birkhäuser, (2000).
5. David, H. A., *Order Statistics*, Second edition, New York: John Wiley & Sons, (1981).
6. Arnold, B. C., Balakrishnan, N. and Nagaraja, H. N., *A First Course in Order Statistics*, New York: John Wiley & Sons, (1992).
7. Lehmann, E. L., The Power of Rank Tests, *Annals of Mathematical Statistics*, **24**, 23-42, (1953).
8. Davies, R. B., Rank Tests for 'Lehmann Alternative', *Journal of the American Statistical Association*, **66**, 879-883, (1971).
9. Lehmann, E. L., *Nonparametrics: Statistical Methods Based on Ranks*, New York: McGraw-Hill, (1975).
10. Gibbons, J. D. and Chakraborti, S., *Nonparametric Statistical Inference*, Third edition, New York: Marcel Dekker, (1992).
11. Johnson, N. L., Kotz, S. and Balakrishnan, N., *Continuous Univariate Distributions, Vol. 1*, Second edition, New York: John Wiley & Sons, (1994).
12. Balakrishnan, N., and Aggarwala, R., *Progressive Censoring: Theory, Methods and Applications*, Boston: Birkhäuser, (2000).

CHAPTER 8

TOTAL TIME ON TEST PROCESSES AND THEIR APPLICATION TO MAINTENANCE PROBLEM

Tadashi Dohi

Faculty of Engineering, Hiroshima University
1-4-1 Kagamiyama, Higashi-Hiroshima 739-8527, Japan
E-mail: dohi@gal.sys.hiroshima-u.ac.jp

Naoto Kaio

Faculty of Economic Sciences, Hiroshima Shudo University
1-1-1 Ozukahigashi, Asaminami-ku, Hiroshima 731-3195, Japan
E-mail: kaio@shudo-u.ac.jp

Shunji Osaki

Faculty of Mathematical and Information Sciences, Nanzan University
Seirei-cho, Seto 489-0863, Japan
E-mail: shunji@it.nanzan-u.ac.jp

One of the most important contributions in reliability theory by Richard E. Barlow is the concept of the *total time on test (TTT) processes.* In fact, the TTT is a very useful statistical device to study aging properties of the underlying lifetime distribution and at the same time can be applied to solve geometrically some stochastic maintenance problems. This article concerns the TTT processes and surveys many theoretical results on them since the seminal work by Barlow and Campo (1975). A comprehensive bibliography on the TTT processes and their applications is also provided.

1. Introduction

As well known, Professor Richard E. Barlow's research interests are wide-ranging. In fact, his work includes numerous papers and books on maintenance optimization, failure data analysis and Bayesian statistics[1,2,3]. One of his most important contributions in reliability theory is the concept of the total time on test (TTT) processes. The TTT is a very useful statistical device to study aging properties of the underlying lifetime distribution[4,5] and at the same time can be applied to solve geometrically some stochastic maintenance problems. Inspite of its importance, only a few textbooks on reliability engineering and practice do treat the TTT concept. Since the seminal contribution by Barlow and Campo[4], many interesting results on the TTT processes have been developed in the literature. This article concerns the TTT processes and surveys many significant results on them.

Suppose that the lifetime X is a continuous random variable and obeys the following exponential distribution:

$$F(x) = 1 - \exp\Big\{-x/\theta\Big\}, \quad x \geq 0 \tag{1}$$

with parameter θ (> 0). Define n (> 0) order statistics $0 = x_0 \leq x_1 \leq \cdots \leq x_n$ sampled from $F(x)$. When r ($\leq n$)-th order statistics is observed, the maximum likelihood estimator of the parameter θ is given by

$$\hat{\theta}_{r,n} = \frac{1}{r}\Big\{\sum_{i=1}^{r} x_i + (n-r)x_r\Big\}. \tag{2}$$

This estimator is also the MVUE (minimum variance and unbiased estimate) for θ. Define

$$\begin{aligned} T(x_r) &= r\hat{\theta}_{r,n} \\ &= nx_1 + (n-1)(x_2 - x_1) + \cdots + (n-r+1)(x_r - x_{r-1}). \end{aligned} \tag{3}$$

The functuion $T(x_r)$ is the sum of all observed complete and incomplete failure data up to the rth failure. Hereafter, we focus on the statistics $T(x_r)$, which is calld the total time on test (TTT) statistics, and its related quantities.

This article is organized as follows. First, we formally define the scaled TTT transform of continuous probability distribution function and describe some aging and stochastic ordering properties. Second, we define the scaled TTT statistics as the empirical counterpart of the scaled TTT transform. The asymptotic behavior of the scaled TTT statistics is examined. Third, a graphical method to determine the optimal maintenance policy is explained. In general, the determination of the optimal maintenance schedule strongly depends on the underlying failure or repair time distribution. More precisely, one often needs to select an appropriate distribution and estimates for model parameters from the data using standard statistical estimation techniques. However, in the graphical method based on the scaled TTT statistics, the probability distribution function does not need to be specified subjectively and resulting estimator of the optimal maintenance policy will be optimal asymptotically. A comprehensive bibliography on the TTT processes and their applications is also provided.

2. Scaled TTT Transform

2.1. *Definition*

Let $F(x)$ be a probability distribution function, that is, $F(0) = 0$ and $\lim_{x\to\infty} F(x) = 1$. In the following discussion, we use the notation;

$$F^{-1}(p) = \inf\{x : F(x) \geq p\}, \quad p \in [0,1). \tag{4}$$

We use "increasing" in place of "nondecreasing" and "decreasing" in place of "nonincreasing". Let

$$H_F^{-1}(p) = \int_0^{F^{-1}(p)} \overline{F}(x)dx, \quad p \in [0,1], \tag{5}$$

where $\overline{F}(\cdot) = 1 - F(\cdot)$ is the survivor function. Barlow and Campo[4] called H_F^{-1} the *total time on test process*. If there exists a finite mean $\int_0^\infty \overline{F}(x)dx = \mu\ (> 0)$, then

$$H_F^{-1}(1) = \int_0^{F^{-1}(1)} \overline{F}(x)dx = \mu. \tag{6}$$

Further, we define the *scaled total time on test transform* (scaled TTT transform) by

$$\phi(p) = H_F^{-1}(p)/H_F^{-1}(1) = H_F^{-1}(p)/\mu. \tag{7}$$

From the definition above, it can be seen that the function $\phi(p), p \in [0,1]$ is equivalent to the *equilibrium distribution* of the probability distribution function $F(x)$, if $F(x)$ is non-arithmetic[1,2]. Then the curve $\mathcal{T} = (p, \phi(p)) \in [0,1] \times [0,1]$ is called the *scaled TTT curve*.

2.2. *Some Aging Properties*

Let $H_F^{-1}(\cdot)$ (equivalently $F^{-1}(\cdot)$) be continuous at $p \in (0,1)$. Then, it is seen that

$$\frac{d}{dp} H_F^{-1}(p) \mid_{p=F(x)} = \frac{1}{r(x)}, \tag{8}$$

where $r(x) = (dF(x)/dx)/\overline{F}(x)$ is the failure rate (hazard rate) function of $F(x)$, if it exists. Thus, if $r(x)$ is increasing, $H_F^{-1}(p)$, say, $\phi(p)$ is concave in $p \in [0, 1]$, while $r(x)$ is decreasing then $H_F^{-1}(p)$ is convex in $p \in [0, 1]$. In the sequel, consider the relationship between some aging properties of the probability distribution function and the corresponding scaled TTT transform. First, we define the classes of probability distributions to be considered.

Definition 1: The probability distribution function $F(x)$ is IFR (Increasing Failure Rate) [DFR (Decreasing Failure Rate)] if $\overline{F}(y+x)/\overline{F}(y)$ is decreasing [increasing] in $y \in [0, \infty)$ for each $x > 0$ [or if $F(\cdot)$ is degenerate].

Definition 2: The probability distribution function $F(x)$ is IFRA (Increading Failure Rate Average) [DFRA (Decreasing Failure Rate Average)] if the function $-(1/x)\log \overline{F}(x)$ is increasing [decreasing] in $x \geq 0$ or if $F(\cdot)$ is degenerate at 0.

Definition 3: The probability distribution function $F(x)$ is NBU (New Better Than Used) [NWU (New Worse Than Used)] if $\overline{F}(x+y) \leq [\geq]\overline{F}(x)\overline{F}(y)$ for $x > 0$, $y > 0$.

Definition 4: The probability distribution function $F(x)$ with finite mean μ is NBUE (New Better Than Used in Expectation) [NWUE (New Worse Than Used in Expectation)] if $\int_x^\infty \overline{F}(u)du \leq [\geq]\overline{F}(x)\mu$ for $x > 0$.

Definition 5: The probability distribution function $F(x)$ is DMRL (Decreasing Mean Residual Life) [IMRL (Increasing Mean Residual Life)] if the mean residual life $(1/\overline{F}(x)) \int_x^\infty \overline{F}(u)du$ is decreasing [increasing] in $x > 0$.

Definition 6: The probability distribution function $F(x)$ is HNBUE (Harmonic New Better Than Used in Expectation) [HNWUE (Harmonic New Worse Than Used in Expectation)] if $\int_x^\infty \overline{F}(u)du \leq [\geq]\mu \exp(-x/\mu)$.

Figure 1 illustrates the implications among the classes of probability distributions defined above.

The following theorem presents the characterization of the distribution functions based on the scaled TTT trasnform.

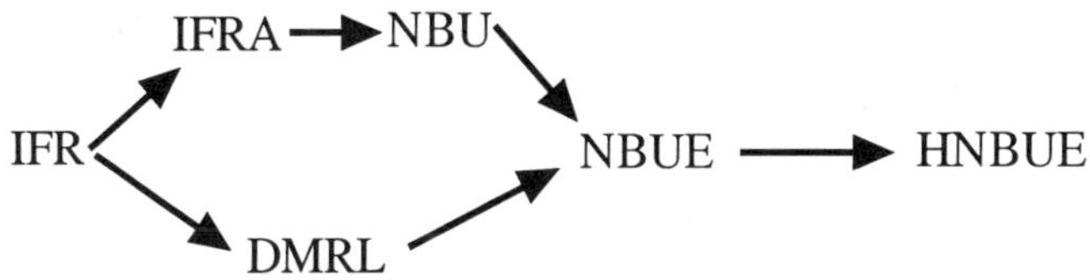

Fig. 1. Implications among notions of aging.

Theorem 7:

(i) The probability distribution function $F(x)$ is IFR [DFR] if and only if $\phi(p)$ is concave [convex] in $p \in [0,1]$.
(ii) If the probability distribution function $F(x)$ is IFRA [DFRA], then $\phi(p)/p$ is decreasing [increasing] in $p \in [0,1]$.
(iii) The probability distribution function $F(x)$ is NBUE [NWUE] if and only if $\phi(p) \geq [\leq]\ p$ for $p \in [0,1]$.
(iv) The probability distribution function $F(x)$ is DMRL [IMRL] if and only if $(1-\phi(p))/(1-p)$ is decreasing [increasing] in $p \in [0,1]$.
(v) The strictly increasing probability distribution function $F(x)$ is HNBUE [HNWUE] if and only if $\phi(p) \leq [\geq]\ 1 - \exp\{-F^{-1}(p)/\mu\}$ for $p \in [0,1]$.

The result (i) was independently proved by Barlow and Campo[4] and Lee and Thompson[6]. The result (ii) is also due to Barlow and Campo[4]. The NBUE (NWUE) characterization in (iii) was made by Bergman[7]. Langberg, León and Proschan[8] gave more detailed results on IFRA (DFRA) and NBU (NWU). Both results on DMRL (IMRL) and HNBUE (HNWUE) were obtained by Klefsjö[9].

2.3. *Stochastic Ordering*

Next, we consider the relationship between the TTT transform and the stochastic ordering (dominance) of the probability distribution functions. In fact, the concept of stochastic ordering is needed to compare two arbitrary probability distribution functions.

Definition 8: For two non-negative real valued random variables X and Y, define the coresponding survivor functions $\overline{F}_X(t)$ and $\overline{F}_Y(t)$, $t \geq 0$, respectively. Then the random variable X is said to be *smaller* than Y in the sense of *stochastic ordering*, *i.e.* $X \prec Y$, if $\overline{F}_X(t) \leq \overline{F}_Y(t)$ for $t \geq 0$.

Definition 9: The random variable X is said to be *smaller* than Y in the sense of *convex ordering, i.e.* $X \prec_c Y$, if $\int_t^\infty \overline{F}_X(u)du \le \int_t^\infty \overline{F}_Y(u)du$ for $t \ge 0$.

Definition 10: The random variable X is said to be *smaller* than Y in the sense of *star-shaped ordering, i.e.* $X \prec_* Y$, if $F_Y^{-1}(F_X(t))/t$ is increasing in $t \ge 0$.

Definition 11: The random variable X is said to be *smaller* than Y in the sense of *DMRL ordering, i.e.* $X \prec_{\text{DMRL}} Y$, if $v_X(F_X^{-1}(p)) \,/ v_Y(F_Y^{-1}(p))$ is decreasing in $p \in [0,1]$, where $v_j(t) = \int_t^\infty \overline{F_j}(u)du \,/ \overline{F_j}(t)$, $j = X, Y$.

For the stochastic ordering, see Shaked and Shanthikumar[10] and Szekli[11].

Theorem 12: Let $H_{F_X}^{-1}(p)$ and $H_{F_Y}^{-1}(p)$ be the TTT transforms of the distribution functions $F_X(t)$ and $F_Y(t)$ for $t \ge 0$, respectively.

(i) If $X \prec_c Y$ then $H_{F_X}^{-1}(p) \prec_c H_{F_Y}^{-1}(p)$ for $p \in [0,1]$.
(ii) If $X \prec_* Y$ then $H_{F_X}^{-1}(p) \prec_* H_{F_Y}^{-1}(p)$ for $p \in [0,1]$.
(iii) $X \prec_{\text{DMRL}} Y$ if and only if $(1-\phi_X(p))/(1-\phi_Y(p))$ is decreasing in $p \in [0,1]$.

The results in (i) and (ii) were given by Barlow[12]. These results are valid even for $\phi_X(p) = H_{F_X}^{-1}(p)/H_{F_X}^{-1}(1)$ and $\phi_Y(p) = H_{F_Y}^{-1}(p)/H_{F_Y}^{-1}(1)$, if $H_{F_X}^{-1}(1) = H_{F_Y}^{-1}(1)$. On the other hand, Kochar[13] proved the partial ordering result in (iii). For more detailed results on the DMRL (IMRL) distributions, see Bergman and Klefsjö[14].

3. Scaled TTT Statistics

3.1. *Definition*

Suppose that the random variable X has to be estimated from an ordered complete sample $0 = x_0 \le x_1 \le x_2 \le \cdots \le x_n$ from an absolutely continuous distribution $F(\cdot)$, which is unknown. At the moment, we restrict our attention to the complete sample. The case with incomplete (censored) sample will be discussed in the latter part. Define the empirical distribution for the order statistics $x_1, x_2, \cdots, x_n$ by

$$F_n(u) = \begin{cases} 0, & u < x_1, \\ i/n, & x_i \le u < x_{i+1}, \quad i = 1, \cdots, n-1 \\ 1, & u \ge x_n. \end{cases} \tag{9}$$

If there exists the inverse function $F_n^{-1}(u) = \inf\{x \geq 0; F_n(x) > u\}$, it can be seen that

$$\int_0^{F_n^{-1}(r/n)} \overline{F}_n(u)du = \sum_{j=1}^{r}\Big(1 - \frac{j-1}{n}\Big)(x_j - x_{j-1}) = \frac{T(x_r)}{n}, \tag{10}$$

where $T(x_r)$ is given in Eq. (2). From the well-known property on the empirical distribution, it turns out that

$$\lim_{n\to\infty}\lim_{r/n\to p}\int_0^{F_n^{-1}(r/n)} \overline{F}_n(u)du = \int_0^{F^{-1}(p)} \overline{F}(u)du \tag{11}$$

uniformly in $p \in [0,1]$. For the sample mean $\mu_n = \sum_{j=1}^{n} x_j/n$, the statistics $\phi_{r,n} = T(x_r)/\mu_r$, $r = 1, 2, \cdots, n$, is calld the *scaled TTT statistics* and is a numerical counter part of the scaled TTT transform of the distribution function $F(x)$. In a fashion similar to the scaled TTT curve, we define the scaled TTT plot $\mathcal{T}_n$, which yields by connecting the points $(i/n, \phi_{i,n})$, $i = 0, 1, \cdots, n$ with the line segments.

Finally, we give an asymptotic convergence property of the scaled TTT statistics. Let $\phi_{j,n} = \{\phi_n(p), p \in [0,1]\} = \{H_n^{-1}(p)/H_n^{-1}(1), p \in [0,1]\}$ be the scaled TTT process. Further, we define the standardized TTT process $\{S_n(p), p \in [0,1]\}$, that is,

$$S_n(p) = \sqrt{n}\Big\{\frac{H_n^{-1}(p)}{\sum_{j=1}^{n} x_j} - \phi(t)\Big\} \tag{12}$$

for $(j-1)/n \leq p \leq j/n$ and $1 \leq j \leq n$, where $S_n(0) = S_n(1) = 0$. First, by an integration by parts, we get

$$\phi(t) = \int_0^{F^{-1}(p)} \overline{F}(u)du = \int_0^{p} F^{-1}(u)du + (1-p)F^{-1}(p). \tag{13}$$

On the other hand, it can be seen that

$$\frac{H_n^{-1}(j/n)}{H_n^{-1}(1)} = \int_0^{j/n} \frac{F_n^{-1}(u)}{\sum_{j=1}^{n} x_j} d\nu_n(u) + (1 - j/n)\frac{x_j}{\sum_{j=1}^{n} x_j} \to \int_0^{p} \frac{F^{-1}(u)}{\mu}du + \frac{(1-p)}{\mu}F^{-1}(p) \tag{14}$$

with probability one and uniformly in $p \in [0,1]$ as $n \to \infty$, where the function $\nu_n(u)$ puts mass $1/n$ at $u = j/n$, $j = 1, 2, \cdots, n$. Barlow and Campo[4] showed

$$S_n(j/n) = \sqrt{n}\Big\{\frac{H_n^{-1}(j/n)}{\sum_{j=1}^n x_j} - \phi(t)\Big\}$$
$$\approx \int_0^{j/n} \sqrt{n}\Big\{\frac{x_{[nu]}}{\sum_{j=1}^n x_j} - \frac{F^{-1}(u)}{\mu}\Big\}d\nu_n(u)$$
$$+(1-j/n)\sqrt{n}\Big\{\frac{x_j}{\sum_{j=1}^n x_j} - \frac{F^{-1}(j/n)}{\mu}\Big\}, \qquad (15)$$

where $[nu]$ means the greatest integer in nu

Define the function

$$P(t) = -\frac{dg(t)/dt}{\mu}U(t) + \frac{g(t)}{\mu^2}\int_0^\infty U(F(x))dx, \qquad (16)$$

where $g(t) = F^{-1}(t)$, $dg(t)/dt = 1/(dF(x)/dx) \mid_{x=F^{-1}(t)}$ and $\{U(p), p \in [0,1]\}$ is the Brownian bridge process.

The following results given by Barlow and Campo[4] present the asymptotic normality of the scaled TTT process.

Theorem 13: If $F(x)$ is absolutely continuous, then in the sense of weak convergence,

$$\lim_{n\to\infty} \sqrt{n}\Big\{\frac{H_n^{-1}(p)}{\sum_{j=1}^n x_j} - \phi(t)\Big\} = \int_0^p P(u)du + (1-p)P(p), \qquad (17)$$

where

$$P(p) = -\frac{dg(p)/dp}{\mu}U(p) + \frac{g(p)}{\mu^2}\int_0^\infty U(F(x))dx. \qquad (18)$$

Theorem 14:

$$\lim_{n\to\infty} \sqrt{n}\Big\{\frac{H_n^{-1}(p)}{\sum_{j=1}^n x_j} - \phi(t)\Big\} = U(p). \qquad (19)$$

3.2. *Other Related Topics*

In practice we often encounter the cases where the complete sample data can not be necessarily obtained. In other words, the incomplete data such as grouped data, truncated data and censored data should be handled for the scaled TTT plotting. If the censored data is available, we can use the well-known Kaplan-Meier estimator[15] of the probability distribution function instead of the empirical distribution. Consider the time-t censored data. Let $x_1 < x_2 < \cdots < x_n$ denote the recorded functioning times, either until failure or to censoring, ordered according to size. Let J_t denote the set of all indices j where $x_j \leq t$ and x_j represents a failure time. Define the number of units functioning and in observation immediately before time x_j by n_j, $j = 1, 2, \cdots, n$. Then the Kaplan-Meier estimator or the product limit estimator of the survivor function $F(x)$ is given by

$$R_n(x) = \Pi_{j \in J_x} \frac{n_j - 1}{n_j}. \tag{20}$$

Replacing $\overline{F}_n(u)$ in Eq. (10) by $R_n(u)$ in Eq. (20), we obtain the improved scaled TTT plot for the censored data. As an alternative method to handle the censored sample, Kim and Proschan[16] introduced the piecewise exponential estimator of the survivor function. Using this, Westberg and Klefsjö[17] developed a TTT-plotting for censored data based on the piecewise exponential estimator. Recently, Sun and Kececioglu[18] proposed a new method called the mean-order-number method for the TTT plotting with a censored data.

To analyze the failure data, the probability distribution with bathtub failure rate plays an central role for characterizing hardware failure phenomena. Barlow[19] analyzed a retrospective failure data using the TTT plot. Aarset[20] and Xie[21] proposed and illustrated the use of the usual TTT plot for identifying bathtub failure rates. Mudholkar and Srivastava[22] focused on a simple generalization of the Weibull distribution called the exponentiated-Weibull family, and showed that it could be used to test not only exponentiality but also goodness-of-fit of the Weibull distribution. For the detailed survey on the statistical test based on the scaled TTT plot, see Klefsjö[23] and Hoyland and Rausand[24].

More generally, the scaled TTT plot is useful to identify some specified stochastic processes. Barlow and Davis[25] characterized a superposition of non-homogeneous Poisson processes by the scaled TTT plot and analyzed the times between failures for engines placed in the tractor. Klefsjö[26] introduced a new plot called the *double TTT plot* to identify the non-homogeneous Poisson processes versus a homogeneous Poisson process. Klefsjö and Kumar[27] gave the goodness-of-fit tests for a special class of non-homogeneous Poisson process called the *power-law process*, based on the scaled TTT-plot. Also, some statisticians derived some theoretical results on the scaled TTT statistics. Ebrahimi and Spizzichino[28] characterized the lifetimes with Schur-constant density in terms of the joint density function of the TTT statistics. Nappo and Spizzichino[29] studied the ordering properties of the TTT-plot of lifetimes with Schur joint densities. Csörgö and Yu[30] showed strong uniform consistency of the TTT plot for a sequence of dependent random variables to the scaled TTT transform of the unimensional marginal distribution. Recently, Hayakawa[31,32] discussed the relationship between the total time on test statistics and l_1-isotropy and applied it to characterize the properties of mixtures of exponential distributions.

4. Application to Maintenance Problem

In this section, we consider the maintenance optimization models and apply the TTT concept to determine the optimal maintenance policies. First, Bergman[34,35] found a graphical method to determine the optimal age replacement time[36,37] based on the scaled TTT statistics. Latter, Bergman[38] and Bergman and Klefsjö[39,40,41] applied this idea to some age replacement-like maintenance problems with different cost criteria. Klefsjö[42] and Jinlin[43] considered burn-in problems with the TTT approach. The minimal repair problems by the TTT approach were considered by Klefsjö[26,44]. Kumar and Westberg[45] extended the original Bergman's approach[35] to an age replacement problem based on the proportional hazards model. Reineke, Pohl and Murdock[46] reconsidered the age replacement model with highly censored data. On the other hand, Dohi *et al.*[47,48,49,50,51,52,53,54,55] applied the scaled TTT approach to the different repair-limit replacement problems[56,57,58,59,60,61,62,63,64]. Also, in the literature[65,66,67], the same authors developed an alternative graphical method based on the Lorenz transform to solve the other kinds of repair-limit problems. In the recent year, the authors used the scaled TTT statistics to determine the optimal power saving strategy for a portable personal computer[68]. Also, in the literature[69], the optimal checkpointing and rollback strategies were analyzed with the same graphical method. The preventive maintenance problems for an operational software system were formulated by Dohi, Goševa-Popstojanova and Trivedi[70,71,72,73]. They applied the scaled TTT statistics to estimation problems for rejuvenating the telecommunication billing applications.

In this section, we will give an example on the repair limit replacement mopdel under the earning rate criterion, respectively, discussed in the literature[50]. Consider a single-unit production system which is repairable. When the unit fails, the repair is started immediately. If the repair is completed up to a time (repair time limit) $t_0 \in [0, \infty)$, then the unit is installed at that time. The repair is assumed to be perfect. The mean failure time is $1/\lambda (> 0)$. On the other hand, if the repair time is greater than t_0, *i.e.* the repair is not completed at the time t_0, then the failed unit is scrapped and is replaced immediately by a new spare unit. Similarly, the time required for replacement is negligible.

Define the following notation:

$G(t)$, $g(t)$, $1/\lambda_g$: c.d.f., p.d.f. and the mean for the repair time for each unit.

e_0: the net earning rate per unit time made by the production of the working system.

e_1: the repair cost per unit time

e_2: the replacement cost.

Then the expected earning rate in the steady state is, from the familiar renewal reward argument[74],

$$C(t_0) = \lim_{t \to \infty} \frac{\text{E[total profit on } (0, t]]}{t} = \frac{V(t_0)}{T(t_0)}, \tag{21}$$

where the mean time length of one cycle $T(t_0)$ and the expected total profit for one cycle $V(t_0)$ are

$$T(t_0) = \frac{1}{\lambda} + \int_0^{t_0} \overline{G}(t)dt \tag{22}$$

and

$$V(t_0) = \frac{e_0}{\lambda} - e_1 \int_0^{t_0} \overline{G}(t)dt - e_2\overline{G}(t_0), \tag{23}$$

respectively. Of our interest is to obtain the optimal repair time limit $t_0{}^*$ which maximizes $C(t_0)$.

Define the numerator of the derivative of $C(t_0)$ with respect to t_0, divided by $\overline{G}(t_0)$, as $q(t_0)$, *i.e.*

$$q(t_0) = (e_2 r_g(t_0) - e_1)T(t_0) - V(t_0), \tag{24}$$

where

$$r_g(t) = g(t)/\overline{G}(t) \tag{25}$$

is called the instantaneous repair rate and is equivalent to the failure rate of the distribution $G(t)$. We also assume that $r_g(t)$ is differentiable with respect to t.

Theorem 15: If $q(\infty) < 0$ or $q(0) > 0$, then there exists at least one optimal repair time limit t_0^* $(0 \le t_0^* < \infty$ or $0 < t_0{}^* \le \infty)$ which maximizes the expected earning rate in the steady state.

Theorem 16: (1) Suppose that the repair time distribution $G(t)$ is strictly DFR.

(i) If $q(0) > 0$ and $q(\infty) < 0$, then there exists a finite and unique optimal repair time limit $t_0{}^*$ $(0 < t_0{}^* < \infty)$ satisfying $q(t_0^*) = 0$ and the corresponding maximum expected earning rate is

$$C(t_0^*) = e_2 r_g(t_0{}^*) - e_1. \tag{26}$$

(ii) If $q(0) \le 0$, then the optimal repair time limit is $t_0{}^* = 0$, *i.e.* it is optimal that the repair is not carried out but only the replacement should be executed. The corresponding maximum expected earning rate is given by $C(0) = e_0 - e_2\lambda$.

(iii) If $q(\infty) \ge 0$, then the optimal repair time limit is $t_0{}^* \to \infty$, *i.e.* it is optimal that the replacement for the failed unit is not executed but only the repair should be carried out. The corresponding maximum expected earning rate is given by $C(\infty) = (e_0\lambda_g - e_1\lambda)/(\lambda + \lambda_g)$.

(2) Suppose that the repair time distribution is IFR. If $V(0)T(\infty) \ge V(\infty)T(0)$, then $t_0{}^* = 0$, otherwise, $t_0{}^* \to \infty$.

Denoting the scaled TTT transform of the repair time distribution by $\phi(p)$, the following result can be easily obtained from a simple algibraic manipulation.

Theorem 17: Obtaining the optimal repair time limit $t_0{}^*$ which maximize the expected earning rate in the steady state $C(t_0)$ is equivalent to obtaining p^* $(0 \le p^* \le 1)$ which minimizes

$$\frac{\phi(p) - (-\lambda_g/\lambda)}{p - (1 - \frac{e_0+e_1}{e_2\lambda})}. \tag{27}$$

Next, we treat the scaled TTT transform $\phi(p)$ of the repair time distribution $G(t)$. In the plane $(x, y) = (p, \phi(p))$, we define the following three points.

$$\mathrm{B} = (x_\mathrm{B}, y_\mathrm{B}) \equiv \left(1 - \frac{e_0 + e_1}{e_2 \lambda}, -\lambda_g/\lambda\right), \tag{28}$$

$$\mathrm{Z} = (x_\mathrm{Z}, y_\mathrm{Z}) \equiv (-\frac{\lambda_g}{\lambda \phi'(0)}, -\frac{\lambda_g}{\lambda}) \tag{29}$$

and

$$\mathrm{I} = (x_\mathrm{I}, y_\mathrm{I}) \equiv (1 - \frac{1 + \lambda_g/\lambda}{\phi'(1)}, -\frac{\lambda_g}{\lambda}), \tag{30}$$

where the point Z is the intersection of the tangent line for the $y = \phi(p)$ at the point O and the straight line $y = -\lambda_g/\lambda$, and the point I is the intersection of the tangent line at the point U and $y = -\lambda_g/\lambda$.

Theorem 18: (1) Suppose that the scaled TTT transform of the repair time distribution $\phi(p)$ is strictly convex in p.

(i) If $x_\mathrm{Z} < x_\mathrm{B} < x_\mathrm{I}$, then there exists a unique solution $p^*(0 < p^* < 1)$ maximizing the expected earning rate in Eq. (21), where p^* is given by the coordinate $x = p$ in the point of contact $(p, \phi(p))$ for $y = \phi(p)$ from the point B. The corresponding maximum expected earning rate is given by Eq. (26).

(ii) If $x_\mathrm{B} \le x_\mathrm{Z}$, then the optimal repair limit policy is $p^* = 0$ $(t_0{}^* = 0)$ and the corresponding maximum expected earning rate is given by $C(0) = (e_0/\lambda - e_2)\lambda$.

(iii) If $x_\mathrm{B} \ge x_\mathrm{I}$, then the optimal repair limit policy is $p^* = 1$ $(t_0{}^* \to \infty)$ and the corresponding maximum earning rate is given by $C(\infty) = (e_0/\lambda - e_1/\lambda_g)\lambda\lambda_g/(\lambda + \lambda_g)$.

(2) Suppose that the scaled TTT transform of the repair time distribution $\phi(p)$ is concave in p. Then the optimal repair limit policy is given by $p^* = 0$ if $x_\mathrm{B} \le -\lambda_g/\lambda$, otherwise, $p^* = 1$.

Figure 2 depicts a schematic illustration on the determination of the optimal repair time limit. From the results above, we can obtain the optimal policy graphically when the repair time distribution has a monotone aging property (*i.e.* IFR or DFR). On the other hand, if the underlying distribution function has the other aging property except IFR and DFR, it will not be easy to calculate the optimal policy analytically. In general,

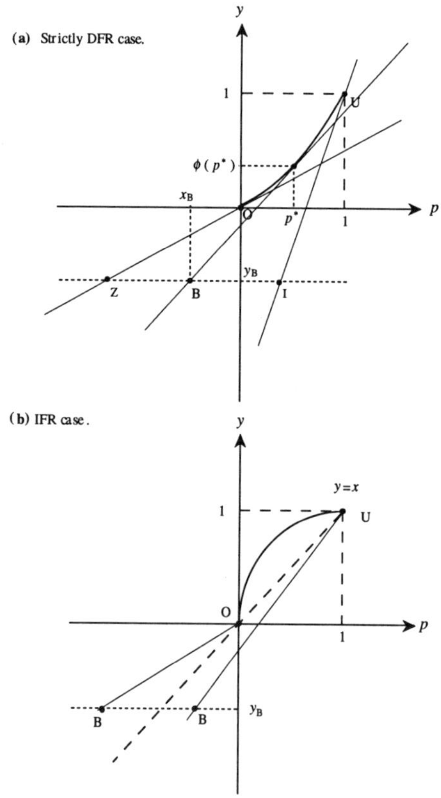

Fig. 2. Schematic illustration for the graphical determination of the optimal repair-time limit.

the precise investigation of aging properties from empirical data is rather troublesome and it is not always possible to determine whether the hypothesis of being IFR or DFR is rejected. Hence, we should consider the case where the distribution function belongs to more general classes and should analyze such a case using the graphical methods proposed here.

As an empirical counterpart, we define the scaled TTT plot of the failure time distribution. Suppose an ordered complete sample $0 = x_0 \leq x_1 \leq x_2 \leq \cdots \leq x_n$ from the underlying distribution function G, which is unknown. The scaled TTT statistics based on this sample is defined by $\phi_{j,n}$, $j = 1, 2, \cdots, n$. Applying the empirical distribution function for

$i = 0, 1, 2, \cdots, n-1$,

$$G_n(x) = \begin{cases} \frac{i}{n} & \text{for} \quad x_i \leq x < x_{i+1}, \\ 1 & \text{for} \quad\quad x_n \leq x, \end{cases} \tag{31}$$

plotting the point $(j/n, \phi_{j,n})$, $j = 1, 2, \cdots, n$ and connecting them by line segments yield a curve called the scaled TTT plot.

Theorem 19: Suppose that the optimal repair time limit has to be estimated from an ordered complete sample $0 = x_0 \leq x_1 \leq x_2 \leq \cdots \leq x_n$ of repair times from an absolutely continuous repair time distribution G, which is unknown. Then the estimator of the optimal repair time limit which maximizes the expected earning rate in the steady state is given by x_{i^*}, where

$$i^* = \left\{ i \mid \min_{0 \leq i \leq n} \frac{v_i - y_{\mathrm{B}}}{i/n - x_{\mathrm{B}}} \right\}. \tag{32}$$

Note that the estimators derived above are consistent estimators, that is, the estimate x_{i^*} approaches to the real optimal solution $t_0{}^*$ as the number of data increases. This statistical property is very powerful and is useful to estimate the optimal maintenance schedule accurately when one can collect a sufficiently large number of data.

Example 1: The repair limit replacement problem is examined. The parameters are the following; $\theta = 0.800$, $\beta = 0.600$, $\lambda = 1.300$, $e_0 = 1.400$, $e_1 = 0.700$, $e_2 = 1.200$. The output result is displayed in Figure 3. In this case, the minimum slope is attained by the tangent passing through point $\mathrm{B} = (-0.346, -0.639)$ to the point $(0.590, 0.323)$ and the optimal repair time limit and the maximum expected earning rate are $t_0{}^* = 0.661$ and $C(t_0{}^*) = 0.272$, respectively.

Example 2: We show the other example for the repair limit replacement problem. It is assumed that the repair-time data, which is generated by the 40 random Weibull number with same parameters as those in Example 1, is observed. The other model parameters are $\lambda = 0.091$, $e_0 = 3.142$, $e_1 = 5.682$ and $e_3 = 80.284$. In Figure 4, we estimate the optimal repair-time limit $\hat{t_0}^* = 5.400$.

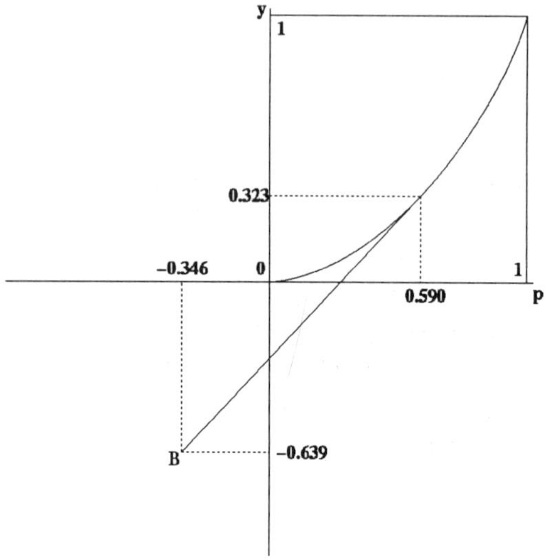

Fig. 3. Determination of the optimal repair-time limit.

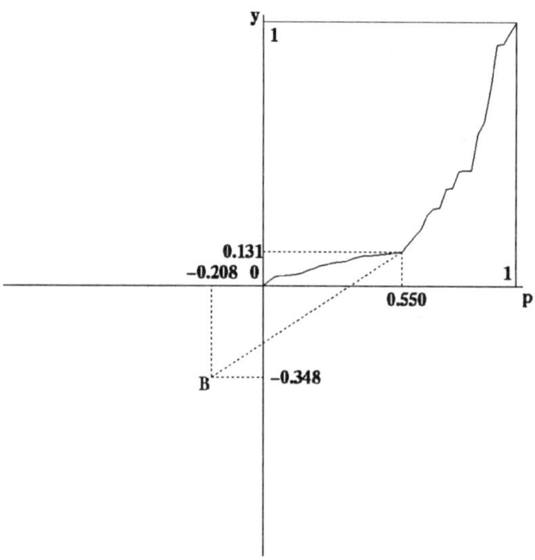

Fig. 4. Estimation of the optimal repair-time limit.

Acknowledgments

This work was supported by a Grant-in-Aid for Scientific Research from the Ministry of Education, Sports, Science and Culture of Japan under Grant No. 13780367, the Research Program 1999 under the Institute for Advanced Studies of the Hiroshima Shudo University, Hiroshima 731-3195, Japan and Nanzan University Pache Research Subsidy I-A. The authors are grateful to Prof. Richard E. Barlow who stimulated their interests in reliability engineering through his famous books co-authored with Frank Proschan.

References

1. R. E. Barlow and F. Proschan, *Mathematical Theory of Reliability* (John Wiley & Sons, New York, 1965).
2. R. E. Barlow and F. Proschan, *Statistical Theory of Reliability and Life Testing* (Holt, Rinehart and Winston, New York, 1975).
3. R. E. Barlow, *Engineering Reliability* (SIAM, Philadelphia, 1998).
4. R. E. Barlow and R. Campo, Total time on test processes and applications to failure data analysis, in *Reliability and Fault Tree Analysis*, Eds. R. E. Barlow, J. Fussell and N. D. Singpurwalla (SIAM, Philadelphia, 1975), pp. 451–481.
5. M. Shaked and J. G. Shanthikumar, Reliability and maintainability, in *Stochastic Models, Handbooks in Operations Research and Management Science Vol. 2*, Eds. D. P. Heyman and M. J. Sobel (North-Holland, Amsterdam 1990), pp. 653–713
6. L. Lee and W. A. Thompson, Failure rate – a unified approach, *Journal of Applied Probability*, **13**, 176–182 (1976).
7. B. Bergman, Crossings in the total time on test plot, *Scandinavian Journal of Statistics*, **4**, 171–177 (1977).
8. N. A. Langberg, R. V. León and F. Proschan, Characterization of non-parametric classes of life distributions, *Annals of Probability*, **8**, 1163-1170 (1980).
9. B. Klefsjö, On aging properties and total time on test transforms, *Scandinavian Journal of Statistics*, **9**, 37–41 (1982).
10. M. Shaked and J. G. Shanthikumar, *Stochastic Orders and Their Applications* (Academic Press, London, 1994).
11. R. Szekli, *Stochastic Ordering and Dependence in Applied Probability, Lecture Note in Statistics, Vol. 97* (Springer-Verlag, New York, 1995).
12. R. E. Barlow, Geometry of the total time on test transform, *Naval Research Logistics Quarterly*, **26**, 393–402 (1979).
13. S. C. Kochar, On extensions of DMRL and related partial orderings of life distributions, *Communications in Statistics – Stochastic Models*, **5**, 235–245

(1989).

14. B. Bergman and B. Klefsjö, A family of test statistics for detecting monotone mean residual life, *Journal of Statistical Planning and Inference*, **21**, 161–178 (1989).
15. E. L. Kaplan and P. Meier, Non-parametric estimation from incomplete observations, *Journal of Americal Statistical Association*, **53**, 457–481 (1958).
16. J. S. Kim and F. Proschan, Piecewise exponential estimator of the survivor function, *IEEE Transactions on Reliability*, **40**, 134–139 (1991).
17. U. Westberg and B. Klefsjö, TTT-plotting for censored data based on the piecewise exponential estimator, *International Journal of Reliability, Quality and Safety Engineering*, **1**, 1–13 (1994).
18. F.-B. Sun and D. B. Kececioglu, A new method for obtaining the TTT plot for a censored sample, in *Proceedings of the 1999 Annual Reliability and Maintainability Symposium* (IEEE Reliability Society, Piscataway, 1999), pp. 112–117.
19. R. E. Barlow, Analysis of retrospective failure data using computer graphics, in *Proceedings of the 1978 Annual Reliability and Maintainability Symposium*, 113–116 (1978).
20. M. V. Aarset, How to identify a bathtub hazard rate, *IEEE Transactions on Reliability*, **36**, 106–108 (1987).
21. M. Xie, Some total time on test quantities useful for testing constant against bathtub-shaped failure rate distributions, Scandinavian Journal of Statistics, **16**, 137-144 (1989).
22. G. S. Mudholkar and D. K. Srivastava, Exponential Weibull family for analyzing bathtub failure-rate data, *IEEE Transactions on Reliability*, **42**, 299–302 (1993).
23. B. Klefsjö, Some tests against aging based on the total time on test transform, *Communications in Statistics A – Theory and Methods*, **12**, 907–927 (1983).
24. A. Hoyland and M. Rausand, *System Reliability Theory – Models and Statistical Methods* (John Wiley & Sons, New York, 1994).
25. R. E. Barlow and B. Davis, Analysis of time between failures for repairable components, *Technical Paper, Operations Research Center*, University of California, Berkeley (1977).
26. B. Klefsjö, TTT-transforms – a useful tool when analysing different reliability problems, *Reliability Engineering*, **15**, 231–241 (1986).
27. B. Klefsjö and U. Kumar, Goodness-of-fit tests for the power-law process based on the TTT-plot, *IEEE Transactions on Reliability*, **41**, 593–598 (1992).
28. N. Ebrahimi and F. Spizzichino, Some results on normalized total time on test and spacings, *Statistics & Probability Letters*, **36**, 231-243 (1997).
29. G. Nappo and F. Spizzichino, Ordering properties of the TTT-plot of lifetimes with Schur joint densities, *Statistics & Probability Letters*, **39**, 195-203 (1998).

30. M. Csörgö and H. Yu, Estimation of total time on test transforms for stationary observations, *Stochastic Processes and Their Applications*, **68**, 229-253 (1997).
31. Y. Hayakawa, The total time on test statistics and l_1-isotropy, *Research Report Series of School of Matrhematical and Computing Sciences, Victoria University of Wellington*, 98-32 (1998).
32. Y. Hayakawa, Characterisation properties of mixtures of exponential distributions, in *Proceedings of the 1st Western Pacific and 3rd Australia-Japan Workshop on Stochastic Models in Engineering, Technology and Management*, Eds. R. J. Wilson, S. Osaki and M. J. Faddy (Technology Management Centre, The University of Queensland, Brisbane, 1999), pp. 127–136.
33. W. D. Kaigh, Total time on test function principal components, *Statistics & Probability Letters*, **44**, 337-341 (1999).
34. B. Bergman, Some graphical methods for maintenance planning, in *Proceedings of the 1977 Annual Reliability and Maintainability Symposium* (IEEE Reliability Society Press, Piscataway, 1977), pp. 468–471.
35. B. Bergman, On age replacement and the total time on test concept, *Scandinavian Journal of Statistics*, **6**, 161–168 (1979).
36. R. E. Barlow and L. C. Hunter, Optimum preventive maintenance policies, *Operations Research*, **8**, 90-100 (1960).
37. R. E. Barlow and L. C. Hunter, Reliability analysis of a one-unit system, *Operations Research*, **9**, 200-208 (1961).
38. B. Bergman, On the decision to replace a unit early or late – a graphical solution, *Microelectronics and Reliability*, **20**, 895–896 (1980).
39. B. Bergman and B. Klefsjö, A graphical method applicable to age replacement problems, *IEEE Transactions on Reliability*, **31**, 478–481 (1982).
40. B. Bergman and B. Klefsjö, TTT transforms and age replacement witrh discounted costs, *Naval Research Logistics Quarterly*, **30**, 631-639 (1983).
41. B. Bergman and B. Klefsjö, The total time on test concept and its use in reliability theory, *Operations Research*, **32**, 596–606 (1984).
42. B. Klefsjö, TTT-plotting – a tool for both theoretical and practical problems, *Journal of Statistical Planning and Inference*, **29**, 99–110 (1991).
43. L. Jinlin, A model of reliability screening and its solution, in *Proceedings of the First China-Japan International Symposium on Industrial Management*, (Eds. W. Huafang, W. Xiaolu and W. Gang, International Academic Publishers, Beijing), 43–46 (1991).
44. U. Westberg and B. Klefsjö, Applications of the piecewise exponential estimator for the maintenance policy block replacement with minimal repair, *IAPQR Transactions*, **20**, 197–210 (1995).
45. D. Kumar and U. Westberg, Maintenance scheduling under age replacement policy using proportional hazards model and TTT-plotting, *European Journal of Operetional Research*, **99**, 507–515 (1997).
46. D. M. Reineke, E. A. Pohl and W. P. Murdock, Survival analysis and maintenance policies for a series system with highly censored data, in *Proceedings*

of the 1998 Annual Reliability and Maintainability Symposium (IEEE Reliability Society Press, Piscataway, 1998), pp. 182–188.
47. T. Dohi, N. Kaio and S. Osaki, Solution procedure for a repair limit problem using the TTT concept, *IMA Journal of Mathematics Applied in Business and Industry*, **6**, 101-111 (1995).
48. H. Koshimae, T. Dohi, N. Kaio and S. Osaki, Graphical / statistical approach to repair limit replacement problem, *Journal of the Operations Research Society of Japan*, **39**, 230-246 (1996).
49. T. Dohi, N. Matsushima, N. Kaio and S. Osaki, Nonparametric repair limit replacement policies with imperfect repair, *European Journal of Operational Research*, **96**, 260-273 (1996).
50. T. Dohi, T. Aoki, N. Kaio and S. Osaki, Nonparametric preventive maintenance optimization models under earning rate criteria, *IIE Transactions on Quality and Reliability Engineering*, **30**, 1099–1108 (1998).
51. T. Dohi, T. Danjou, N. Kaio and S. Osaki, The statistical estimation for delivery schedule of spare units, *Proceedings of 6th ISSAT International Conference on Reliability and Quality in Design*, 296–300 (2000).
52. T. Dohi, A. Ashioka, N. Kaio and S. Osaki, Optimizing the repair-time limit replacement schedule with discounting and imperfect repair, *Journal of Quality in Maintenance Engineering*, **7**, 71–84 (2001).
53. T. Dohi, N. Kaio and S. Osaki, A new graphical method to estimate the optimal repair-time limit with incomplete repair and discounting, *Computers & Mathematics with Applications*, accepted for publication.
54. T. Dohi, H. Koshimae, N. Kaio and S. Osaki, Geometrical interpretations of repair cost limit replacement policies, *International Journal of Reliability, Quality and Safety Engineering*, **4**, 309–333 (1997).
55. T. Dohi, N. Kaio and S. Osaki, A graphical method to repair-cost limit replacement policies with imperfect repair, *Mathematical and Computer Modelling*, **31**, 99–106 (2000).
56. N. A. J. Hastings, The repair limit replacement method, *Operational Research Quarterly*, **20**, 337-349 (1969).
57. T. Nakagawa and S. Osaki, The optimum repair limit replacement policies, *Operational Research Quarterly*, **25**, 311-317 (1974).
58. K. Okumoto and S. Osaki, Repair limit replacement policies with lead time, *Zeitschrift für Operations Research*, **20**, 133-142 (1976).
59. T. Nakagawa, Optimum preventive maintenance and repair limit policies maximizing the expected earning rate, *R.A.I.R.O. Recherche opèrationnelle/Operations Research*, **11**, 103-109 (1977).
60. T. Nakagawa and S. Osaki, Optimum ordering policies with lead time for an operating unit, *R.A.I.R.O. Recherche Opèrationnelle/Operations Research*, **12**, 383-393 (1978).
61. D. G. Nguyen and D. N. P. Murthy, A note on the repair limit replacement policy, *Journal of Operational Research Society*, **31**, 1103-1104 (1980).
62. D. G. Nguyen and D. N. P. Murthy, Optimal repair limit replacement poli-

cies with imperfect repair, *Journal of Operational Research Society*, **32**, 409-416 (1981).
63. N. Kaio and S. Osaki, Optimum repair limit policies with a time constraint, *International Journal of Systems Science*, **13**, 1345-1350 (1982).
64. N. Kaio and S. Osaki, Optimum repair limit policies with cost constraint, *Microelectronics and Reliability*, **21**, 597–599 (1981).
65. T. Dohi, K. Takeita and S. Osaki, Graphical methods for determining/estimating optimal repair-limit replacement policies, *International Journal of Reliability, Quality and Safety Engineering*, **7**, 43–60 (2000).
66. T. Dohi, F. S. Othman, N. Kaio and S. Osaki, The Lorenz transform approach to the optimal repair-cost limit replacement policy with imperfect repair, *R.A.I.R.O. Recherche Opėrationnelle/Operations Research*, **35**, 21–36 (2001).
67. T. Dohi, N. Kaio and S. Osaki, Determination of optimal repair-cost limit on the Lorenz curve, *Journal of the Operations Research Society of Japan*, accepted for publication.
68. T. Dohi, N. Kaio and S. Osaki, Nonparametric approach to power saving strategies for a portable personal computer, *Electronics and Communications in Japan*, **69**, 80-90 (1997).
69. T. Dohi, N. Kaio and S. Osaki, Optimal checkpointing and rollback strategies with media failures: statistical estimation algorithms, in *Proceedings of 1999 Pacific Rim International Symposium on Dependable Computing* (IEEE Computer Society Press, Los Alamitos, 1999), pp. 161–168.
70. T. Dohi, K. Goševa-Popstojanova and K. S. Trivedi, Analysis of software cost models with rejuvenation, in *Proceedings of 5th IEEE International Symposium on High Assurance Systems Engineering* (IEEE Computer Society Press, Los Alamitos, 2000), pp. 25–34.
71. T. Dohi, K. Goševa-Popstojanova and K. S. Trivedi, Statistical nonparametric algorithms to estimate the optimal software rejuvenation schedule, in *Proceedings of 2000 Pacific Rim International Symposium on Dependable Computing* (IEEE Computer Society Press, Los Alamitos, 2000), pp. 77–84.
72. T. Dohi, K. Goševa-Popstojanova and K. S. Trivedi, Estimating software rejuvenation schedule in high assurance systems, *The Computer Journal*, accepted for publication.
73. T. Dohi, K. Goševa-Popstojanova, K. Vaidyanathan, K. S. Trivedi and S. Osaki, Software rejuvenation – modeling and applications, *Springer Handbook of Reliability*, Ed. H. Pham (Springer, New York, 2001), accepted for publication.
74. S. M. Ross, *Applied Probability Models with Optimization Applications* (Holden-Day, San Francisco, 1970).

PART 2

AGEING PROPERTIES

CHAPTER 9

NONMONOTONIC FAILURE RATES AND MEAN RESIDUAL LIFE FUNCTIONS

Ramesh C. Gupta

Department of Mathematics and Statistics, University of Maine
Orono, ME 04469-5752, U.S.A.
E-mail: RCGUPTA@MAINE.maine.edu

This paper deals with the identification of the shape of the failure rate and the mean residual life function. In this regards we review Glaser's method and present some examples where the expressions for the failure rates are complicated and Glaser's methods can be applied. Glaser's methods are then extended to accomodate more than one turning point and in particular to the case of roller coaster failure rates. The shape of the mean residual life function, in comparison to the failure rate, is then studied for one or more than one turning points. Several illustrative examples are provided.

1. Introduction

Survival and failure time data are frequently modeled by increasing or decreasing failure rate. This may be inappropriate when the course of disease is such that the mortality reaches a peak after some finite period and then declines. In such a case the failure rate is upside down bathtub shaped and the data is analyzed with appropriate models like lognormal, inverse Gaussian, log logistic and Burr type XII distributions having non-monotonic failure rates. In addition to the bathtub and the upside down bathtub shaped failure rates, Wong (1988, 1989, 1991) presents situations where the hazard rate curve is of the roller coaster shape and remarks that the bathtub does not hold water any more. His articles suggest some plausible physical reasons for the formation of the roller coaster type. He also remarks that believing in an erroneous hazard rate curve can lead us into making wrong decisions and worst of all spending a great deal of effort in developing the wrong reliability methods.

Since most of the failure rates have complex expressions because of the integral in the denominator, the determination of the monotonicity is not straightforward. To alleviate this difficulty, Glaser (1980) presented a method to determine the monotonicity of the failure rate with one turning point.Glaser's method uses the density function instead of the failure rate, which, in many cases, is much simpler. Glaser's method can be used in many complicated examples such as power quadratic exponential family having generalized Rayleigh, half normal, gamma, Maxwell Boltzman, classical Rayleigh and chi square as special cases. Other examples include inverse Gaussian, lognormal and weighted inverse Gaussian distributions, see Gupta and Akman (1995b). Glaser's method can be used to determine the monotonicity in certain mixture models having two components, see Gurland and Sethuraman (1994,1995) and Al-Hussaini and Abd-El Hakim (1989).

There are certain situations, as in the case of roller coaster failure rates described above, where there is more than one turning point of the failure rate. For such situations, Gupta and Warren (2001) extended Glaser's techniques and discussed some examples..

In this paper, we will review Glaser's method and present several examples where the expressions for the failure rates are complicated and Glaser's methods can be applied. Glaser's methods are then extended to accomodate more than one turning point of the failure rate and in particular the case of roller coaster failure rate. The shape of the mean residual life function, in comparison to the failure rate, is then studied for one or more than one turning point.

The organization of the paper is as follows: In section 2, we present some definitions and Glaser's (1980) procedure for the determination of the shape of the failure rate. In section 3, we discuss some examples, including the case of power quadratic exponential family and skew normal distribution, to illustrate the procedure. Section 4 contains the extension to more than one turning point of the failure rate and its application to the mixed gamma situation. Finally,in section 5, the reciprocity of the shape of the mean residual life function (MRLF) and the failure rate is examined and some examples are discussed. Also the case of multiple turning points of the MRLF is studied and the location of the turning points of the MRLF is investigated in relation to the location of the turning point of the failure rate.

2. Background and Glaser's Procedure

Let T be a nonnegative random variable denoting the life length of a component having distribution function $F(t)$ with $F(0) = 0$ and the probability density function (pdf) $f(t)$. Then the failure rate of T is given by $r(t) = f(t)/R(t)$, where $R(t) = 1 - F(t)$ is the survival (reliability) function of T. We also assume that $f(t)$ is continuous and twice differentiable on $(0,\infty)$.

Let $h : R_+ \to R_+$ be a real valued differentiable function. Then $h(t)$ is said to be

(1) increasing if $h'(t) > 0$ for all t and is denoted by I.

(2) decreasing if $h'(t) < 0$ for all t and is denoted by D.

(3) bathtub shaped if $h'(t) < 0$ for $t \in (0, t_0), h'(t_0) = 0, h'(t) > 0$ for $t > t_0$ and is denoted by B.

(4) upside down bathtub shaped if $h'(t) > 0$ fort $\in (0, t_0), h'(t_0) = 0, h'(t) < 0$ for $t > t_0$ and is denoted by U.

(5) upside down bathtub and then bathtub if there exist t_1 and t_2 such that $h'(t) > 0$ for $t \in (0, t_1), h'(t_1) = 0, h'(t) < 0$ for $t \in (t_1, t_2), h'(t_2) = 0, h'(t) > 0$ for $t > t_2$ and is denoted by UB.

(6) bathtub and then upside down bathtub if there exist t_1 and t_2 such that $h'(t) < 0$ for $t \in (0, t_1), h'(t_1) = 0, h'(t) > 0$ for $t \in (t_1, t_2), h'(t_2) = 0$, $h'(t) < 0$ for $t > t_2$ and is denoted by BU.

(7) increasing (not strictly) if $h'(t) > 0$ for all t and $h'(y) = 0$ for some $y > 0$ and is denoted by I*.

(8) decreasing (not strictly) if $h'(t) < 0$ for all t and $h'(y) = 0$ for some $y > 0$ and is denoted by D*.

For some definitions given above, see Barlow and Proschan (1975). Also see Barlow et al. (1963) for some properties of probability distributions with monotone hazard rate.

In order to determine the monotonicity of the failure rates, we proceed as follows:

Define

$$\eta(t) = -f'(t)/f(t) \tag{1}$$

This function contains useful information about $r(t)$ and is simpler because it does not involе $R(t)$. In particular the shape of $\eta(t)$ (I,D,B etc.) often determines the shape of the failure rate. The relations between $r(t)$

and $\eta(t)$ are given by

$$\frac{d}{dt}\ln r(t) = r(t) - \eta(t) \tag{2}$$

and

$$[\frac{1}{r(t)}]' = \frac{\eta(t)}{r(t)} - 1. \tag{3}$$

The above equations also suggest that the turning point of $r(t)$ is a solution of the equation $\eta(t) = r(t)$. Also it can be verified that $\lim_{t\to\infty} r(t) = \lim_{t\to\infty} \eta(t)$.

Remark 1: Lillo et al. (2001) have used the function $\eta(t)$ to define the shifted likelihood ratio order between two random variables.

We now present the following result due to Glaser (1980) which helps us to determine the shape of the failure rates of the first four types described above.

Theorem 2: (a) If $\eta(t) \in$ I, then r(t) $\in$ I (IFR).

(b) If $\eta(t) \in$ D, then $r(t) \in$ D (DFR)

(c) If $\eta(t) \in B$ and (i) if there exist a y_0 such that $r'(y_0) = 0$, then $r(t) \in B$, (ii) otherwise $r(t) \in$ I.

(d) If $\eta(t) \in$ U *and* (i) if there exists a y_0 such that $r'(y_0) = 0,$ *then* $r(t) \in U$, (ii) otherwise $r(t) \in D$.

In the last two cases, determining the existence of y_0 leaves us with the original difficulty of evaluating the derivative of $r(t)$. However, we can simplify this problem in many situations with the following lemma.

Lemma 3: *Let* $\in = \lim_{t\to 0} f(t)$ *and* $\delta = \lim_{t\to 0} g(t)\eta(t)$, *where* $g(t) = 1/r(t)$.

1. Suppose $\eta(t) \in B$, *then*

(a) If either $\in = 0$ *or* $\delta < 1$, *then* $r(t) \in I$.

(b) If either $\in = \infty$ *or* $\delta > 1$, *then* $r(t) \in B$.

2. Suppose $\eta(t) \in U$, *then*

(a) If either $\in = 0$ *or* $\delta < 1$, *then* $r(t) \in U$.

(b) If either $\in = 0$ *or* $\delta > 1$, *then* $r(t) \in D$.

3. Some Examples

3.1. *Lognormal Distribution*

$$f(x) = \frac{1}{x\sigma\sqrt{2\pi}} \exp\{-\frac{1}{2\sigma^2}(\ln x - \mu)^2\}, x > 0, \sigma > 0. \tag{4}$$

$$r(t) = \frac{(1/\sqrt{2\pi}x\sigma)\exp\{-(\ln at)^2/2\sigma^2\}}{1 - \Phi[\ln at/\sigma]}, \tag{5}$$

where $a = e^{-\mu}$.

Even though the expression of $r(t)$ is quite complicated, it can be easily shown by using Glaser's procedure that $r(t)$ is of the type U. For the estimation of the change point, see Gupta et al. (1997).

3.2. *Inverse Gaussian Distribution*

$$f(x) = (\lambda/2\pi x^3)\exp[-\lambda(x-\mu)^2/2\mu^2 x], x, \lambda, \mu > 0. \tag{6}$$

$$r(t) = \frac{(\lambda/2\pi t^3)^{1/2}\exp[-\lambda(t-\mu)^2/2\mu^2 t]}{\Phi(\sqrt{\lambda/t}(1-t/\mu)) - e^{2\lambda/\mu}\Phi(-\sqrt{\lambda/t}(1+t/\mu))} \tag{7}$$

In this case also, it can be seen that $r(t)$ is of the type U, see Chhikara and Folks (1977).

3.3. *Mixture Inverse Gaussian Distribution*

$$f_p(x) = (1-p)f(x) + pf^*(x), \tag{8}$$

where $f(x)$ is given by (6) and $f^*(x)$ is the length biased version of $f(x)$ given by

$$f^*(x) = \frac{xf(x)}{\mu}, \tag{9}$$

see Jorgensen et al. (1991). The failure rate is given by

$$r_p(t) = \frac{(\lambda/2\pi t)^{1/2}\exp[-\lambda(t-\mu)^2/2\mu^2 t](1-p+pt/\mu)}{\Phi(-\alpha(t)) - (1-2p)e^{2\lambda/\mu}\Phi(\beta(t))}, \tag{10}$$

where $\alpha(t) = \sqrt{\lambda/t}(t/\mu - 1)$ and $\beta(t) = -\sqrt{\lambda/t}(t/\mu + 1)$.

Using Glaser's method, it can be shown that $r_p(t)$ is of the type U. For details see Gupta and Akman (1995b).

3.4. *Skew Normal Distribution*

The pdf is given by

$$f(z;\lambda) = 2\phi(z)\Phi(\lambda z), -\infty < z, \lambda < \infty, \tag{11}$$

see Azzalini (1985). Here $\phi(z)$ is the pdf of a standard normal distribution and $\Phi(z)$ is the corresponding cumulative distribution function. The introduction of the parameter λ makes the distribution skewed which for $\lambda = 0$, reduces to the usual standard normal distribution. The failure rate is given by

$$r(t) = \frac{2\phi(t)\Phi(\lambda t)}{\Phi(t) + 2T(t;\lambda)}, \tag{12}$$

where $T(z;\lambda)$ is given by

$$T(z;\lambda) = \int_z^\infty \int_0^{\lambda t} \phi(u)\phi(t)dudt, \tag{13}$$

see Gupta and Brown (2001). Since the expression for $r(t)$ is quite complex, we use Glaser's procedure as follows:

It can be verified that

$$\eta(t) = \frac{-\phi'(t;\lambda)}{\phi(t;\lambda)}. \tag{14}$$

This gives

$$\begin{aligned} -\eta'(t) &= \frac{d^2}{dt^2}[\ln 2 + \ln\phi(t) + \ln\Phi(\lambda t)] \\ &= -1 + \frac{-\lambda^2\phi(\lambda t)}{\Phi(\lambda t)}[\frac{\phi(\lambda t)}{\Phi(\lambda t)} + \lambda t] \end{aligned} \tag{15}$$

We shall now show that

$$\frac{\phi(\lambda t)}{\Phi(\lambda t)} + \lambda t > 0. \tag{16}$$

Case I

If $\lambda t > 0$, the expression given in (16) is positive.

Case II

If $\lambda t < 0$, let $u = -\lambda t$. Then

$$\phi(\lambda t) = \phi(-\lambda t) = \phi(u)$$

and

$$\Phi(\lambda t) = 1 - \Phi(-\lambda t) = 1 - \Phi(u).$$

Thus

$$\frac{\phi(\lambda t)}{\Phi(\lambda t)} + \lambda t = \frac{\phi(u)}{1 - \Phi(u)} - u$$
$$= h(u) - u > 0,$$

where $h(u)$ is the failure rate of a standard normal distribution. Thus $\eta'(t) > 0$ and the failure rate of a skew normal distribution is increasing. This in turn implies that the failure rate of a standard normal distribution is increasing, a well known result.

3.5. *Power Quadratic Exponential Family*

In this case, the pdf is given by

$$f(x; \alpha, \beta, \gamma) = c(\alpha, \beta, \gamma)x^{\gamma} \exp(-\alpha x - \beta x^2), \tag{17}$$

where $c(\alpha, \beta, \gamma)$ is the normalizing constant and the parameter space is given by

$$\Omega = \Omega_1 \cup \Omega_2,$$

where

$$\Omega_1 = \{(\alpha, \beta, \gamma) | -\infty < \alpha < \infty, \beta > 0, \gamma > -1\}$$

and

$$\Omega_2 = \{(\alpha, \beta, \gamma) | \alpha > 0, \beta = 0, \gamma > -1.$$

The special cases of the above family are:

Generalized Rayleigh ($\alpha = 0$), Half normal ($\gamma = 0$), $\Gamma(\beta = 0)$, Maxwell-Boltzman ($\gamma = 2$), Classical Rayleigh ($\gamma = 1$) and Chi square with $(\gamma + 1)$ degrees of freedom ($\beta = 1/2$), see Glaser (1980) and Pham-Gia (1994).

Before proceeding further, we shall try to find the failure rate of the above family of distributions.

It can be verified that

$$(r(t))^{-1} = \int_0^{\infty} (1 + \frac{u}{t})^{\gamma} \exp[-\{(\alpha + 2\beta t)u + \beta u^2\}]du$$
$$= \int_1^{\infty} t y^{\gamma} \exp[-(\alpha + 2\beta t)ty + (\alpha + 2\beta t)t - \beta t^2 (y-1)^2 dy.$$

Assuming that $\beta \neq 0 (\beta = 0$ gives gamma distribution), the above expression can be written as

$$(r(t))^{-1} = \sqrt{2\pi} t \exp(\alpha t + \beta t^2 + \alpha^2/4\beta) \int_1^\infty \frac{1}{\sqrt{2\pi}} y^\gamma e^{-(y-\mu)^2/2\sigma^2} dy, \quad (18)$$

where $\mu = \alpha/2\beta t$ and $\sigma^2 = 1/2\beta t^2$. Thus the above equation can be written as

$$(r(t))^{-1} = \sqrt{2\pi} t \exp(\alpha t + \beta t^2 + \alpha^2/4\beta) I(\gamma), \quad (19)$$

where

$$I(\gamma) = \frac{1}{\sqrt{2\pi}} \int_1^\infty y^\gamma e^{-(y-\mu)^2/2\sigma^2} dy \quad (20)$$
$$= E(X^\gamma) - \int_{-\infty}^1 \frac{1}{\sqrt{2\pi}} e^{-(y-\mu)^2/2\sigma^2} dy,$$

where X has a normal distribution with mean μ and variance σ^2.

In order to evaluate the integral in (20), we proceed as follows:

Let

$$E_{-\infty}^z(X^n) = \int_{-\infty}^z x^n \frac{1}{\sqrt{2\pi}} e^{-(x-\mu)^2/2\sigma^2} dx. \quad (21)$$

Then

$$E_{-\infty}^z(X^n) = -\sigma^2 z^{n-1} f_N(z) + (n-1)\sigma^2 E_{-\infty}^z(X^{n-2}) + \mu E_{-\infty}^z(X^{n-1}), (22)$$

where

$$f_N(z) = \frac{1}{\sqrt{2\pi}\sigma} e^{-(z-\mu)^2/2\sigma^2}, \quad (23)$$

see equation (3.4) of Winkler et al. (1972). Using the above procedure $I(\gamma)$ can be calculated.

4. Some Special Cases

4.1. *Maxwell-Boltzman Distribution* ($\gamma = 2$)

In this case

$$I(2) = E(X^2) - [-\sigma(1+\mu)\phi(\frac{1-\mu}{\sigma}) + (\mu^2+\sigma^2)\Phi(\frac{1-\mu}{\sigma})] \quad (24)$$
$$= (\mu^2+\sigma^2) + \sigma(1+\mu)\phi(\frac{1-\mu}{\sigma}) - (\mu^2+\sigma^2)\Phi(\frac{1-\mu}{\sigma})$$
$$= \Phi(\frac{\mu-1}{\sigma}) + \sigma(1+\mu)\phi(\frac{1-\mu}{\sigma}).$$

Thus

$$(r(t))^{-1} \tag{25}$$
$$= \sqrt{2\pi} t e^{\alpha t+\beta t^2+\alpha^2/4\beta}[(\mu^2+\sigma^2)\Phi(\frac{\mu-1}{\sigma})+\sigma(1+\mu)\phi(\frac{1-\mu}{\sigma})],$$

where $\mu = \alpha/2\beta t$ and $\sigma^2 = 1/2\beta t^2$.

4.2. *Classical Rayleigh* ($\gamma = 1$)

In this case

$$I(1) = E(X) - [-\sigma\phi(\frac{1-\mu}{\sigma}) + \mu\Phi(\frac{1-\mu}{\sigma})]. \tag{26}$$

Thus

$$(r(t))^{-1} = \sqrt{2\pi} t e^{\alpha t+\beta t^2+\alpha^2/4\beta}[\mu\Phi(\frac{\mu-1}{\sigma}) + \sigma\phi(\frac{1-\mu}{\sigma})], \tag{27}$$

where $\mu = \alpha/2\beta t$ and $\sigma^2 = 1/2\beta t^2$.

It is clear from the complicated form of the failure rate that straight forward methods will not be helpful in determining its monotonicity.However, Glaser's procedure will be applicable as follows:

4.3. *Monotonicity of the Failure Rate*

For the power quadratic exponential family, we have

$$\eta(t) = -\frac{f'(t)}{f(t)} = \alpha + 2\beta t - \gamma/t \tag{28}$$

Case I $\beta > 0$

(a) If $\gamma \geq 0, \eta'(t) > 0$ and hence $r(t)$ is of the type I

(b) If $\gamma < 0, r(t)$ is of the type B and the turning point is given by $t_0 = (-\gamma/2\beta)^{1/2}$. Note that in this case $\lim_{t\to 0} f(t) = \infty$.

Case II $\beta = 0$

In this case $\eta'(t) = \gamma/t^2$ and hence $r(t) \in I$ if $\gamma > 0$ and $r(t) \in D$ if $\gamma < 0$. For $\gamma = 0$, we have a constant failure rate and hence an exponential distribution.

5. Extension of Glaser's Result

In this section, we seek to expand our understanding of the relationship between $\eta(t)$ and $r(t)$ so that we can determine precisely the shape of $r(t)$.

Suppose that η' has n zeros, $z_1, z_2, ..., z_n$ (n finite) with

$$0 < z_1 < z_2 < ... < z_n < \infty$$

The following theorem shows the location of the critical points of $r(t)$.

Theorem 4: Let $f(t)$ be a twice differentiable function on (0,∞). Then the equation $r'(t) = 0$ has atmost one solution on the (closed) interval $[z_{k-1}, z_k]$ for k$\in 1, 2, \ldots, n$.

For proof, see Gupta and Warren (2001).

Remark 5: If $\eta(t)$ has k critical points, then $r(t)$ has atmost k critical points.

Remark 6: If $\eta(t)$ has two critical points and $r(t)$ also has two critical points, one must be in the interval (z_0, z_1) and the other must be in (z_1, z_2).

We now present the following result dealing with the common zeros of $\eta'(t)$ and $r'(t)$.

Theorem 7: (1) Suppose $\eta'(t)$ and $r'(t)$ have a common zero, say z_k.Then z_k is the unique zero of $r'(t)$ on $[z_{k-1}, z_{k+1}]$.

(2) Suppose $\eta'(t)$ and $r'(t)$ have m common zeros.Then $r'(t) has$ atmost $n - m$ zeros.

(3) Suppose $\eta(t) \in B(U)$ on the interval (z_{k-1}, z_{k+1}). Then z_k is a common critical point of $\eta(t)$ and $r(t)$ if and only if $r(t) \in D^*(I^*)$ on (z_{k-1}, z_{k+1}).

For proof, see Gupta and Warren (2001).

The following theorem deals with two critical points of $\eta(t)$.

Theorem 8: (1) Suppose $\eta(t) \in UB(BU)$ with critical points z_1 and z_2. Also $r'(y_0) = 0$ for some $y_0 \in (0, z_1)$. Then there is a $y_1 \in (z_1, z_2)$ such that $r'(y_1) = 0$. Furthermore $r(t) \in UB(BU)$.

(2) Suppose $\eta(t) \in UB(BU)$ with critical points z_1 and z_2 and there exists a $y_0 \neq z_1$(a unique critical point of $r(t)$). Then $y_0 \in (z_1, z_2)$ and $r(t) \in B(U)$.

(3) (a) Suppose $\eta(t) \in UB$ with z_1 known and $r(t)$ increases in a neighborhood of zero. Then $r'(z_1) < 0 \Rightarrow r(t) \in UB, r'(z_1) = 0 \Rightarrow r(t) \in I^*$ and $r'(z_1) > 0 \Rightarrow r(t) \in I$.

(b) If $\eta(t) \in UB$ with z_1 known and $r(t)$ decreases in a neighborhood of zero, then $r(t) \in B$.

(c) If $\eta(t) \in BU$ with z_1 known and $r(t)$ decreases in a neighborhood of zero, then $r'(z_1) < 0 \Rightarrow r(t) \in D, r'(z_1) = 0 \Rightarrow r(t) \in D^*$ and $r'(z_1) > 0 \Rightarrow r(t) \in BU$.

(4) If $\eta(t) \in BU$ with z_1 known and $r(t)$ increases in a neighborhood of zero, then $r(t) \in U$.

For proof, see Gupta and Warren (2001).

Finally, we present the following generalization of Glaser due to Gupta and Warren (2001) when the shape of $\eta(t)$ is known and the number of critical points of $r(t)$ are known.

Theorem 9: (1) Suppose $\eta(t) \in UB$. Then

a) If $r'(t)$ has no zeros, then $r(t) \in I$.

b) If $r'(t)$ has one zero, then $r(t) \in I^*$ or B.

c) If $r'(t)$ has two zeros, then $r(t) \in UB$.

(2) Suppose $\eta(t) \in BU$. Then

a) If $r'(t)$ has no zeros, then $r(t) \in D$.

b) If $r'(t)$ has one zero, then $r(t) \in D^*$or U.

c) If $r'(t)$ has two zeros, then $r(t) \in BU$.

We now present the following example of mixture of two gamma distributions where the results of this section will be applicable.

6. Mixture of Gamma Distributions

Let

$$f(x) = pf_1(x) + (1-p)f_2(x), \tag{29}$$

where

$$f_i(x) = \frac{1}{\beta^{\alpha_1}\Gamma(\alpha_i)} x^{\alpha_i - 1} e^{-x/\beta}, x, \alpha_1, \alpha_2, \beta > 0, i = 1, 2. \tag{30}$$

Using Glaser's method, all except case (8) of the following can be established.

(1) If $\alpha_1 \leq 1$ and $\alpha_2 \leq 1$, both have DFR, so the mixture has DFR.

(2) If $\alpha_1 = \alpha_2 = 1$, both have exponential distribution and hence the mixture has DFR.

(3) If $\alpha_1 > 1$ and $\alpha_2 < 1$, then $r(t) \in B$.

(4) If $1 < \alpha_1 < 2$ and $\alpha_2 = 1$, $r(t) \in IFR$.

(5) If $\alpha_1 > 1$ and $\alpha_2 = 1$, then $r(t) \in B$.

(6) If $\alpha_1 > 1, \alpha_2 > 1$ and $0 < \alpha_1 - \alpha_2 < 1$, then $r(t) \in IFR$.

(7) If $\alpha_1 > 1, \alpha_2 > 1, \alpha_1 - \alpha_2 > 1$ and $\alpha_2 - 1 > (\alpha_1 - \alpha_2 - 1)^2/4$, then $r(t) \in IFR$

(8) If $\alpha_1 > 1, \alpha_2 > 1, \alpha_1 - \alpha_2 > 1.$ and $4(\alpha_2 - 1) < (\alpha_1 - \alpha_2 - 1)^2$, then $\eta'(t)$ has two real roots and Glaser,s method is not applicable and Glaser's conjecture that $r(t) \in IFR$ is not correct.

For example, let $\alpha_1 = 7, \alpha_2 = 2$ and $p = .5$. In this case $r(1) = .2122$ and $r(2) = .2017$. For details see Gupta and Warren (2001).

As has been noted before that in this case $\eta'(t)$ has two real roots. Also it can be verified that $\eta(t) \in UB$ and $r(t)$ increases in the neighborhood of zero.

The above facts imply that $r(t) \in I$ or I^* or UB. But $r(t)$ has a net decrease over (1,2). This eliminates I and I^* as candidates. Hence $r(t) \in UB$.

For a general discussion of this case, the reader is referred to Gupta and Warren (2001).

7. Mean Residual Life Function and its Reciprocity with the Failure Rate

The mean residual life function (MRLF) of a random variable T is defined as

$$\begin{aligned} \mu(t) &= E(T - t|T > t) \\ &= \frac{\int_t^\infty R(x)dx}{R(t)}. \end{aligned} \tag{31}$$

As in the case of failure rate, the MRLF also determines the distribution function uniquely : see Gupta (1981), and is related to the failure rate $r(t)$ by

$$r(t) = \frac{1 + \mu'(t)}{\mu(t)}. \tag{32}$$

It is well known that if $r(t) \in IFR$, then $\mu(t) \in DMRL$ (decreasing mean residual life) class and if $r(t) \in DFR$, then $\mu(t) \in IMRL$ (increasing mean residual life) class .But, the converse is not necessarily true. For counter examples, see Bryson and Siddiqui (1969). However, if $\mu(t)$ is decreasing and convex, then the corresponding $r(t)$ is increasing, see Kupka and Loo (1988).

In order to study the limiting behavior of $r(t)$ and $\mu(t)$, by applying L'Hopital rule to (5.1), Calabria and Pulcini (1987) derived the relationship

$$\lim_{t\to\infty} \mu(t) = \lim_{t\to\infty} \frac{1}{r(t)}, \tag{33}$$

provided the latter limit exists. They then used (32) to conclude that $\lim_{t\to\infty}\mu'(t) = 0$ or equivalently $\lim_{t\to\infty}r(t)\mu(t) = 1$. Unfortunately, one can't infer this from (33) unless one assumes that $\lim_{t\to\infty} r(t)$ is finite and strictly positive. For counter examples, see Bradley and Gupta (2001).

Now the question arises: Does type U (B) failure rate imply type B (U) MRLF? We present the following counter example.

Example 10:

$$\mu(t) = \frac{1}{1+2.3t^2}, t > 0. \tag{34}$$

The corresponding failure rate is given by

$$r(t) = \frac{(1+2.3t^2) - 4.6t}{1+2.3t^2}, t > 0 \tag{35}$$

It can be verified that $r(t) \in B$ while $\mu(t)$ is decreasing, see Muth (1977).

We now present the following result due to Gupta and Akman (1995a) which characterizes the shape of the MRLF when $r(t) \in B(U)$.

Theorem 11: a) Suppose r(t)$\in B$ and $\mu = \mu(0)$. Then
i) If $\mu r(0) \leq 1$, then $\mu(t) \in D$
ii) If $\mu r(0) > 1$, then $\mu(t) \in U$.
b) Suppose $r(t) \in U$ and $\mu = \mu(0)$. Then
i) If $\mu r(0) \geq 1$, then $\mu(t) \in I$
ii) If $\mu r(0) < 1$, then $\mu(t) \in B$.

The following are some examples which illustrate the above result.

Example 12: In Example 10
r(t)$\in B$ and $\mu r(0) \leq 1$. Hence $\mu(t) \in D$.

$$\mu(t) = \frac{1}{4(1-t)(1+2t)}, 0 \le t \le 1. \tag{36}$$

The corresponding $r(t)$ is given by

$$r(t) = \frac{5-4t}{(1-t)(1+2t)}. \tag{37}$$

In this case $r(t) \in B$ and $\mu r(0) > 1$. Hence $\mu(t) \in U$.

Example 13:

$$\mu(t) = 1 + t^2, 0 \le t \le \pi/2 \tag{38}$$

The corresponding $r(t)$ is given by

$$r(t) = \frac{1+2t}{1+t^2} \tag{39}$$

In this case $r(t) \in U$ and $\mu r(0) \ge 1$. Hence $\mu(t) \in I$.

Example 14: Inverse Gaussian Distribution

The expression for $r(t)$ is given by equation (7).

The MRLF is given by

$$\mu(t) = \frac{(\mu - t)\Phi(\sqrt{\lambda/t})(1 - t/\mu) + (\mu - t)e^{2\lambda/\mu}\Phi(-\sqrt{\lambda/t})(1 + t/\mu)}{\Phi(\sqrt{\lambda/t}(1 - t/\mu)) - e^{2\lambda/\mu}\Phi(-\sqrt{\lambda/t}(1 + t/\mu))} \tag{40}$$

In this case $r(t) \in U$ and $\mu r(0) < 1$. Hence $\mu(t) \in B$.

We now present the following result which determines the location of the turning point of $\mu(t)$ when $\mu(t) \in U(B)$.

Theorem 15: Suppose $r(t) \in U$ with $r'(t^*) = 0$ and r(0)$\mu(0) < 1$. Then there exists a unique point k^* such that $\mu'(t) < 0$ for $t < k^*, \mu'(k^*) = 0$ and $\mu'(t) > 0$ for $t > k^*$. In such a case $k^* < t^*$.

Proof: Define

$$S(t) = \mu'(t)R(t). \tag{41}$$

Then

$$\begin{aligned} S(t) &= r(t)\mu(t)R(t) - R(t) \qquad (42) \\ &= r(t)\int_t^\infty R(x)dx - R(t). \end{aligned}$$

The above equation gives

$$\begin{aligned} S(t^*) &> \int_{t^*}^{\infty} r(x)R(x)dx - R(t^*) \\ &= R(t^*) - R(t^*) \\ &= 0 \end{aligned} \tag{43}$$

Also $S(0) = r(0)\mu(0) - 1 < 0$. Therefore, there exists atleast one point $k^*, 0 < k^* < t^*$ such that $S(k^*) = 0$ We shall now show that such a k^* is unique.

Suppose on the contrary $S(k_1^*) = S(k_2^*) = 0$ and $k_1^* < k_2^* < t^*$. This means that there is a point t_0 such that $k_1^* < t_0 < k_2^*$ and $S'(t_0) = 0$. This implies that $r'(t_0) = 0$, which is not true. Hence k^* is unique.

For $t > t^*$

$$\begin{aligned} S(t) &= r(t)\int_{t}^{\infty} R(x)dx - R(t) \\ &> \int_{t}^{\infty} r(x)R(x)dx - R(t) \\ &= R(t) - R(t) \\ &= 0 \end{aligned} \tag{44}$$

Thus

$$\begin{aligned} \mu'(t) &< 0 \text{ if } t < k^* \\ &= 0 \text{ if } t = k^* \\ &> 0 \text{ if } k^* < t^*. \end{aligned} \tag{45}$$

This completes the proof, see alsot Tang et al. (1999). □

In a similar manner we can prove the following theorem

Theorem 16: Suppose r(t)$\in B$ with $r(t^*) = 0$ and r(0)$\mu(0) > 1$. Then there exists a unique point k^* such that $\mu'(t) > 0$ if $t < k^*, \mu'(k^*) = 0$ and $\mu'(t) < 0$ if $t > k^*$.

Remark 17: Another version of the above two theorems has been given by Gupta and Akman (1995b).

Finally, we present the following theorem dealing with the multiple turning points of the failure rate

Theorem 18: Suppose F has a roller coaster failure rate of the type UB....(BU....) with consecutive change points $0 = t_0^* < t_1^* < t_2^* < ... < t_n^* < \infty$.Also $\mu'(t_{i-1}^*)\mu'(t_i^*) < 0, i = 1, 2, ..., n-1$. Then $\mu(t)$ is of the type BU....(UB...) with change points $k_i^*, i = 1, 2, ..., n$ such that $t_{i-1}^* < k_i^* < t_i^*, i = 1, 2, ..., n$.

The proof follows on the same lines as in Theorem 15. Also see Tang et al. for an induction proof.

References

1. Al-Hussaini, E.K.and Abd-El-Hakim, N.S. (1989). Failure rate of the inverse Gaussian-Weibull mixture model. *Annals of the Institute of Statistical Mathematics*, 41(3),617–622.
2. Barlow, R.E., Marshall, A.W. and Proschan, F.(1963). Properties of probability distributions with monotone hazard rate. *Annals of Mathematical Statistics*, 34, 375–389.
3. Barlow, R.E. and Proschan, F. (1975). *Statistical Theory of Reliability and Life Testing.* Rinehart and Winston, Inc., New York.
4. Azzalini, A. (1985). A class of distributions which includes the normal ones. *Scandinavian Journal of Statistics*, 12, 171–178.
5. Bradley, D.M. and Gupta, R.C.(2001). The mean residual life and its limiting behaviour. Submitted for publication.
6. Bryson, M.C. and Siddiqui, M.M. (1969). Some criteria for aging. *Journal of the American Statistical Association*, 64, 1472–83.
7. Calabria, R. and Pulcini, G. (1987). On the asymptotic behaviour of the mean residual life function. *Reliability Engineering*, 19,165-170.
8. Chhikara, R.S. and Folks, J.L. (1977). The inverse Gaussian distribution as a lifetime model. *Technometrics,* 19(4), 461–468.
9. Glaser, R.E. (1980). Bathtub and related failure rate characterizations. *Journal of the American Statistical Association,* 75, 667–672.
10. Gupta, R.C. (1981). On the mean residual life function in survival studies. Statistical Distributions in Scientific Work 5, *D-Reidel Publishing Co.* Boston, 327-334.
11. Gupta, R.C. and Akman, O. (1995a). Mean residual life function for certain types of non-monotonic ageing. *Communications in Statistics; Stochastic Models,* 11(1), 219–225.
12. Gupta, R.C. and Akman,O (1995b). On the reliability studies of a weighted inverse Gaussian model. *Journal of Statistical Planning and Inference,* 48,69–83.
13. Gupta, R.C.,Kannan, N. and Raychaudhari, A. (1997). Analysis of log normal survival data. *Mathematical Biosciences,* 139,103–115.
14. Gupta, R.C. and Warren, R. (2001). Determination of change points of non monotonic failure rates. To be published in *Communications in Statistics* (Rao's volume).

15. Gupta, R.C. and Brown, N. (2001). Reliability studies of the skew-normal distribution and its application to a strength-stress model. To be published in *Communications in Statistics* (Maine Conference).
16. Gurland, J. and Sethuraman, J. (1994). Reversal of increasing failure rates when pooling failure data. *Technometrics,* 36(4), 416–418.
17. Gurland, J. and Sethuraman, J. (1995). How pooling data may reverse increasing failure rate. *Journal of the American Statistical Association*, 90 (432), 1416–1423.
18. Jorgensen, B.,Seshadri,V. and Whitmore, G.A. (1991). On the mixture of the inverse Gaussian distribution with its complementary reciprocal. *Scandinavian Journal of Statistics,* 18,77–89.
19. Kupka, J. and Loo, S. (1988). The hazard and vitality measures of ageing. *Journal of Applied Probability,* 26, 532–542.
20. Lillo, R.E.,Nanda, A.K. and Shaked, M. (2001). Preservation of some likelihood ratio stochastic orders by order staistics. Statistics and Probability letters, 51, 111–119.
21. Muth, E.J. (1977). Reliability models with positive memory derived from the mean residual life function. In *Theory and Applications of Reliability,* eds.C.P.Tsokos and I.N,Shimi, Academic Press, 401–434.
22. Pham-Gia, T.(1994). The hazard rate of the power-quadratic exponential family of distributions. *Statistics and Probability Letters*, 20, 375–382.
23. Tang, L.C., Lu,Y. and Chew, E.P. (1999). Mean residual lifetime distributions. *IEEE Transactions on Reliability,* 48(1),73–78.
24. Winkler, R.L.,Roodman, G.M. and Britney, R.R. (1972). The determination of partial moments. *Management Science,* 19(3),290–296.
25. Wong, K.L. (1988). The bathtub does not hold water any more. *Quality and Reliability Engineering International*, 4, 279–282.
26. Wong, K.L. (1989). The roller-coaster curve is in. Quality and Reliability Engineering International, 5, 29–36.
27. Wong, K.L. (1991). The physical basis for the roller-coaster hazard rate curve for electronics. *Quality and Reliability Engineering International*, 7, 489–495.

CHAPTER 10

THE FAILURE RATE AND THE MEAN RESIDUAL LIFETIME OF MIXTURES

M. S. Finkelstein

Department of Mathematical Statistics, University of the Free State
P.O. Box 339 Bloemfontein, Republic of South Africa
E-mail: msf@wwg3.uovs.ac.za

It is well known that mixtures of decreasing failure rate distributions (distributions with increasing mean residual life function) also have the decreasing failure rate (increasing mean residual lifetime function). A Bayes explanation of this fact was given by Professor **R. Barlow.** It turns out that very often mixtures of increasing failure rate (decreasing mean residual lifetime function) distributions can decrease (increase) or show even more complicated patterns of dependence on time. For studying this and other relevant effects several direct models of mixing are considered. It is shown that for the proportional hazard model of mixing an inverse problem can be solved, which means that given an arbitrary shape of the mixture failure rate and a mixing distribution, the failure rate for a governing distribution can be uniquely obtained. Some examples are considered where this operation can be performed explicitly. Asymptotic properties of the mixture failure rate and of the corresponding mixture mean residual lifetime function are studied. Possible generalizations are discussed.

1. Introduction

In most practical situations the population of lifetimes is not homogeneous. This means that not all of the items in the population do have exactly the same distribution: there are usually lifetimes, which are different from the majority. For example, for electronic components, most of the population might be exponential, with long lives, while the certain percentage often has an exponential distribution with short lives and the certain percentage can be characterized by lives of the intermediate duration. Thus, the mixing procedure arises naturally when pooling from heterogeneous populations. Another origin of mixing can be found when modeling an impact of a random environment on reliability characteristics of various objects. The observed failure rate, which turns to be the mixture failure rate in this case, estimated via standard statistical methods, does not "reveal" the governing and the mixing distributions. To say more, obtaining the mixture failure rate in this direct way can often present certain difficulties, as the corresponding data can be scarce. On the other hand, it is clear that the proper modeling can add some valuable information for estimating the mixture failure rate and for analyzing and comparing its shape with the shape of the failure rate of the governing cumulative distribution function (c.d.f.).

The origin of mixing in practice can be "physical" when, for instance, a number of devices of different (heterogeneous) types, performing the same function and not distinguishable in operation, are chosen in accordance with some probability law [1]. On the other hand, mixing in a subjective approach is suggested by an unknown parameter and is performed via some prior distribution. Thus, the mixing procedure in the Bayesian framework turns out to be quite natural. One can say that mixing arises when the initial c.d.f. is indexed in some way by a random parameter which is the case in different applications.

The mean residual lifetime (MRL) function, similar to the corresponding failure rate, under certain assumptions also defines the corresponding c.d.f. The MRL function can be very helpful in describing the aging properties of distributions. It is worth noting that the mixing procedures based on the MRL function were less investigated in the literature than those based on the failure rate.

Mixtures of decreasing failure rate (DFR) distributions are always DFR. A Bayes explanation of this preservation of aging properties under mixing result was given by Professor R. Barlow [2], whereas the formal proof was obtained in Barlow and Proschan [3]. As usually, we use the term 'decreasing' (increasing) for 'non-increasing' (non-decreasing). The corresponding preservation property for the MRL mixtures was discussed in Klefsjo [4]. It is well known that this property does not exist for general mixtures of distributions with increasing failure rates (IFR), however, under additional assumptions it can still take place [5,6]. It turns out that very often the mixture failure rate of IFR distributions is ultimately decreasing, which means that the operation of mixing can change dramatically the corresponding pattern of aging. This fact is rather surprising and should be taken into account in various applications due to its practical importance. Monotonic properties of failure rates of mixtures were also studied in references [1,6,7,8,9,10] to name a few. Block *et al* [8] and Block and Joe [9] considered the corresponding limiting behavior as $t \to \infty$. The MRL mixtures were investigated in [11,12] In Badia *et al* [13] some useful inequalities for the MRL mixtures were obtained. General properties of the MRL function were discussed in references [14,15,16].

Our goal in this paper is to obtain some results on the shapes of the mixture failure rate and of the mixture MRL function. In Section 2 the main definitions and properties, describing the mixture failure rate and the mixture MRL function, are given. Section 3 is devoted to the mixture failure rate modeling via the conditional probability density of the mixing distribution. An asymptotic convergence to the failure rate of the strongest population is discussed as well. In Section 4 a direct model of the MRL mixing is studied. The corresponding approach is based on transforming this model into a model based on the failure rate mixing. In Section 5 the asymptotic comparison of the shapes of the mixture failure rate and the mixture MRL function is performed. Section 6 is devoted to the inverse problem in the mixture failure rate modeling: given the mixture failure rate, obtain the failure rate of a governing distribution. Some relevant remarks are discussed in the last section.

2. Basic Notions

Consider a lifetime random variable (r.v.) $T \geq 0$ with a c.d.f. $F(t)$. We shall call $F(t)$ the governing c.d.f. Let $F(t)$ be indexed by a parameter $\hat{\theta}$, so that $P(T \leq t|\hat{\theta} = \theta) = F(t, \theta)$ and that the probability density function $f(t, \theta)$ exists. Then the corresponding failure rate $\lambda(t, \theta)$ can be defined in a usual way as $f(t, \theta)/\bar{F}(t, \theta)$. Let $\hat{\theta}$ be interpreted as a non-negative r.v. with a positive support: $[a, b)$; $a \geq 0, b \leq \infty$ and the probability density function $\pi(\theta)$. A mixture c.d.f. is defined by

$$F_m(t) = \int_a^b F(t, \theta)\pi(\theta)d\theta, \tag{1}$$

whereas the mixture failure rate in accordance with this definition is

$$\lambda_m(t) = \frac{\int_a^b f(t,\theta)\pi(\theta)d\theta}{\int_a^b \bar{F}(t,\theta)\pi(\theta)d\theta}, \tag{2}$$

where $\bar{F}(t,\theta) \equiv 1 - F(t,\theta)$.

Following Block *et all* [9] and Lynn and Singpurwalla [18], consider the conditional probability density function $\pi(\theta|t)$ of θ given $T \geq t$:

$$\pi(\theta|t) = \frac{\pi(\theta)\bar{F}(t,\theta)}{\int_a^b \bar{F}(t,\theta)\pi(\theta)d\theta}. \tag{3}$$

With the help of $\pi(\theta|t)$ the mixture failure rate $\lambda_m(t)$ can be written as:

$$\lambda_m(t) = \int_a^b \lambda(t,\theta)\pi(\theta|t)d\theta. \tag{4}$$

Denote by $E[\theta|t]$ the conditional expectation of $\hat{\theta}$ given $T \geq t$, where for the sake of notation convenience the "hat" sign in $E[\hat{\theta}|t]$ is omitted :

$$E[\theta|t] = \int_a^b \theta\,\pi(\theta|t)d\theta.$$

It was proved in Finkelstein and Esaulova [5] that the derivative $E'[t|\theta]$ with respect to t can be obtained by the following relation:

$$E'[\theta|t] = \int_a^b \theta\pi'(\theta)d\theta = \lambda_m(t)E[\theta|t] - \frac{\int_a^b \theta\, f(t,\theta)\pi(\theta)d\theta}{\int_a^b \bar{F}(t,\theta)\pi(\theta)d\theta}. \tag{5}$$

Relation (5) will be used in the next section while considering specific cases.

Denote by $m(t)$ the MRL as a function of t (defined by the governing distribution $F(t)$):

$$m(t) = \frac{\int_t^\infty \bar{F}(u)du}{\bar{F}(t)} \tag{6}$$

and assume that $E[T] = m(0) = \int_0^\infty \bar{F}(u)du < \infty$. The following useful inverse relationship shows that $m(t)$ uniquely determines $F(t)$ [14] :

$$\bar{F}(t) = \frac{m(0)}{m(t)} exp\left\{ -\int_0^t \frac{1}{m(u)} du \right\}. \tag{7}$$

It follows from this equation that the accumulated failure rate is also uniquely determined by the MRL function $m(t)$:

$$\int_0^x \lambda(u)du = lnm(x) - lnm(0) + \int_0^x \frac{1}{m(u)} du. \tag{8}$$

From definition (6), assuming that $m(t)$ is differentiable:

$$m'(t) = \frac{\lambda(t) \int_t^\infty \bar{F}(u)du - \bar{F}(t)}{\bar{F}(t)} \equiv \frac{d(t)}{\bar{F}(t)} = \lambda(t)m(t) - 1 > -1, \tag{9}$$

where $d(t)$ is the notation for the numerator. From equation (9):

$$\lambda(t) = \frac{m'(t) + 1}{m(t)}. \tag{10}$$

This simple relation will play an important role in what follows. It is clear that equation (7) could be used for constructing distribution functions with a specified shape of $m(t)$. Differentiable functions $m(t)$ should meet the following characterization conditions:

a. $m(t) > 0, \ t \in [0, \infty)$
b. $m(0) < \infty$
c. $m'(t) > -1, \ t \in [0, \infty)$
d. $lnm(t) + \int_0^t (m(u))^{-1} du \to \infty, \quad t \to \infty.$

Assumptions 'a' and 'b' are quite natural, if we want to describe the properties of the "proper" [12] MRL function; the third condition is obtained from relation (9), and the last one is just the statement that the corresponding cumulative failure rate $\Lambda(t) = \int_0^\infty \lambda(u)du$ should tend to infinity as $t \to \infty$.

The MRL, being one of the major functions characterizing the aging properties of lifetime random variables, can also constitute a convenient and reasonable in applications model of mixing, though we think that this approach did not receive the proper attention in the literature. Taking into account definitions (1) and (6), let:

$$m(t,\theta) = \frac{\int_t^\infty \bar{F}(u,\theta)du}{\bar{F}(t,\theta)}, \tag{11}$$

$$m_m(t) = \frac{\int_t^\infty \int_a^b \bar{F}(u,\theta)\pi(\theta)d\theta du}{\int_a^b \bar{F}(t,\theta)\pi(\theta)d\theta} = \frac{\int_t^\infty \bar{F}_m(u)du}{\bar{F}_m(t)}. \tag{12}$$

In accordance with condition 'a', assume that $m(0,\theta) \equiv \int_0^\infty \bar{F}(u,\theta)du < \infty$ for $\forall\theta \in [a,b]$ and that the mixture MRL function also meets this condition:

$$m_m(0) \equiv \int_0^\infty \bar{F}_m(u)du < \infty, \tag{13}$$

which creates certain restrictions, as the first moment defined by $F(t,\theta)$ can, for instance, exist for $\forall\theta \in [0,\infty)$, while relation (13) does not hold. The simplest example illustrating this situation is when $F(t,\theta)$ is exponential ($\bar{F}(t,\theta) = exp\{-\theta u\}$) and $\pi(\theta)$ is the probability density function of the gamma distribution. Note that we are not "constructing" $F(t,\theta)$ via $m(t,\theta)$ and $F_m(t)$ via $m_m(t)$. It means that the last two characterization conditions are just properties of $m(t,\theta)$ and $m_m(t)$, respectively.

Taking into account relations (3) and (11), it is easy to transform (12) by changing the order of integration, which is due to our assumptions, a proper operation for the case under consideration:

$$m_m(t) = \frac{\int_t^\infty \int_a^b \bar{F}(u,\theta)\pi(\theta)d\theta du}{\int_0^\infty \bar{F}(t,\theta)\pi(\theta)d\theta} = \int_0^\infty m(t,\theta)\pi(\theta|t)d\theta. \tag{14}$$

Thus the mixture MRL function is written in the same form as the mixture failure rate in (4)! Relation (14) enables to analyze the shape of $m_m(t)$. It can be also done directly via (12) or the corresponding mixture failure rate $\lambda_m(t)$, because in some situations it is more convenient and natural to define $\lambda_m(t)$ from the beginning. This approach will be considered later.

If $m(t, \theta)$ is an increasing function, then $m(t)$ is also increasing. Thus, the corresponding closure property of the operation of mixing exists for this case [4]. As it was stated in the previous section, the shape of the MRL function can be different from the one of the governing distribution for other cases.

3. Models for the Failure Rate

Assume that the family $\{\lambda(t, \theta)\}, \theta \in [a, b]$ is such that given $\Delta\theta > 0$, $(\theta + \Delta\theta \in [a, b])$, the value $\varepsilon > 0$ can be found:

$$|\lambda(t, \theta + \Delta\theta) - \lambda(t, \theta)| \geq \varepsilon, \quad \forall t \in [0, \infty]. \tag{15}$$

In other words, condition (15) means that there are no crossings between the functions of this family with different values of θ and that they can be "separated". All models to be considered later meet this condition. Relation (15) holds, for instance, for $\lambda(t, \theta) = \theta\phi(t)$, $\phi(t) > 0$, where $\phi(t)$ is an increasing or decreasing function and $lim_{t \to \infty}\phi(t) = c > 0$. If $c = 0$, than (15) does not hold (one cannot separate different functions at the infinity). We are not considering the formally possible but practically unrealistic case when $|\lambda(t, \theta) - g(t)| \to 0$ as $t \to \infty$ uniformly in $\theta \in [a, b]$, where $g(t)$ is a continuous function. It is clear that the operation of mixing will trivially result then in the asymptotic mixture failure rate $g(t)$.

Condition (15) leads to the hazard rate ordering [17] (assume for definiteness that larger values of θ lead to larger values of $\lambda(t, \theta)$ for a fixed t):

$$\lambda(t, \theta_1) > \lambda(t, \theta_2), \quad \theta_1 > \theta_2, \ \forall t \in [0, \infty), \tag{16}$$

which leads to the corresponding stochastic ordering:

$$\bar{F}(t, \theta_1) < \bar{F}(t, \theta_2), \quad \theta_1 > \theta_2, \ \forall t \in [0, \infty). \tag{17}$$

Consider the conditional c.d.f.:

$$\Pi(\theta|T \geq t) = P(\hat{\theta} < \theta|T \geq t) = \int_a^\theta \pi(u|t)du. \tag{18}$$

Intuitively it is clear that under given assumptions the following weak convergence to a degenerate distribution with a mass in a (convergence in distribution) takes place as $t \to \infty$:

$$\Pi(\theta|T \geq t) \to \begin{cases} 1, & \theta \in [a.b], \\ 0, & \theta \notin [a,b]. \end{cases} \tag{19}$$

Indeed, as time increases, the proportion of items with smaller values of the failure rate in the survived population increases (weaker populations are dying out first). This means that for each internal point of the support interval:

$$lim_{t\to\infty}\Pi(\theta|T \geq t) = lim_{t\to\infty}\left(1 - \int_{\theta}^{b} \pi(u|t)dt\right) = 1,$$

which defines weak convergence in the continuity points. The strict proof of this fact was obtained in Finkelstein and Esaulova [5] for a specific model of $\lambda(t,\theta)$ but it can be easily generalized to the described above family of failure rates. As a is a discontinuity point, relation

$$lim_{t\to\infty}(\lambda_m(t) - \lambda(t,a)) = 0 \tag{20}$$

is not necessarily valid (see also Block *et al* [8]). We shall return to this topic after defining several specific models of mixing:

$$\lambda(t,\theta) = \lambda(t) + \theta, \tag{21}$$

$$\lambda(t,\theta) = \theta\lambda(t), \tag{22}$$

$$\lambda(t,\theta) = \theta\lambda(\theta t), \tag{23}$$

$$\lambda(t,\theta) = \alpha(t) + \theta\lambda(t), \tag{24}$$

$$\lambda(t,\theta) = \lambda(\theta t), \tag{25}$$

where, as previously, $\lambda(t)$ denotes the failure rate of the governing distribution $F(t)$ and $\alpha(t)$ is a continuous function.

Model (21) defines the additive failure rate model, relation (22)—the proportional hazard model, relation (23)—the accelerated life model. These models are widely used in practice for describing an impact of external covariates on the lifetime distribution of an object. We shall focus the study on model (24), which is a generalization of (21) and (22).

Let $\lambda(t)$ be an ultimately increasing function. It is clear that ordering (16) holds for models (21), (22), (24) (whereas $\lambda(t)$ should be increasing in $[0,\infty)$ for models (23) and (25) to meet the ordering condition). As models

(21) and (22) are the specific cases of model (24), it is sufficient to consider the last named one. It can be easily seen directly by applying relation (5) to model (24) that

$$E'[\theta|t] = \lambda(t)\left([E[\theta|t]]^2 - \int_a^b \theta^2\pi(\theta|t)d\theta\right) = -\lambda(t)Var(\theta|t) < 0, \quad (26)$$

which means that the conditional expectation $E[\theta|t]$ is decreasing in $t \in [0,\infty)$ (weaker populations are dying out first). This fact can be also obtained from general considerations which were used for interpreting weak convergence (19), whereas (26) gives the concrete expression for the specific model. Weak convergence (19) suggests (it was proved for the proportional hazard model (22) [5] and can be easily generalized to model (24)) that the following convergence to the Dirac δ–function takes place:

$$\pi(\theta|t) \to \delta(a), \quad t \to \infty$$

to be understood in terms of the generalized functions:

$$lim_{t\to\infty} \int_a^b G(\theta)\pi(\theta|t)d\theta = G(a), \quad (27)$$

where $G(\theta)$ is a continuous function. By substituting $G(\theta) = \theta$ in (27) the following asymptotic result can be obtained:

$$lim_{t\to\infty} E[\theta|t] = a, \quad (28)$$

which is intuitively evident, as weaker populations are dying out first!

It turns out that relation (28) is not sufficient for concluding that convergence (20) takes place. Furthermore, from equation (4) it follows that

$$lim_{t\to\infty}\lambda_m(t) = \alpha(t) + lim_{t\to\infty}\lambda(t)E[\theta|t], \quad (29)$$

$$lim_{t\to\infty}(\lambda_m(t) - \lambda(t,a)) = lim_{t\to\infty}\lambda(t)(E[\theta|t] - a) \quad (30)$$

and the limiting behavior depends on asymptotic properties of $\lambda(t)$ and $E[\theta|t]$. Therefore, for convergence to the stronger population (20), $E[\theta|t]-a$ should decrease to 0 sharper than $\lambda(t)$ increases.

For illustrating the possible asymptotic behavior consider the following examples [5] for the model (22) and the infinite support: $\theta \in [0,\infty)$:

Example 1: Let $F(t,\theta)$ be a specific type of the Weibull distribution with linear increasing failure rate $\lambda(t,\theta) = 2\theta t(\lambda(t) = 2t)$ and assume that $\pi(\theta)$ is a gamma probability density function:

$$\pi(\theta) = \frac{\vartheta^{\beta}\theta^{\beta-1}exp\{-\theta\vartheta\}}{\Gamma(\beta)},\ \beta,\vartheta > 0,\ \theta \geq 0.$$

Using relation (2), the mixture failure rate can be easily obtained via direct integration:

$$\lambda_m(t) = \frac{2\beta t}{\vartheta + t^2}.$$

This expression is equal to zero at $t = 0$ and tends to zero as $t \to \infty$ with a single maximum at $t = \sqrt{\theta}$. Hence, the mixture of IFR distributions has a decreasing (tending to zero!) failure rate for sufficiently large t and this is rather surprising. Furthermore, the same result asymptotically holds for arbitrary Weibull distributions with increasing failure. It is easy to derive that the conditional expectation for this case is

$$E[\theta|t] = \frac{2\beta}{\vartheta + t^2}$$

and that it decreases to 0 sharper than $\lambda(t) = 2\theta t$ increases.

Example 2: Consider the truncated extreme value distribution defined in a following way:

$$F(t,k,\theta) = 1 - exp\{-\theta\, k(exp\{t\} - 1)\},\ t \geq 0,$$

$$\lambda(t,\theta) = \theta\lambda(t) = \theta\, kexp\{t\},$$

where $k > 0$ is a constant.

Let for simplicity $\pi(\theta)$ be an exponential probability density function with parameter ϑ. Then by direct computations via relation (2):

$$\lambda_m(t) = \frac{k\, exp\{t\}}{kexp\{t\} - k + \vartheta} = 1 + \frac{k - \vartheta}{kexp\{t\} - k + \vartheta}.$$

For analyzing the shape of $\lambda_m(t)$ it is convenient to write it in the following way:

$$\lambda_m(t) = \frac{1}{1 + d/z},$$

where $z = kexp\{t\}$ and $d = -k + \vartheta$. Thus $\lambda_m(0) = k/\vartheta$. When $d < 0$ ($k > \vartheta$), the mixture failure rate is monotonically decreasing asymptotically converging to 1. For $d > 0$ ($k < \vartheta$) it is monotonically increasing asymptotically converging to 1. When $d = 0 (k = \vartheta)$, the mixture failure rate is equal to 1: $\lambda_m(t) \equiv 1, \forall t \in [0, \infty)$.

This result is even more surprising than in the first example. The initial failure rate is increasing extremely sharply and still for some values of parameters the mixture failure rate is decreasing! The conditional expectation for this case is

$$E[\theta|t] = \frac{1}{kexp\{t\} - k + \vartheta} = \frac{1}{z + d}$$

and $\lambda(t)(E[\theta|t] - 0)$ monotonically converges to 1, as $t \to \infty$.

It follows from (30) that the mixture failure rate converges to the failure rate of the strongest population if and only if

$$lim_{t\to\infty}\lambda(t)(E[\theta|t] - a) = 0, \quad a \geq 0. \tag{31}$$

If, for instance, $\lambda(t)$ is an increasing bounded function (e.g., as in the case of the gamma distribution), thus having a limit value $0 < b < \infty$ as $t \to \infty$, then it follows from (31) that the mixture failure rate automatically converges to the failure rate of the strongest population (see also [9]):

$$lim_{t\to\infty}\lambda_m(t) = ba.$$

Specifically, when $a = 0$ the mixture failure rate converges to 0.

It is clear that, as it was already stated, the ordering condition (16) holds for models (23) and (25), if $\lambda(t)$ is an increasing function. The weak convergence result (19) also takes place, but obtaining conditions similar to relation (31) is more complicated and can be considered as a special topic for future studies.

4. Models for the MRL Function

Similar to the previous section, it makes sense to define the following models for the direct mixing procedure via the MRL function:

$$m(t, \theta) = \frac{m(t)}{\theta}, \tag{32}$$

$$m(t, \theta) = m(t/\theta), \tag{33}$$

where, as previously, $m(t)$ denotes the MRL function defined in (6) by the governing distribution $F(t)$. We shall consider model (32), as it leads to the more natural, speaking for itself results. It will be seen later that it is more convenient for this setting to divide by θ than to multiply by it.

As follows from results of Section 2, the baseline $m(t)$ should satisfy characterization conditions 'a'–'d'. The MRL function $m(t,\theta)$ should satisfy these conditions for $\forall\theta \in [a,b)$ as well. Conditions 'a' and 'b' are trivially satisfied whereas condition 'c' implies that

$$m'(t,\theta) = \frac{m'(t)}{\theta} > -1 \rightarrow m'(t) > -\theta. \tag{34}$$

Considering (34) and (9) results in a conclusion that for increasing baseline $m(t)$ equation (32) defines the MRL function for $\forall\theta \in (0,\infty)$, whereas for decreasing and non-monotonic baseline $m(t)$ it defines the MRL function for $\forall\theta \in [1,\infty)$.

It is clearly seen from the corresponding closure property that the mixture MRL function $m_m(t)$ for increasing $m(t)$ is also increasing. Thus, it makes sense to consider only the model of mixing (24). Assume that $m(t)$ is an ultimately decreasing function and let the corresponding support interval be $[1,\infty)$. It is convenient to transform this model to the equivalent failure rate setting. It is worth noting that the case of decreasing $m(t)$ is the most important in practical applications. From equation (10):

$$\lambda(t,\theta) = \frac{m'(t,\theta)+1}{m(t,\theta)} = \frac{m'(t)}{m(t)} + \frac{\theta}{m(t)} = m'(t)r(t) + \theta r(t), \tag{35}$$

where $r(t) \equiv (m(t))^{-1}$. We can see that equation (35) actually defines the failure model (24). It can be written even in a more speaking for itself form:

$$\lambda(t,\theta) = r(t)m'(t) + r(t) + (\theta-1)r(t) = \lambda(t) + (\theta-1)r(t). \tag{36}$$

Therefore, using definition (4) for the mixture failure rate:

$$\lambda_m(t) = \int_1^\infty \lambda(t,\theta)\pi(\theta|t)d\theta = \lambda(t) + r(t)(E[\theta|t]-1) \tag{37}$$

and the baseline failure rate $\lambda(t)$ plays the role of the failure rate of the strongest population in this setting. Similar to (30) and (31):

$$lim_{t\to\infty}(\lambda_m(t) - \lambda(t)) = 0, \tag{38}$$

if and only if

$$lim_{t\to\infty}r(t)(E[\theta|t]-1)=0. \tag{39}$$

Applying the L'Hospital rule and using (9) and equation similar to (26), condition (39) can be written as:

$$lim_{t\to\infty}r(t)(E[\theta|t]-1)=lim_{t\to\infty}-\frac{Var(\theta|t)}{(m(t)\lambda(t)-1)m(t)}=0. \tag{40}$$

Therefore, if:

a. *The function $r(t)$ is ultimately increasing*
b. *Condition (39) takes place.*

Then the mixture failure rate converges to the failure rate of a strongest population, as defined by relation (38).

If $\lambda(t)$ is ultimately increasing and condition (32) takes place, then it is easy to see the following convergence to a stronger population:

$$lim_{t\to\infty}(r_m(t)-r(t))=0, \tag{41}$$

where $r_m(t)\equiv(m_m(t))^{-1}$.
Indeed, as it was stated before, the increasing (ultimately increasing) failure rate implies the decreasing (ultimately decreasing) MRL function. This means that $r(t)$ is increasing (ultimately increasing). The rest comes from the previous result, noting that vanishing as $t\to\infty$ perturbation in the increasing failure rate $\lambda(t)$ (as in (37) under condition (38)) leads to a vanishing perturbation in $(m(t))^{-1}$.

5. Asymptotic Comparison

Let $\lambda(t)(\lambda_m(t))$ be an ultimately increasing function. It is interesting to compare its asymptotic behavior with $r(t)(r_m(t))$. We shall obtain the corresponding comparison result for functions $\lambda(t)$ and $r(t)$ (the result is the same for mixtures).

It is well known [11] that for the stated assumption:

$$lim_{t\to\infty}\lambda(t)=lim_{t\to\infty}r(t)=c. \tag{42}$$

If $c<\infty$, then it is nothing to say more about asymptotic comparison. Consider the case $c=\infty$. It is reasonable to rewrite equation (10) as

$$\lambda(t)=-\frac{r'(t)}{r(t)}+r(t). \tag{43}$$

Using this expression it is easy to see that the function $r(t)$ is asymptotically equivalent to $\lambda(t)$ in the following sense:

$$|\lambda(t) - r(t)| \to 0, \quad t \to \infty \tag{44}$$

if and only if

$$\left|\frac{r'(t)}{r(t)}\right| = \left|\frac{m'(t)}{m(t)}\right| \to 0, \quad t \to \infty. \tag{45}$$

If, for instance, $r(t) = \beta t^{\beta-1}$; $0 < \beta < \infty$, $t \in [a, \infty), a > 0$, then (45) is valid. Thus, $|\lambda(t) - \beta t^{\beta-1}| \to 0$ as $t \to \infty$, and $\beta t^{\beta-1}$ is the failure rate for the Weibull distribution of the form: $\bar{F}(t) = exp\{-t^\beta\}$. For the sharply increasing functions $r(t)$, such as $r(t) = exp\{t\}$ or $r(t) = exp\{t^2\}$, relation (25) does not hold while the weaker version:

$$\lambda(t) = r(t)(1 + o(1)), \quad t \to \infty \tag{46}$$

certainly takes place.

Many of the standard distributions have failure rates, which are polynomials or ratios of polynomials. Asymptotic result (44)–(45) can be easily generalized on these classes of functions by assuming that $r(t)$ is a regularly varying function. The theory of regular varying functions was studied in references [18,19,20]. The regularly varying $r(t)$ can be written as $r(t) = t^b l(t)(1 + o(1))$, where $-\infty < b < \infty$,$t \to \infty$ and $l(t)$ is a slowly varying function: $l(kt)/l(t) \to 1$ as $t \to \infty$ for all $k > 0$.

If $r(t) = t^{\beta-1}l(t)(1 + o(1))$, $-\infty < \beta < \infty, \beta \neq 0$, $t \to \infty$ is a regularly varying function, where $l(t)$ is a slowly varying function and $r'(t)$ is ultimately monotone, then relation (45) takes place. The proof of this result comes from the Monotone Density Theorem [19], according to which for the ultimately monotone $r'(t)$:

$$r'(t) = t^{\beta-2}\tilde{l}(t)(1 + o(1)), \quad t \to \infty,$$

where $\tilde{l}(t)$ is a slowly varying function. Using expressions for regularly varying $r(t)$ and $r'(t)$, the following relation can be obtained:

$$\left|\frac{r'(t)}{r(t)}\right| = t^{-1}\hat{l}(t)(1 + o(1)), \quad t \to \infty,$$

where $\hat{l}(t)$ is a slowly varying function. Due to this fact: $t^{-1}\hat{l}(t) \to 0$ as $t \to \infty$.

6. Inverse Problem

In the process of analyzing the shape of the mixture failure rate the following inverse problem may be of interest: Given the mixture failure rate and the mixing distribution, obtain the failure rate of the governing distribution. It can be shown that this problem can be explicitly solved [10] for specific cases of mixing (21), (22) and (24). In this section the proportional hazard model (22) will be considered. What is the reason for solving the stated inverse problem? It appears, that in some situations we want to model a certain shape (a bathtub shape, for example) of the mixture failure rate, because it explains our understanding of the lifetime random variable to be modeled. Then another reasonable question should be answered: what is the shape of the failure rate of the governing distribution function, which after the operation of mixing results in the desired mixture failure rate shape? Is this shape realistic in concrete applications? The answer to these questions can be provided by the solution of the inverse problem.

The mixture failure rate $\lambda_m(t)$ should be equal to some arbitrary continuous function $g(t)$, such that $\int_0^\infty g(t)dt = \infty,\ \ g(t) > 0$. Thus, the corresponding equation for solving the inverse problem is $\lambda_m(t) = g(t)$. Let, specifically the interval of support $[a,b]$ be infinite: $a = 0, b = \infty$. For the model under consideration:

$$\lambda_m(t) = \int_0^\infty \lambda(t,\theta)\pi(\theta|t)d\theta = \lambda(t)E[\theta|t]. \tag{47}$$

Thus, taking into account the definition of $E[\theta|t]$, the equation for solving the inverse problem can be written as the equation for obtaining the cumulative failure rate $\Lambda(t) = \int_0^t \lambda(u)du$ [10]:

$$\Lambda'(t)\frac{\int_0^\infty \theta exp\{-\theta\Lambda(t)\}\pi(\theta)d\theta}{\int_0^\infty exp\{-\theta\Lambda(t)\}\pi(\theta)d\theta} = g(t), \tag{48}$$

Denote by $\pi^*(t)$ the Laplace transform of $\pi(\theta)$ for $t \geq 0$:

$$\int_0^\infty exp\{-\theta t\}\pi(\theta)d\theta \equiv \pi^*(t).$$

Taking into account that $\Lambda(t)$ is monotonically increasing:

$$\int_0^\infty exp\{-\theta\Lambda(t)\}\pi(\theta)d\theta \equiv \pi^*(\Lambda(t)) \equiv S(t). \tag{49}$$

Differentiating both sides of (49):

$$S'(t) = -\Lambda'(t)\int_0^\infty \theta\, exp\{-\theta\Lambda(t)\}\pi(\theta)d\theta. \tag{50}$$

Combining relations (48) with (49) and (50), a simple expected result can be obtained in terms of the Laplace transform:

$$\pi^*(\Lambda(t)) = S(t) = exp\left\{-\int_0^t g(u)du\right\}, \tag{51}$$

which leads to

$$\Lambda(t) = (\pi^*)^{-1}\left(exp\left\{-\int_0^t g(u)du\right\}\right). \tag{52}$$

Equation (51) always has a unique solution, as $\Lambda(t)$ is a monotonically increasing function and $\pi^*(t)$ $(\pi^*(\Lambda(t))$ is a survival function (monotonically decreasing from 1 at $t = 0$ $(\Lambda(0) = 0)$ to 0 as $t \to \infty(\Lambda(t) \to \infty))$. The solution can be easily obtained explicitly for mixing distributions with a "nice" Laplace transform $\pi^*(t)$.

Example 3: Let, as in Example 1, $\pi(\theta)$ be the gamma probability density function:

$$\pi(\theta) = exp\{-\vartheta\theta\}\frac{\theta^{n-1}\vartheta^n}{\Gamma(n)}, \quad n > 0$$

$$\pi^*(t) = \int_0^\infty exp\{-\theta t\}\frac{exp\{-\vartheta\theta\}\theta^{n-1}}{\Gamma(n)}\vartheta^n d\theta = \frac{\vartheta^n}{(t+\vartheta)^n} \quad .$$

Thus,

$$\pi^*(\Lambda(t)) = \frac{\vartheta^n}{(\Lambda(t)+\vartheta)^n},$$

$$\Lambda(t) = \vartheta exp\left\{\frac{1}{n}\int_0^t g(u)du\right\} - \vartheta.$$

Eventually:

$$\lambda(t) = \Lambda'(t) = \frac{\vartheta}{n}g(t)exp\left\{\frac{1}{n}\int_0^t g(u)du\right\},$$

which completes the solution of the inverse problem.

7. Conclusions and Outlook

The change in the pattern of the mixture failure rate (MRL function) behavior compared with the pattern of the failure rate (MRL function), which generates this mixture, is usually non-desirable because one should expect in modeling the aging properties of a mixture be similar to the corresponding properties of the governing distribution. The subjective assumption on the prior distribution for a parameter of interest can lead to the same consequences, but in this case the impact can be even more dramatic, because it is the **assumption** that the parameter is random had caused the mentioned change. Therefore, it is important for practical applications that the effect of mixing should be understood and modeled properly.

Several types of mixing models were considered. It turned out that the model of mixing via the failure rate (24) (the model of mixing via the MRL function (32)) lead to explicit asymptotic results on the shape of the mixture failure rate (MRL function). It is clear from general considerations of Section 3 that the scale transformation $\lambda(\theta t)(m(\theta t))$ leads to the similar results for the monotonically increasing $\lambda(t)$(monotonically decreasing $m(t)$), but the proof of this fact is not so straightforward. It is worth noting that the direct model of mixing (32) was analyzed using the transformation, which allowed considering the problem in terms of the corresponding failure rate.

In many instances the mixture failure rate $\lambda_m(t)$ exhibits a tendency to decrease at least for substantially large t and eventually converges to 0. Some surprising examples show that this takes place even when $\lambda(t,\theta)$ is sharply increasing as $t \to \infty$. Simplifying the situation, one can say that the mixture failure rate tends to increase, but at the same time another process due to the effect of dying out of the weaker populations takes place.

Interaction between these two tendencies can result in different shapes of $\lambda_m(t)$.

It was shown that the inverse problem of mixing could be efficiently solved for model (22) (the generalization to model (24) is trivial). This means that under given assumptions the failure rate of the baseline distribution can be obtained in such a way that the operation of mixing will result in a given (!) shape of the mixture failure rate. For instance, as follows from Example 2, the constant failure rate of the mixture $\lambda_m(t) = const$ can result for some values of parameters ($k = \vartheta$) in the truncated extreme value governing distribution with a sharply increasing failure rate! It is clear that due to the mentioned above transformation the inverse problem can be also solved for model (32).

References

1. N. J. Lynn and N.D. Singpurwalla, Comment: "Burn-in" makes us feel good, *Statistical Science*, **12**, 13 (1997).
2. R. E. Barlow, A Bayes explanation of an apparent failure rate paradox, *IEEE Trans. Reliability*, **34**, 107 (1985)
3. R. E. Barlow and F. Proschan, *Statistical Theory of Reliability: Probability Models* (Holt, Rinehart & Winston, New York, 1975).
4. B. Klefsjo, The NBUE and HNBUE classes of life distributions, *Naval Research Logistics Quartely*, **29**, 331 (1982).
5. M. S. Finkelstein and V. Esaulova, Modeling a failure rate for a mixture of distribution functions. *Probability in the Engineering and Informational Sciences*, **15**, 383 (2001).
6. J. D. Lynch, On conditions for mixtures of increasing failure rate distributions to have an increasing failure rate, *Probability in the Engineering and Informational Sciences*, **15**, 33 (1999)
7. H. W. Block and T. H. Savits, Burn-in, *Statistical Science*, **12**, 1 (1997).
8. H. W. Block, J. Mi and , T. H. Savits, Burn-in and mixed populations, *J. Appl. Prob.* **30**, 692 (1993).
9. H. W. Block and H. Joe, Tail behavior of the failure rate functions of mixtures, *Lifetime Data* Analyses, **3**, 269 (1997).
10. M. S. Finkelstein and V. Esaulova, On inverse problem in mixture failure rate modeling, *Applied Stochastic Models in Business and Industry*, **17**, 221 (2001)
11. R. C. Gupta and H. O. Akman, Mean residual life function for certain types of non-monotonic aging, *Stochastic* models, **11**, 219 (1995).
12. H. Zahedi, Proportional mean remaining life model, *Journal of Statistical Planning and* Inference, **29**, 221 (1991).
13. F. G. Badia, M. D. Berrade, C. A. Campos and M. A. Navascues, On the

behavior of aging characteristics in mixed populations, *Probability in the Engineering and Informational* Sciences, **15**, 83 (2001).
14. J. Beirlant, M. Broniatowski, J. L. Teugels and P. Vynckier, The mean residual life function at great age: Application to tail estimation, *Journal of Statistical Planning and* Inference, **45**, 21 (1995).
15. M. C. Bhattachrjee, The class of mean residual lives and some consequences, *Siam J. Alg. Disc.* Math. **3**, 56 (1982).
16. J. Mi, Bathtub failure rate and upside-down bathtub mean residual life, *IEEE Transactions on Reliability*, **44**, 388 (1995).
17. S. M. Ross, *Stochastic Processes* (John Wiley & Sons, New York,1996).
18. J. Beirlant, J. L Teugels and P. Vynckier, *Practical Analysis of Extreme Values* (Univ. Press, Leuven, 1996).
19. N. H. Bingham, C. M. Goldie and J. L. Teugels, *Regular Variation* (Univ. Press, Cambridge, 1987).
20. L. de Haan, Equivalence classes of regularly varying functions. *Stochastic process. Appl.* **2**, 243 (1974).

CHAPTER 11

ON SOME DISCRETE NOTIONS OF AGING

Cyril Bracquemond* and Olivier Gaudoin†
Institut National Polytechnique de Grenoble, Laboratoire IMAG-LMC
BP 53, 38 041 Grenoble Cedex 9, France
*E-mail: * Cyril.Bracquemond@imag.fr,*
E-mail: † Olivier.Gaudoin@imag.fr

Dilip Roy
Department of Business Administration, Burdwan University
Golapbag, Burdwan 713104, India
E-mail: dilip_roy@yahoo.com

Min Xie
Industrial and Systems Engineering Department, National University of Singapore 10 Kent Ridge Crescent, Singapore 119260
E-mail: mxie@nus.edu.sg

This chapter is dedicated to the study of basic notions of aging, such as IFR (Increasing Failure Rate) IFRA (Increasing Failure Rate in Average) and NBU (New Better than Used), when system lifetimes are discrete random variables. As stated by Shaked, Shantikumar and Valdez-Torres, several different definitions are possible for these notions, which are equivalent in the continuous case but not in the discrete case. It is shown that the problem lies in the usual definition of failure rate for discrete distributions. An alternative definition, also known as the second rate of failure, is discussed to solve the above-mentioned problem.

1. Introduction : Discrete Time Reliability

Discrete distributions are used in reliability when lifetime measurements are taken in discrete time. This is the case for example when an equipment operates on demand and the observation consists in the number of demands successfully completed before failure, or when continuous lifetimes are grouped.

Let the random variable K be a discrete system lifetime. K is defined over the set of positive integers $\mathbb{N}^*$. Let $p(k) = P(K = k)$ be the probability that the system fails at time k. The reliability function is given by the probability that the system is still alive at time k :

$$R(k) = P(K > k) = \sum_{i=k+1}^{\infty} p(i) \tag{1}$$

The failure rate of a discrete distribution has been defined by Barlow, Marshall and Proschan[1] as :

$$\begin{aligned} \forall k \in \mathbb{N}^*, \qquad \lambda(k) &= P(K = k | K \geq k) = \frac{P(K = k)}{P(K \geq k)} \\ &= \frac{p(k)}{R(k-1)} = \frac{R(k-1) - R(k)}{R(k-1)} \end{aligned} \tag{2}$$

Then, the probability function and reliability function can be written as :

$$\forall k \in \mathbb{N}^*, \qquad p(k) = \lambda(k) \prod_{i=1}^{k-1} (1 - \lambda(i)) \tag{3}$$

$$\forall k \in \mathbb{N}^*, \qquad R(k) = \prod_{i=1}^{k} (1 - \lambda(i)). \tag{4}$$

The cumulative hazard function is :

$$\forall k \in \mathbb{N}^*, \qquad \Lambda(k) = \sum_{i=1}^{k} \lambda(i) \tag{5}$$

The residual reliability function at time k is :

$$\forall i \in \mathbb{N}^*, \qquad R(i|k) = P(K > k + i | K > k) = \frac{R(k+i)}{R(k)} \tag{6}$$

Some short discussions on discrete reliability appeared in Barlow and Proschan[2], Cox[3], Kalbfleisch and Prentice[4] and Lawless[5]. More precise concepts of discrete reliability theory have been settled by Salvia and Bollinger[6]. The first discrete reliability distribution has been defined by Nakagawa and

Osaki[7] to be the discrete analogous of the continuous Weibull distribution. Roy and Gupta[8] examined classification of discrete life distributions and in this process they introduced the concept of second rate of failure to maintain analogy with the continuous aging classes. Roy[9] extended the discrete classification system to multivariate set up. More recent work deals with characterization of discrete reliability models (Gupta, Gupta and Tripathi[10], Roy and Gupta[11]) and discrete mean residual life (Tang, Lu and Chew[12], Ghai and Mi[13]).

This chapter is dedicated to notions of aging for discrete random variables. In section 2, the basic definitions and results for aging notions in continuous case are recalled. In section 3, similar definitions for aging notions in discrete time are discussed according to Roy and Gupta[8], Shaked, Shantikumar and Valdez-Torres[14]. It appears that several different definitions are possible for these notions, which are equivalent in the continuous case but not in the discrete case. In section 4, it is shown that the problem lies in the usual definition of failure rate for discrete distributions. An alternative definition is discussed in section 5. With this definition, all the definitions of discrete aging notions are equivalent. Then, it is possible to ensure the IFRA closure theorem for discrete distributions.

2. Notions of Aging in Continuous Time

The characterization of classes of lifetime distributions based on notions of aging, as proposed in Barlow and Proschan[15], has been a very fruitful research field in reliability theory. The basic definitions are recalled here for systems with continuous lifetimes.

Let T be a random lifetime with a continuous distribution on $\mathbb{R}^+$. The reliability, failure rate, residual reliability and cumulative hazard functions are defined as :

$$\forall t \geq 0, \qquad R(t) = P(T > t)$$

$$\forall t \geq 0, \qquad \lambda(t) = \lim_{dt \to 0} \frac{1}{dt} P(t < T \leq t + dt | T > t) = -\frac{R'(t)}{R(t)}$$

$$\forall t \geq 0, \forall s \geq 0, \quad R(s|t) = P(T > s + t | T > t) = \frac{R(s+t)}{R(t)}$$

$$\forall t \geq 0, \qquad \Lambda(t) = \int_0^t \lambda(u) du = -\ln R(t)$$

2.1. *Increasing Failure Rate (IFR)*

T is said to be IFR if and only if, equivalently (theorem 4.1. in Barlow and Proschan[2]) :

- IFR1 : $\lambda(t)$ is an increasing function.
- IFR2 : $\forall s \geq 0$, $R(s|t)$ is a decreasing function in t.
- IFR3 : R is a Pólya frequency function of order 2 (PF_2), i.e. : $\forall s_1, s_2, t_1, t_2 \in \mathbb{R}^+$, $s_1 < s_2$, $t_1 < t_2$, $\begin{vmatrix} R(s_1 - t_1) & R(s_1 - t_2) \\ R(s_2 - t_1) & R(s_2 - t_2) \end{vmatrix} \geq 0.$
- IFR4 : $\ln R(t)$ is a concave function.

2.2. *Increasing Failure Rate in Average (IFRA)*

T is said to be IFRA if and only if, equivalently :

- IFRA1 : $[R(t)]^{1/t}$ is a decreasing function.
- IFRA2 : $\dfrac{\Lambda(t)}{t}$ is an increasing function.

The equivalence between IFRA1 and IFRA2 is immediate since $\Lambda(t) = -\ln R(t)$. This notion has been introduced by Birnbaum, Esary and Marshall [16] in order to find the smallest class of lifetimes distributions of coherent systems with IFR components.

An important result is the *IFRA closure theorem* (theorem 2.6 p.85 in Barlow and Proschan[15]) :

Proposition 1: *A coherent system of independent components with continuous IFRA lifetimes distributions has itself an IFRA lifetime distribution.*

2.3. *New Better than Used (NBU)*

T is said to be NBU (New Better Than Used) if and only if, equivalently :

- NBU1 : $R(s|t) \leq R(s) \quad \forall s \geq 0, \forall t \geq 0.$
- NBU2 : $\Lambda(t+s) \geq \Lambda(t) + \Lambda(s) \quad \forall t \geq 0, \forall s \geq 0.$

The equivalence between NBU1 and NBU2 comes also from the fact that $\Lambda(t) = -\ln R(t)$.

2.4. *Relationships between Basic Aging otions*

In continuous time, it is easy to show that (Barlow and Proschan[15] p.159):

$$IFR \Rightarrow IFRA \Rightarrow NBU. \tag{7}$$

Similar results are valid to describe the dual negative aging notions DFR (Decreasing Failure Rate), DFRA (Decreasing Failure Rate in Average), and NWU (New Worse than Used).

3. Notions of Aging in Discrete Time

In this section, the discrete counterpart of the above continuous notions of aging are defined, according to Shaked, Shantikumar and Valdez-Torres[14].

3.1. *Increasing Failure Rate (IFR)*

The definitions corresponding to (2.1) of Increasing Failure Rate in discrete time are :

- IFR1 : $\{\lambda(k)\}_{k\geq 1}$ is an increasing sequence.
- IFR2 : $\forall i \in \mathbb{N}^*$, $\{R(i|k)\}_{k\geq 1}$ is a decreasing sequence in k.
- IFR3 : $\forall j_1, j_2, k_1, k_2 \in \mathbb{N}^*$, $j_1 < j_2$, $k_1 < k_2$, $\begin{vmatrix} R(j_1-k_1) & R(j_1-k_2) \\ R(j_2-k_1) & R(j_2-k_2) \end{vmatrix} \geq 0$
- IFR4 : $\{\ln R(k)\}_{k\geq 1}$ is a concave sequence.

As in the continuous case, Barlow, Marshall and Proschan[1] proved that these four definitions are equivalent.

3.2. *Increasing Failure Rate in Average (IFRA)*

The definitions corresponding to (2.2) of Increasing Failure Rate in Average in discrete time are :

- IFRA1 : $\left\{R(k)^{\frac{1}{k}}\right\}_{k\geq 1}$ is a decreasing sequence.
- IFRA2 : $\left\{\dfrac{\Lambda(k)}{k}\right\}_{k\geq 1}$ is an increasing sequence.

Roy and Gupta[8] have noticed that both IFRA definitions are not equivalent. More precisely in their Result 3.4 they ensured the following proposition:

Proposition 2: *IFRA1 $\Rightarrow$ IFRA2 but IFRA2 $\not\Rightarrow$ IFRA1.*

In the continuous case, the equivalence between both IFRA definitions is immediately proved by the fact that $-\ln R(t) = \Lambda(t) = \int_0^t \lambda(u)du$. But in the discrete case, $H(k) = -\ln R(k) \neq \Lambda(k) = \sum_{i=1}^{k} \lambda(i)$. Lawless[5] had noticed this fact and have just concluded that the term "cumulative hazard rate" was not appropriate for H in discrete time. Going further, we can think that this can be the reason of the non-equivalence of both IFRA definitions in the discrete case.

The relationship between IFR and IFRA1 aging notions in discrete time is given in Roy and Gupta[8] in their Result 3.3. We present the same proposition with an alternative proof for the first part:

Proposition 3: *IFR $\Rightarrow$ IFRA1 but IFRA1 $\not\Rightarrow$ IFR.*

Proof: 1. IFR $\Rightarrow$ IFRA1
Using an analogy with the continuous case, we will use the concept of *star-shaped sequence* in order to prove this implication. In continuous time, a function $\varphi(x)$ defined on $\mathbb{R}^+$ is star-shaped if and only if, equivalently, $\forall x \in \mathbb{R}^+$, $\forall \alpha \in [0,1]$, $\varphi(\alpha x) \leq \alpha\varphi(x)$, or $\dfrac{\varphi(x)}{x}$ is an increasing function. Moreover, every convex continuous function passing through the origin is star-shaped (Barlow-Proschan [15]).

We present here the equivalent of these definitions in discrete time.

Definition 4: A sequence $\{\varphi(k)\}_{k\geq 1}$ is star-shaped if and only if, equivalently :

- $\forall (j,k) \in \mathbb{N}^{*2}$, $j \leq k \Rightarrow \varphi(j) \leq \dfrac{j}{k}\varphi(k)$
- $\left\{\dfrac{\varphi(k)}{k}\right\}_{k\geq 1}$ is an increasing sequence.

Lemma 5: *Every convex sequence passing through the origin is star-shaped.*

Proof: (of the lemma). A sequence $\{\varphi(k)\}_{k\geq 1}$ is a convex sequence if and only if $\forall(k_1,k_2,k_3)\in\mathbb{N}^{*3}$, :

$$k_1<k_2<k_3\Rightarrow\frac{\varphi(k_2)-\varphi(k_1)}{k_2-k_1}\leq\frac{\varphi(k_3)-\varphi(k_1)}{k_3-k_1}\leq\frac{\varphi(k_3)-\varphi(k_2)}{k_3-k_2}\qquad(8)$$

It is easy to prove that this is equivalent to :

$$\forall(u,v,w)\in\mathbb{N}^{*3},\ (u+v)\varphi(w+u)-v\varphi(w)-u\varphi(w+u+v)\leq 0\qquad(9)$$

φ passing through the origin, $\varphi(w)=0$ for $w=0$. Thus :

$$\forall(u,v)\in\mathbb{N}^{*2},\ (u+v)\varphi(u)-u\varphi(u+v)\leq 0\qquad(10)$$

Let $j=u$ and $k=u+v$. Then :

$$\forall(j,k)\in\mathbb{N}^{*2},\ j<k\Rightarrow k\varphi(j)-j\varphi(k)\leq 0\,,\ \Rightarrow\varphi(j)\leq\frac{j}{k}\varphi(k)\qquad(11)$$

So $\{\varphi(k)\}_{k\geq 1}$ is star-shaped. □

Now it is easy to prove that IFR $\Rightarrow$ IFRA1 :

$$\begin{aligned}\text{K IFR}\ &\Rightarrow\{\ln R(k)\}_{k\geq 1}\ \text{ is concave}\\&\Rightarrow\{-\ln R(k)\}_{k\geq 1}\ \text{ is a convex sequence (passing through the origin)}\\&\Rightarrow\{-\ln R(k)\}_{k\geq 1}\ \text{ is star-shaped}\\&\Rightarrow\left\{-\frac{\ln R(k)}{k}\right\}_{k\geq 1}\ \text{ is increasing}\\&\Rightarrow\left\{R(k)^{\frac{1}{k}}\right\}\ \text{ is decreasing}\\&\Rightarrow\ \text{K is IFRA1}\end{aligned}$$

2. IFRA1 $\not\Rightarrow$ IFR
Consider the sequence $\{\lambda(k)\}_{k\geq 1}$ defined by :

$$\lambda(1)=0.1,\ \lambda(2)=0.4,\ \text{and}\ \forall k\geq 3\quad\lambda(k)=0.3.$$

We have $R(1)=0.9$, $R(2)=0.54$, and $\forall k\geq 3\quad R(k)=0.54*0.7^{k-2}$.

It is easy to show that $\dfrac{R(k+1)^{1/(k+1)}}{R(k)^{1/k}}=\left[\dfrac{49}{54}\right]^{\frac{1}{k(k+1)}}\leq 1.$

So the sequence $\{R(k)^{\frac{1}{k}}\}_{k\geq 1}$ is decreasing and the distribution is IFRA1 but not IFR. □

3.3. *New Better than Used (NBU)*

The definitions corresponding to (2.3) of New Better than Used in discrete time are :

- NBU1 : $\forall k \in \mathbb{N}^*, \forall i \in \mathbb{N}^*, \quad R(i|k) \leq R(i)$,
- NBU2 : $\forall k \in \mathbb{N}^*, \forall i \in \mathbb{N}^*, \quad \Lambda(k+i) \geq \Lambda(k) + \Lambda(i)$.

Proposition 6: *NBU1 $\not\Rightarrow$ NBU2 and NBU2 $\not\Rightarrow$ NBU1.*

Proof: The proposition is proved with 2 counterexamples, proposed by Shaked *et al.*[14]. Consider the failure rate defined by :

$\lambda(1) = 0.5$, $\lambda(2) = 0.8$, $\lambda(3) = 0.8$, $\lambda(4) = 0.5$, $\lambda(5) = 0.9$, $\lambda(6) = 0.6$, and $\forall k \geq 7, \lambda(k) = 0.99$.

The distribution is NBU1 but not NBU2.

Consider the failure rate defined by :

$\lambda(1) = 0.4$, $\lambda(2) = 0.6$, $\lambda(3) = 0.5$, $\lambda(4) = 0.5$, and $\forall k \geq 5, \lambda(k) = 0.99$.

The distribution is NBU2 but not NBU1. □

3.4. *Relationships between Basic Aging Notions*

It is easy to prove that IFRA1 $\Rightarrow$ NBU1 and IFRA2 $\Rightarrow$ NBU2. So the analogous of (7) in discrete time is, as shown by Roy and Gupta[8] and Shaked *et al* [14] :

$$\begin{array}{ccccc} IFR & \Rightarrow & IFRA1 & \Rightarrow & NBU1 \\ & & \Downarrow & & \\ & & IFRA2 & \Rightarrow & NBU2 \end{array} \tag{12}$$

The main result of this section is that several different definitions are possible for IFRA and NBU notions of aging, which are equivalent in the continuous case but not in the discrete case. This is a serious problem regarding the understanding of these notions and the use of discrete reliability in practice. In next section, it is shown that the problem lies in the usual definition of failure rate for discrete distributions.

4. Some Problems with the Usual Definition of Discrete Failure Rate

The definition of the failure rate as in (2) is different from that of its continuous counterpart in many aspects.

First of all, the failure rate is a (conditional) probability in the discrete case, but not in the continuous case. This fact may add to the common confusion between failure rate and failure probability.

Moreover, the failure rate is bounded (by 1) in the discrete case, but it is a conditional probability rate in the continuous case. Most of the continuous reliability models have unbounded failure rates, for example the Weibull distribution. So it is not possible to have an exact discrete counterpart of the continuous Weibull distribution.

Furthermore, $\lambda(k)$ cannot be a convex function. This is a serious problem when the interpretation of failure rate for actual data is concerned. The function $\lambda(k)$ can for example never grow linearly, not to say exponentially, that is the case for components in the wear-out lifetime period. In practice, failure rate is related to the wear in engineering, and it is understood that it might increase exponentially in many cases.

Another important problem is that the failure rate defined by (2) is not additive for series systems. More precisely, let us consider a system made up of n independent components in series. Let R_i and λ_i be respectively the reliability and failure rate of component i. The system reliability is $R(k) = \prod_{i=1}^{n} R_i(k)$. Then, the system failure rate is :

$$\lambda(k) = \frac{R(k-1) - R(k)}{R(k-1)} = 1 - \frac{R(k)}{R(k-1)} = 1 - \frac{\prod_{i=1}^{n} R_i(k)}{\prod_{i=1}^{n} R_i(k-1)}$$

$$= 1 - \prod_{i=1}^{n} \frac{R_i(k)}{R_i(k-1)} = 1 - \prod_{i=1}^{n}[1 - \lambda_i(k)] \neq \sum_{i=1}^{n} \lambda_i(k) \qquad (13)$$

So the failure rate of a series system is not the sum of the failure rates of its components. It is disturbing that such a well-known property of the failure rate is not true in the discrete case.

As stated in section 3 and by Lawless[5], the cumulative hazard function in continuous time is such that $-\ln R(t) = \Lambda(t)$, while in discrete time, $H(k) = -\ln R(k)$ is not equal to $\Lambda(k) = \sum_{i=1}^{k} \lambda(i)$.

Finally, IFRA1 and IFRA2 notions are equivalent in the continuous case, but not in the discrete case.

Note that if $\Lambda(k)$ is replaced by $H(k)$ in the definition of IFRA 2, both definitions are equivalent and the problem raised by Lawless is immediately solved.

In this section, we have shown that many of the usual properties of the failure rate and cumulative hazard functions in continuous time are no longer true in discrete time. Then, it seems that definition (2) of the failure rate does not provide the proper discrete counterpart of the usual continuous failure rate. In next section, a new definition is discussed, which addresses to the above-mentioned problems.

5. An Alternative Definition of Discrete Failure Rate

As shown in previous section, an important problem of discrete distributions is that $-\ln R(k) \neq \sum_{i=1}^{k} \lambda(i)$. So an alternative definition of discrete failure rate could be the sequence $\{r(k)\}_{k \geq 1}$ such that :

$$H(k) = -\ln R(k) = \sum_{i=1}^{k} r(i) \quad \text{or} \quad R(k) = exp\big[-\sum_{i=1}^{k} r(i)\big].$$

failure (SRF) is: Then :

$$r(k) = -\ln R(k) + \ln R(k-1) = -\ln \frac{R(k)}{R(k-1)} = \ln \frac{R(k-1)}{R(k)} \qquad (14)$$

This function has been introduced in Roy and Gupta[8], who named it the second rate of failure (SRF).

Definition (2) can be understood as the discrete version of the continuous $\lambda(t) = -\frac{R'(t)}{R(t)}$, while (14) can be understood as the discrete version of the continuous $\lambda(t) = -\frac{d}{dt}\ln R(t)$. Both definitions are equivalent in continuous time, but they are not in discrete time.

With definition (14), the failure rate $r(k)$ is not a probability and is not bounded. So failure rate and failure probability are not mixed up, and a failure rate can be convex, can grow linearly or exponentially, can have exactly the same shape as the failure rate of a continuous Weibull distribution, and so on...

With definition (14), the cumulative hazard function is $H(k)$ instead of $\Lambda(k)$. Then, if $\Lambda(k)$ is replaced by $H(k)$ in the definition of IFRA2, both definitions IFRA 1 and IFRA 2 are equivalent, so there is only one IFRA concept. It is easy to see that same thing happens for the NBU notion.

Finally, the failure rate of a system made up of n independent components in series (see Result 3.1 of Roy and Gupta[8]) is :

$$r(k) = \ln\frac{R(k-1)}{R(k)} = \ln\frac{\prod_{i=1}^{n} R_i(k-1)}{\prod_{i=1}^{n} R_i(k)} = \ln\prod_{i=1}^{n}\frac{R_i(k-1)}{R_i(k)}$$
$$= \sum_{i=1}^{n}\ln\frac{R_i(k-1)}{R_i(k)} = \sum_{i=1}^{n} r_i(k) \qquad (15)$$

This is the expected result : the failure rate of a series system of n independent components is the sum of the failure rates of the components.

So with definition (14) for the failure rate of a discrete distribution, all the problems raised in section 4 are solved.

Both failure rates $\lambda(k)$ and $r(k)$ are linked by the simple relationship :

$$r(k) = -\ln\frac{R(k)}{R(k-1)} = -\ln[1-\lambda(k)] \qquad (16)$$

So both functions have the same monotonicity property : $r(k)$ is increasing/decreasing if and only if $\lambda(k)$ is increasing/decreasing.

Furthermore, a limited expansion of the logarithm leads to :

$$r(k) = -\ln[1 - \lambda(k)] = \lambda(k) + o(\lambda(k)) \tag{17}$$

So, for small values of $\lambda(k)$ (which is the practical case for reliable systems), both failure rates are equivalent.

Then $\lambda(k)$ and $r(k)$ are very similar functions. All the work done on $\lambda(k)$ can be used with $r(k)$.

Finally, definition (2) has been used by everybody as the definition of failure rate in discrete time, probably because of its nice probabilistic interpretation as the conditional probability that a system will fail at time k given that it is still alive at time $k-1$. However, this leads to many discrepancies between the discrete and continuous cases. The SRF of a discrete distribution is much more similar to the usual continuous failure rate, and hence can be used in practice without causing any confusion. Moreover, this definition provides only one definition of each usual aging notion in discrete time.

Then, it is possible to show the IFRA closure theorem in discrete time. The equivalent of Proposition 1 in the discrete case is as follows as given in Result 3.5 of Roy and Gupta[8]:

Proposition 7: *A coherent system of independent components with discrete IFRA lifetimes distributions has itself an IFRA lifetime distribution.*

References

1. R.E. Barlow, A.W. Marshall and F. Proschan, Properties of Probability Distributions with Monotone Hazard Rate, *Annals of Mathematical Statistics*, **34**, 375-389, (1963).
2. R.E. Barlow and F. Proschan, *Mathematical Theory of Reliability* (John Wiley and Sons, New York, 1965).
3. D.R. Cox, Regression models with life tables, *Journal of the Royal Statistical Society, Series B*, **34**, 187-226, 1972.
4. J.D. Kalbfleisch and R.L. Prentice, *The statistical analysis of failure time data* (John Wiley and Sons, New York, 1980).
5. J.F. Lawless, *Statistical models and methods for lifetime data* (John Wiley and Sons, New York, 1982).
6. A.A. Salvia and R.C. Bollinger, On discrete hazard functions, *IEEE Transactions on Reliability*, **R-31**, 5, 458-459, (1982).
7. T. Nakagawa and S. Osaki, The discrete Weibull distribution, *IEEE Transactions on Reliability*, **R-24**, 5, 300-301, (1975).

8. D. Roy and R.P. Gupta, Classifications of discrete lives, *Microelectronics and Reliability*, **32**,1459-1473, (1992).
9. D. Roy, On classifications of multivariate life distributions in the discrete setup, *Microelectronics and Reliability*, **37**, 361-366, (1997).
10. P.L. Gupta, R.C. Gupta and R.C. Tripathi, On the monotonic properties of discrete failure rates, *Journal of Statistical Planning and Inference*, **65**, 255-268, (1997).
11. D. Roy and R.P. Gupta, Characterizations and model selections through reliability measures in the discrete case, *Statistics and Probability Letters*, **43**, 197-206, (1999).
12. L.C. Tang, Y. Lu, E.P. Chew, Mean residual life of lifetime distributions, *IEEE Transactions on Reliability*, **R-48**, 1, 73-78, (1999).
13. G.L. Ghai and J. Mi, Mean residual life and its association with failure rate, *IEEE Transactions on Reliability*, **R-48**, 3, 262-266, (1999).
14. M. Shaked, J.G. Shanthikumar and J.B. Valdez-Torres, Discrete hazard rate functions, *Computers and Operations Research*, **22**, 4, 391-402, (1995).
15. R.E. Barlow and F. Proschan, *Statistical Theory of Reliability and Life Testing : Probability Models* (To begin with, Silver Spring, 1981).
16. Z.E. Birnbaum, J.D. Esary and A.W. Marshall, A stochastic characterization of wear-out for components and systems, *Annals of Mathematical Statistics*, **37**, 816-825, (1966).
17. M. Xie, O. Gaudoin and C. Bracquemond, Redefining failure rate function for discrete distributions, *preprint*, Laboratoire LMC, Institut National Polytechnique de Grenoble (2000).

CHAPTER 12

ON GENERALIZED ORDERINGS AND AGEING PROPERTIES WITH THEIR IMPLICATIONS

Taizhong Hu

Department of Statistics and Finance
University of Science and Technology of China
Hefei, Anhui 230026, People's Republic of China
E-mail: thu@ustc.edu.cn

Amarjit Kundu* and Asok K. Nanda†

Department of Mathemetics, Indian Institute of Technology
Kharagpur 721 302, India
*E-mail: *bapai_k@yahoo.com,*
E-mail: †asok@maths.iitkgp.ernet.in

Several concepts of partial orderings between random variables are defined in the literature. They have also been generalized in a number of ways and are shown to be useful in different fields of econometrics, queueing theory, actuarial sciences, applied probability and many other fields. Here we have proved a number of results regarding generalized orderings so that a large number of existing results can be obtained as particular cases. We have given Laplace transform characterization along with quite a few other characterizations of generalized orderings and a number of characterizations for generalized aging properties of random variables.

1. Introduction

Several concepts of partial orderings between random variables have been considered in the literature. They are used in reliability, economics, queues, inventory, actuarial science and stochastic process contexts. They are also used in proving important results in applied probability. Example of applications may be found in Rolski[34], Ross[35], Shaked[39], Stoyan[44] among others. Generalized orderings were discussed in Fagiuoli and Pellerey[12], Fishburn[14], Mukherjee and Chatterjee[24], and O'Brien[31] among others. Recently, Denuit *et al.*[10] used some generalized orderings (s-convex order) in queues and insurance. Kaas *et al.*[19] used generalized orderings in actuarial sciences. They studied s^{th} order stop-loss (s-SL) ordering. The mixture property of s-SL ordering was studied in Hesselager[16]. Different properties of generalized orderings such as moment properties, closure under mixtures were addressed in Nanda *et al.*[29,30]. Nanda[26] used generalized orderings in minimal repair policy where the improvement/deterioration of a system under different generalized orderings is defined and their properties were also studied. By minimal repair we mean that after repair a unit is restored to a condition just prior to failure.

Along with generalized orderings, generalized aging properties also play an important role in reliability. By aging we mean mathematical specification of degradation of equipment over time. For an extensive review of ageing classes and of ageing properties, one may refer to Barlow and Proschan[49]. Averous and Meste[3] (see also Bondesson[4]) gave a complete classification of aging classes. Later, Fagiuoli and Pellerey[12] gave a different type of classification of generalized aging classes in a unified way. Willmot[47] provided bounds to the survival function for higher order equilibrium distributions.

The purpose of this article is to generalize some of the existing results in the literature, which are expected to be useful in different fields. From this general discussion, many known results can be obtained as particular cases of our general results. Such a study is meaningful because it throws an important light on the understanding of the existing results in the literature.

The organization of the paper is as follows.

In Section 2, we give different notations and definitions of generalized orderings and of generalized ageing classes, to be used in the paper.

Section 3 deals with some results relating to generalized orderings given in Definition 1 below. More precisely, we give connections among some generalized orderings, and characterize s-FR orderings in terms of residual lives and of equilibrium distributions, respectively. s-CX and s-CV orderings are also both characterized in terms of equilibrium distributions and of Laplace transform, respectively. Furthermore, we consider the preservation property of s-FR orderings under mixtures of distributions. Here we also give necessary and sufficient conditions under which two distributions ordered in

s-ST ordering sense will be stochastically equal.

The generalized aging properties are discussed in Section 4. More precisely, the s-IFR (s-DFR) (Definition 2 below) aging class is characterized in a number of ways, s-NBUFR (s-NWUFR) (Definition 2) is characterized in terms of ordering between a random variable and its equilibrium random variable, and s-NBUFR and s-NBUCX ageing classes are characterized in terms of Laplace transform. It is well known that all s-IFR (s-DFR) random variables are necessarily $(s+1)$-IFR ($(s+1)$-DFR) for any nonnegative integer s, but the converse is not necessarily true; we also provide a sufficient condition for the converse to hold.

Throughout this paper, the terms "increasing" and "decreasing" mean "non-decreasing" and "non-increasing", respectively. $a/0$ is understood to be ∞ whenever $a > 0$ and $0/0$ is not defined. All expectations and integrals are implicitly assumed to exist whenever they are written, and the random variables under consideration are absolutely continuous and nonnegative. Also, we write $\mathcal{N} = \{0, 1, 2, \ldots\}$, the set of nonnegative integers and $\mathcal{N}^+ = \{1, 2, \ldots\}$, the set of positive integers.

2. Notations, Definitions and Preliminaries

For any nonnegative absolutely continuous random variable X with distribution function $F_X(x)$ and differentiable density function $f_X(x)$, write

$$\overline{T}_0(X, x) = f_X(x)$$

and

$$\overline{T}_s(X, x) = \frac{\int_x^\infty \overline{T}_{s-1}(X, u)du}{\mu_{s-1}(X)}, \quad s \in \mathcal{N}^+, \tag{1}$$

where

$$\mu_s(X) = \int_0^\infty \overline{T}_s(X, x)dx, \quad s \in \mathcal{N}.$$

Clearly, $\mu_1(X) = \mu(X)$, the mean of X. Write $\overline{T}_{-1}(X, x) = \frac{d}{dx} f_X(x)$ and assume that $\lim_{x\to\infty} f_X(x) = 0$.

Note that $\overline{T}_2(X, x)$ is the survival function of the equilibrium distribution of X, which plays an important role in aging concepts,[9] whereas $\overline{T}_s(X, x)$ is the survival function of the equilibrium distribution of a distribution with survival function $\overline{T}_{s-1}(X, x)$, $s = 2, 3, \ldots$. Suppose a person is waiting for a bus at a bus stop and the interarrival times between successive arrivals of bus at the bus stop are independent and identically distributed (*iid*) having the distribution same as X. Then, in the long run, the waiting time of the person will have distribution given by $\overline{T}_2(X, x)$.[48]

Also write, for $s \in \mathcal{N}$,

$$r_s(X, x) = \frac{-\frac{d}{dx}\overline{T}_s(X, x)}{\overline{T}_s(X, x)} = \frac{\overline{T}_{s-1}(X, x)}{\int_x^\infty \overline{T}_{s-1}(X, u)du}. \tag{2}$$

For $s \in \mathcal{N}^+$, (2) represents the failure rate function corresponding to $\overline{T}_s(X, x)$. For $s = 1$, $r_1(X, x)$ is the failure rate function of X, defined as the ratio of the density to the survival function, whereas for $s = 2$, $r_2(X, x)$ is the reciprocal of the mean residual life function $m_X(t) = E(X - t|X > t)$ of X.

It is not very hard to verify that, for $s \in \mathcal{N}$,

$$\frac{\overline{T}_s(X, x)}{\overline{T}_s(X, 0)} = \exp\left\{-\int_0^x r_s(X, u)du\right\}. \tag{3}$$

For $s \in \mathcal{N}^+$, (3) becomes

$$\overline{T}_s(X, x) = \exp\left\{-\int_0^x r_s(X, u)du\right\}.$$

Definition 1: Let X and Y be two nonnegative random variables and $s \in \mathcal{N}$.

(*a*) X is said to be smaller than Y in s-FR ordering ($X \leq_{s-\mathrm{FR}} Y$) if

$$\frac{\overline{T}_s(X, x)}{\overline{T}_s(Y, x)} \text{ is decreasing in } x \geq 0; \tag{4}$$

(*b*) X is said to be smaller than Y in s-ST ordering ($X \leq_{s-\mathrm{ST}} Y$) if

$$\frac{\overline{T}_s(X, x)}{\overline{T}_s(Y, x)} \leq \frac{\overline{T}_s(X, 0)}{\overline{T}_s(Y, 0)}, \quad \forall x \geq 0; \tag{5}$$

(*c*) X is said to be smaller than Y in s-CX ordering ($X \leq_{s-\mathrm{CX}} Y$) if

$$\int_x^\infty \frac{\overline{T}_s(X, u)}{\overline{T}_s(X, 0)} du \leq \int_x^\infty \frac{\overline{T}_s(Y, u)}{\overline{T}_s(Y, 0)} du, \quad \forall x \geq 0; \tag{6}$$

(*d*) X is said to be smaller than Y in s-CV ordering ($X \leq_{s-\mathrm{CV}} Y$) if

$$\int_0^x \frac{\overline{T}_s(X, u)}{\overline{T}_s(X, 0)} du \leq \int_0^x \frac{\overline{T}_s(Y, u)}{\overline{T}_s(Y, 0)} du, \quad \forall x \geq 0. \tag{7}$$

One can verify that (4) can equivalently be written as

$$r_s(X,x) \geq r_s(Y,x) \text{ for all } x \geq 0. \tag{8}$$

For $s \in \mathcal{N}^+$, (5), (6) and (7) reduce respectively to $\overline{T}_s(X,x) \leq \overline{T}_s(Y,x)$,

$$\int_x^\infty \overline{T}_s(X,u)du \leq \int_x^\infty \overline{T}_s(Y,u)du$$

and

$$\int_0^x \overline{T}_s(X,u)du \leq \int_0^x \overline{T}_s(Y,u)du$$

for all $x \geq 0$. One can easily verify that $X \leq_{s-\text{CX}} (\leq_{s-\text{CV}}) Y \Longrightarrow \mu_s(X) \leq \mu_s(Y)$ for $s \in \mathcal{N}^+$. In particular, $X \leq_{\text{icx}} Y \Longrightarrow E(X) \leq E(Y)$.

Definition 2: Let $s \in \mathcal{N}$. The random variable X is said to be

(a) s-IFR (s-DFR) if

$$r_s(X,x) \text{ is increasing (decreasing) in } x. \tag{9}$$

(b) s-NBU (s-NWU) if

$$\overline{T}_s(X,x+t)\overline{T}_s(X,0) \leq (\geq)\overline{T}_s(X,x)\overline{T}_s(X,t), \quad \forall x,t \geq 0. \tag{10}$$

(c) s-NBUFR (s-NWUFR) if

$$r_s(X,x) \geq (\leq) r_s(X,0), \quad \forall x \geq 0. \tag{11}$$

(d) s-NBUCX (s-NWUCX) if, for all $x,t \geq 0$,

$$\overline{T}_s(X,0)\int_x^\infty \overline{T}_s(X,t+u)du \;\leq (\geq)\; \overline{T}_s(X,t)\int_x^\infty \overline{T}_s(X,u)du. \tag{12}$$

For $s \in \mathcal{N}^+$, (10) and (12) becomes

$$\overline{T}_s(X,x+t) \leq (\geq)\, \overline{T}_s(X,x)\cdot \overline{T}_s(X,t)$$

and

$$\int_x^\infty \overline{T}_s(X,t+u)du \leq (\geq)\, \overline{T}_s(X,t)\int_x^\infty \overline{T}_s(X,u)du$$

for all $x,t \geq 0$ respectively. (9) can equivalently be written as

$$\frac{\overline{T}_s(X,x+t)}{\overline{T}_s(X,t)} \text{ is decreasing (increasing) in } t \tag{13}$$

for all $x \geq 0$. This, again can equivalently be written as

$$\overline{T}_s(X, x) \text{ is logconcave (logconvex) in } x \geq 0. \tag{14}$$

For a detailed study of logconcave and concave distributions one may refer to Sengupta and Nanda[38]. They gave sharp bounds to the reliability function under the assumption that the distribution is logconcave/concave along with a lot more interesting results.

It has been observed that the following equivalences hold.

0-FR	$\Leftrightarrow$	lr;	1-FR	$\Leftrightarrow$	hr;
2-FR	$\Leftrightarrow$	mrl;	3-FR	$\Leftrightarrow$	vrl;
0-ST	$\Leftrightarrow$	wlr;	1-ST	$\Leftrightarrow$	st;
2-ST	$\Leftrightarrow$	$hmrl$;	1-CX	$\Leftrightarrow$	icx;
0-IFR	$\Leftrightarrow$	ILR;	1-IFR	$\Leftrightarrow$	IFR;
2-IFR	$\Leftrightarrow$	DMRL;	3-IFR	$\Leftrightarrow$	DVRL;
1-NBU	$\Leftrightarrow$	NBU	1-NBUFR	$\Leftrightarrow$	NBUFR;
2-NBUFR	$\Leftrightarrow$	NBUE.			

For definitions of DMRL and DVRL random variables and their properties one may refer to Launer[22], NBUFR can be obtained in Deshpande *et al.*[9] whereas wlr and vrl orderings can be found in Singh[43]. For an extensive review of different partial orderings and of ageing classes, one may refer to Shaked and Shanthikumar[41] and Barlow and Proschan[49], respectively. It is to be mentioned here that in actuarial language, hazard rate ordering is called *mortality* ordering, whereas increasing convex order (icx) is known as *stop-loss* order.

Denuit *et al.*[10] define s-convex ordering between two random variables as follows.

Definition 3: Let X and Ybe random variables taking values on $[a, b]$, a subinterval of the real line for some $a < b$. If $E[\phi(X)] \leq E[\phi(Y)]$ for all s-convex functions[a] ϕ, provided the expectations exist, then X is said to be smaller than Y in s-convex order and we write $X \leq_{s-\text{convex}} Y$.

If the sth derivative $\phi^{(s)}(\cdot)$ of a function ϕ exists, then ϕ is s-convex if and only if $\phi^{(s)}(x) \geq 0$ for all $x \in [a, b]$. Since every polynomial of degree at most $s - 1$ is an s-convex function, it follows that if $X \leq_{s-\text{convex}} Y$ then

[a]For definition and properties of s-convex functions one may refer to Pečarić *et al.*[32].

$E[X^k] = E[Y^k]$, $k = 1, 2, \ldots, s-1$. In other words, only random variables with identical first $s-1$ moments can be compared in the s-convex order.

Kaas *et al.*[19] define s-Stop Loss (s-SL) order as follows.

Definition 4: For two random variables X and Y, X is said to be smaller than Y in s-stop loss order (or stop loss order of degree s) ($X \leq_{s-\mathrm{SL}} Y$), if

$$E[X^k] \leq E[Y^k], \quad k = 1, 2, \ldots, s-1$$

and, for each $x \geq 0$,

$$E[\{(X-x)_+\}^s] \leq E[\{(Y-x)_+\}^s],$$

where, for any real number x, $x_+ = x$ if $x \geq 0$ and, $= 0$ otherwise.

The s-CX, s-convex and s-SL orderings are in general different types of generalizations of stop-loss order (which is one important tool in actuarial science); each of them reduces to the stop-loss order for a specific value of s; for example, 1-CX $\equiv$ 1-SL $\equiv$ *icx*. If $E(X)$=$E(Y)$, then 2-convex $\equiv$ *icx*. For s-convex order to be defined, one needs $E(X^j) = E(Y^j)$ for $j = 1, 2, \ldots, s-1$; s-SL ordering is defined for random variables X and Y such that $E(X^j) \leq E(Y^j)$ for $j = 1, 2, \ldots, s-1$. But in defining s-CX order, no such conditions are necessary. When $E(X^j) = E(Y^j)$ for $j = 1, 2, \ldots, s-1$, then all the three orderings (s-CX, s-convex and s-SL) become identical. Thus, s-CX ordering discussed in this paper is more general than s-convex and s-SL orders.

3. Some Generalized Ordering Results

3.1. *Connections among the Orderings*

Fagiuoli and Pellerey[12] defined the generalized orderings in a unified way and proved various interrelatons among them. It is known that $X \leq_{(s-1)-\mathrm{FR}} Y \Rightarrow X \leq_{s-\mathrm{FR}} Y$, while the converse is not necessarily true. But, under certain conditions, the converse is also true as is shown in the following theorem. A corresponding aging result is given in Section 4.

Theorem 5: Let $r_s(X, x)/r_s(Y, x)$ be decreasing in x. Then, for $s \in \mathcal{N}^+$,

$$X \leq_{s-\mathrm{FR}} Y \Longrightarrow X \leq_{(s-1)-\mathrm{FR}} Y.$$

Proof: Note that, from (1), (2) and (3),

$$\overline{T}_{s-1}(X,x) = \mu_{s-1}(X) r_s(X,x) \exp\left\{-\int_0^x r_s(X,u)du\right\}.$$

Hence, $X \leq_{(s-1)-\text{FR}} Y$ if and only if

$$\exp\left\{-\int_0^x [r_s(X,u) - r_s(Y,u)]\, du\right\} \cdot \frac{r_s(X,x)}{r_s(Y,x)} \text{ is decreasing in } x,$$

which is true from the hypothesis and by (8). □

Corollary 6: *(a) If $r_X(x)/r_Y(x)$ is decreasing in x, then $X \leq_{\text{hr}} Y \Rightarrow X \leq_{\text{lr}} Y$, where $r_X(\cdot)$ and $r_Y(\cdot)$ are the hazard rate functions of X and Y, respectively.*

(b) Let $m_X(x)$ and $m_Y(x)$ be the mean residual life functions of X and Y, respectively. If $m_X(x)/m_Y(x)$ is increasing in x, then $X \leq_{\text{mrl}} Y \Rightarrow X \leq_{\text{hr}} Y$.

In terms of relative aging of two distributions,[36,37] Corollary 6(a) can be restated as follows: If X ages slower than Y and $X \leq_{\text{hr}} Y$, then $X \leq_{\text{lr}} Y$. Theorem 2(b) of Gupta and Kirmani[15] is a particular case of the above theorem.

Example 7: Suppose X and Y are two independent and exponentially distributed random variables with means λ_1 and λ_2 respectively. It is well known that if $\lambda_1 \leq \lambda_2$, then $X \leq_{\text{lr}} Y$ and hence $X \leq_{s-\text{FR}} Y$ for any $s \in \mathcal{N}$. But, one can also verify that, for any $s \in \mathcal{N}$, $r_s(X,x) = 1/\lambda$ and hence, by Theorem 5, we get, for exponential distribution, all s-FR orderings are identical.

The following lemma will be used in proving the upcoming theorem.

Lemma 8: *For $s \in \mathcal{N}$,*

$$\frac{\overline{T}_s(X,t)}{\overline{T}_s(X,0)} = \frac{r_{s+1}(X,t)}{r_{s+1}(X,0)} \exp\left[-\int_0^t r_{s+1}(X,x)dx\right].$$

Proof: Note that

$$\frac{r_{s+1}(X,t)}{r_{s+1}(X,0)} = \frac{\overline{T}_s(X,t)}{\int_t^\infty \overline{T}_s(X,u)du} \cdot \frac{\mu_s(X)}{\overline{T}_s(X,0)}.$$

Hence,

$$\frac{\overline{T}_s(X,t)}{\overline{T}_s(X,0)} = \frac{r_{s+1}(X,t)}{r_{s+1}(X,0)}\overline{T}_{s+1}(X,t),$$

by using (1). This, along with (3), gives the required result. □

Although $X \leq_{s-\mathrm{FR}} Y \Rightarrow X \leq_{(s+1)-\mathrm{FR}} Y$ and $X \leq_{s-\mathrm{FR}} Y \Rightarrow X \leq_{s-\mathrm{ST}} Y$, it is well known that none of $(s+1)$-FR ordering and s-ST ordering implies the other. The counterexample can be found in Gupta and Kirmani[15]. But, from Theorem 5, if $r_{s+1}(X,t)/r_{s+1}(Y,t)$ is decreasing in t, then $(s+1)$-FR ordering implies s-FR ordering, which, in turn implies s-ST ordering. This is true even under less restrictive condition than the decreasingness of $r_{s+1}(X,t)/r_{s+1}(Y,t)$, and is shown in the following theorem.

Theorem 9: If $r_{s+1}(X,t)/r_{s+1}(Y,t) \leq \mu_s(Y)/\mu_s(X)$, then, for $s \in \mathcal{N}$,

$$X \leq_{(s+1)-\mathrm{FR}} Y \Longrightarrow X \leq_{s-\mathrm{ST}} Y.$$

Proof: On using Lemma 8, we get

$$\begin{aligned}\frac{\overline{T}_s(X,t)}{\overline{T}_s(X,0)} - \frac{\overline{T}_s(Y,t)}{\overline{T}_s(Y,0)} &= \frac{r_{s+1}(X,t)}{r_{s+1}(X,0)}\exp\left\{-\int_0^t r_{s+1}(X,x)dx\right\} \\ &\quad - \frac{r_{s+1}(Y,t)}{r_{s+1}(Y,0)}\exp\left\{-\int_0^t r_{s+1}(Y,x)dx\right\} \\ &\leq \frac{r_{s+1}(Y,t)}{r_{s+1}(Y,0)}\left[\exp\left\{-\int_0^t r_{s+1}(X,x)dx\right\}\right. \\ &\quad \left. - \exp\left\{-\int_0^t r_{s+1}(Y,x)dx\right\}\right] \leq 0.\end{aligned}$$

The first inequality follows from the hypothesis by noting the fact that $r_{s+1}(X,0) = 1/\mu_s(X)$ and $r_{s+1}(Y,0) = 1/\mu_s(Y)$, while the second inequality follows from (8). Hence the result. □

An immediate consequence of Theorem 9 is the following corollary, due to Gupta and Kirmani[15].

Corollary 10: *If $m_Y(t)/E(Y) \leq m_X(t)/E(X)$ for all $t \geq 0$, then*

$$X \leq_{\mathrm{mrl}} Y \Longrightarrow X \leq_{\mathrm{st}} Y,$$

where $m_Z(\cdot)$ is the mean residual life function of the random variable Z.

The following theorem shows that under certain condition $(s+1)$-ST and s-CX orderings are identical. The proof is omitted.

Theorem 11: For two random variables X and Y with $\mu_s(X) = \mu_s(Y)$, and $s \in \mathcal{N}^+$,

$$X \leq_{(s+1)-\mathrm{ST}} Y \Longleftrightarrow X \leq_{s-\mathrm{CX}} Y.$$

Corollary 12: *Let X and Y be two random variables such that $E(X) = E(Y)$. Then $X \leq_{\mathrm{hmrl}} Y$ if and only if $X \leq_{\mathrm{icx}} Y$.*

3.2. *Characterizations in Terms of Residual Lives*

First we give two useful lemmas which will be used later. In Lemma 13, if $z = x$, then it is trivial, but it is to be noted that for any $z \geq x$, the lemma holds.

Lemma 13: *Let $s \in \mathcal{N}^+$. Then $X \leq_{s-\mathrm{FR}} Y$ if and only if, for all $x \leq z$,*

$$\frac{\int_z^\infty \overline{T}_{s-1}(X,u)du}{\overline{T}_{s-1}(X,x)} \leq \frac{\int_z^\infty \overline{T}_{s-1}(Y,u)du}{\overline{T}_{s-1}(Y,x)}.$$

Proof: $X \leq_{s-\mathrm{FR}} Y$ if and only if, for $y \leq z$,

$$\frac{\overline{T}_s(Y,y)}{\overline{T}_s(X,y)} \leq \frac{\overline{T}_s(Y,z)}{\overline{T}_s(X,z)}. \tag{15}$$

Thus, for $x \leq y$,

$$\overline{T}_s(X,y)\overline{T}_s(Y,x) \leq \overline{T}_s(X,x)\overline{T}_s(Y,y). \tag{16}$$

Let $X_{(s)}$ and $Y_{(s)}$ be the random variables having survival functions $\overline{T}_s(X,\cdot)$ and $\overline{T}_s(Y,\cdot)$ respectively. Then (16) can be written as

$$\overline{T}_s(X,y)\left[\overline{T}_s(Y,y) + P\{x \leq Y_{(s)} \leq y\}\right]$$

$$\leq \overline{T}_s(Y,y)\left[\overline{T}_s(X,y) + P\{x \leq X_{(s)} \leq y\}\right].$$

Or equivalently, for all $x \leq y$,

$$\frac{\overline{T}_s(Y,y)}{\overline{T}_s(X,y)} \geq \frac{P\{x \leq Y_{(s)} \leq y\}}{P\{x \leq X_{(s)} \leq y\}} \tag{17}$$

Comparing (15) and (17), we get that, for all $x \le y \le z$,

$$\frac{P\{x \le Y_{(s)} \le y\}}{P\{x \le X_{(s)} \le y\}} \le \frac{\overline{T}_s(Y,z)}{\overline{T}_s(X,z)}.$$

Dividing the numerator and the denominator of the left hand side of the above expression by $y-x$, taking limit as $y \to x$ and using (1), we get the required result. □

Let us write $X_t = [X - t|X > t]$ and $Y_t = [Y - t|Y > t]$, the residual random variables corresponding to X and Y, respectively.

The following lemma which shows the relationship between X and X_t in terms of generalized means, will be used in proving the upcoming theorems.

Lemma 14: *For* $s \in \mathcal{N}^+$,

(*i*) $\overline{T}_s(X_t,x) = \dfrac{\overline{T}_s(X,x+t)}{\overline{T}_s(X,t)}$;

(*ii*) $\mu_s(X_t) = \dfrac{\overline{T}_{s+1}(X,t)\mu_s(X)}{\overline{T}_s(X,t)}$.

Proof: We prove the result by induction on $s \in \mathcal{N}^+$. Let $s = 1$. Then, one can easily verify that

$$\overline{T}_1(X_t,x) = \overline{F}_{X_t}(x) = \frac{\overline{F}_X(x+t)}{\overline{F}_X(t)} = \frac{\overline{T}_1(X,x+t)}{\overline{T}_1(X,t)}, \tag{18}$$

where $\overline{F}_Z(x)$ is the survival function of the random variable Z evaluated at x. Now

$$\mu_1(X_t) = \int_0^\infty \overline{T}_1(X_t,x)dx = \frac{\overline{T}_2(X,t)\mu_1(X)}{\overline{T}_1(X,t)},$$

where the second equality follows from (18) and (1). Hence the result is true for $s = 1$. Let the result be true for $s = s$. Now

$$\overline{T}_{s+1}(X_t,x) = \frac{\int_x^\infty \overline{T}_s(X_t,u)du}{\mu_s(X_t)} = \frac{\int_{x+t}^\infty \overline{T}_s(X,u)du}{\mu_s(X)\overline{T}_{s+1}(X,t)}$$
$$= \frac{\overline{T}_{s+1}(X,x+t)}{\overline{T}_{s+1}(X,t)}.$$

Again

$$\mu_{s+1}(X_t) = \int_0^\infty \overline{T}_{s+1}(X_t,x)dx = \frac{\overline{T}_{s+2}(X,t)\mu_{s+1}(X)}{\overline{T}_{s+1}(X,t)}$$

by (1). Hence the result follows by induction. □

The following theorem characterizes the ordering between two random variables in terms of ordering between their residual lives, which is a generalization of Lemma 2.5(b) of Brown and Shanthikumar[5].

Theorem 15: For $s \in \mathcal{N}^+$,

$$X \leq_{s-\mathrm{FR}} Y \Longleftrightarrow X_t \leq_{(s-1)-\mathrm{CX}} Y_t, \quad \forall t \geq 0.$$

Proof: Suppose $s \geq 2$. Then $X_t \leq_{(s-1)-\mathrm{CX}} Y_t$ if and only if

$$\int_x^\infty \overline{T}_{s-1}(X_t, u)du \leq \int_x^\infty \overline{T}_{s-1}(Y_t, u)du$$

for all $x \geq 0$, which, by Lemma 14, reduces to

$$\frac{\int_x^\infty \overline{T}_{s-1}(X, u+t)du}{\overline{T}_{s-1}(X, t)} \leq \frac{\int_x^\infty \overline{T}_{s-1}(Y, u+t)du}{\overline{T}_{s-1}(Y, t)}.$$

This can equivalently be written as, for all $x \geq 0$,

$$\frac{\int_{x+t}^\infty \overline{T}_{s-1}(X, u)du}{\overline{T}_{s-1}(X, t)} \leq \frac{\int_{x+t}^\infty \overline{T}_{s-1}(Y, u)du}{\overline{T}_{s-1}(Y, t)},$$

which, by Lemma 13, gives the result for $s \geq 2$. Now take $s = 1$. Let X and Y have density functions f and g with survival functions $\overline{F}$ and $\overline{G}$ respectively. Then the density functions of X_t and Y_t are given respectively by $f_{X_t}(x) = f(x+t)/\overline{F}(t)$ and $f_{Y_t}(x) = g(x+t)/\overline{G}(t)$, where $f_Z(x)$ is the density function of the random variable Z. Now one can see that $X_t \leq_{0-\mathrm{CX}} Y_t$ if and only if, for all $x \geq 0$,

$$\frac{f(t)}{\overline{F}(x+t)} \geq \frac{g(t)}{\overline{G}(x+t)},$$

which, by Lemma 13, gives $X \leq_{\mathrm{hr}} Y$. Hence the result. □

Corollary 16: *$X \leq_{\mathrm{mrl}} Y$ if and only if $X_t \leq_{\mathrm{icx}} Y_t$ for all t.* [6]

The following theorem, originally proved by Mukherjee and Chatterjee[24] for $s \in \mathcal{N}^+$, can be obtained easily from Lemma 14. But actually one can prove this result for $s \in \mathcal{N}$. A corresponding aging property is given in Section 4.

Theorem 17: For $s \in \mathcal{N}$,

$$X \leq_{s-\mathrm{FR}} Y \iff X_t \leq_{s-\mathrm{ST}} Y_t, \quad \forall t \geq 0.$$

Corollary 18: *(i) $X \leq_{\mathrm{lr}} Y$ if and only if $X_t \leq_{\mathrm{wlr}} Y_t$ for all $t \geq 0$;*
(ii) $X \leq_{\mathrm{hr}} Y$ if and only if $X_t \leq_{\mathrm{st}} Y_t$ for all $t \geq 0$;
(iii) $X \leq_{\mathrm{mrl}} Y$ if and only if $X_t \leq_{\mathrm{hmrl}} Y_t$ for all $t \geq 0$.

The following theorem is not difficult to prove.

Theorem 19: For $s \in \mathcal{N}$,

$$X \leq_{s-\mathrm{FR}} Y \iff X_t \leq_{s-\mathrm{FR}} Y_t, \quad \forall t \geq 0.$$

Corollary 20: *(i) $X \leq_{\mathrm{lr}} Y$ if and only if $X_t \leq_{\mathrm{lr}} Y_t$ for all $t \geq 0$;*
(ii) $X \leq_{\mathrm{hr}} Y$ if and only if $X_t \leq_{\mathrm{hr}} Y_t$ for all $t \geq 0$;
(iii) $X \leq_{\mathrm{mrl}} Y$ if and only if $X_t \leq_{\mathrm{mrl}} Y_t$ for all $t \geq 0$;
(iv) $X \leq_{\mathrm{vrl}} Y$ if and only if $X_t \leq_{\mathrm{vrl}} Y_t$ for all $t \geq 0$.

Corollary 20(ii) was given in Nanda and Jain[28] in order to prove a weighted distribution result. From Theorems 17 and 19 one can see that s-FR and s-ST orderings between X_t and Y_t are identical.

3.3. *Characterizations in Terms of Equilibrium Distributions*

For a random variable X with survival function $\overline{F}_X$ and mean $\mu(X)$, define another random variable A_X having survival function

$$\overline{F}_{A_X}(x) = \frac{\int_x^\infty \overline{F}_X(t)dt}{\mu(X)}, \quad x \geq 0.$$

$\overline{F}_{A_X}(\cdot)$ is actually the survival function of the equilibrium distribution (also called the stationary renewal excess distribution) of X, and A_X is called equilibrium random variable corresponding to X.

The following lemma is immediate from the definition of A_X.

Lemma 21: *For $s \in \mathcal{N}^+$,*
(a) $\overline{T}_s(A_X, x) = \overline{T}_{s+1}(X, x)$;
(b) $\mu_s(A_X) = \mu_{s+1}(X)$;
(c) $r_s(A_X, x) = r_{s+1}(X, x)$.

The following is an easy-to-prove theorem characterizing s-FR ordering between two random variables in terms of their equilibrium random variables.

Theorem 22: Suppose A_X and A_Y are the equilibrium random variables corresponding to X and Y respectively. Then, for $s \in \mathcal{N}^+$,

$$X \leq_{s-\mathrm{FR}} Y \Longleftrightarrow A_X \leq_{(s-1)-\mathrm{FR}} A_Y.$$

Corollary 23: *(a)* $X \leq_{\mathrm{hr}} Y \Longleftrightarrow A_X \leq_{\mathrm{lr}} A_Y$;
(b) $X \leq_{\mathrm{mrl}} Y \Longleftrightarrow A_X \leq_{\mathrm{hr}} A_Y$;
(c) $X \leq_{\mathrm{vrl}} Y \Longleftrightarrow A_X \leq_{\mathrm{mrl}} A_Y$.

The following theorem characterizes s-ST ordering of two random variables in terms of their equilibrium random variables.

Theorem 24: For $s \in \mathcal{N}^+$,

$$X \leq_{s-\mathrm{ST}} Y \Longleftrightarrow A_X \leq_{(s-1)-\mathrm{ST}} A_Y.$$

Proof: Take $s \geq 2$. We have $X \leq_{s-\mathrm{ST}} Y$ if and only if

$$\overline{T}_s(X, x) \leq \overline{T}_s(Y, x)$$

for all $x \geq 0$. This, by Lemma 21, can be written as

$$\overline{T}_{s-1}(A_X, x) \leq \overline{T}_{s-1}(A_Y, x)$$

for all $x \geq 0$, which gives $A_X \leq_{(s-1)-\mathrm{ST}} A_Y$. For $s = 1$, the result follows by writing the density functions of A_X and A_Y. □

Corollary 25: *(i)* $X \leq_{\mathrm{st}} Y$ *if and only if* $A_X \leq_{\mathrm{wlr}} A_Y$;
(ii) $X \leq_{\mathrm{hmrl}} Y$ *if and only if* $A_X \leq_{\mathrm{st}} A_Y$.

The following theorem can be proved using Lemma 21 on the same line as Theorem 24.

Theorem 26: Let $s \in \mathcal{N}^+$. Then
(a) $X \leq_{s-\mathrm{CX}} Y \Longleftrightarrow A_X \leq_{(s-1)-\mathrm{CX}} A_Y$;
(b) $X \leq_{s-\mathrm{CV}} Y \Longleftrightarrow A_X \leq_{(s-1)-\mathrm{CV}} A_Y$.

A similar type of result for s-SL ordering was given in Kass *et al.*[19].

The following theorem generalizes the well known characterization result of hazard rate ordering.[7]

Theorem 27: Let X and Y be two nonnegative random variables and α and β be two functions such that α/β and β are increasing and β is nonnegative. Then $X \leq_{s-\mathrm{FR}} Y$ if and only if

$$\frac{\int_0^\infty \alpha(x)\overline{T}_{s-1}(X,x)dx}{\int_0^\infty \beta(x)\overline{T}_{s-1}(X,x)dx} \leq \frac{\int_0^\infty \alpha(x)\overline{T}_{s-1}(Y,x)dx}{\int_0^\infty \beta(x)\overline{T}_{s-1}(Y,x)dx} \tag{19}$$

for $s \in \mathcal{N}^+$.

Proof: The necessity of the theorem follows from Capéraà[7] by noting the fact that the functions $\overline{T}_s(X,\cdot)$ and $\overline{T}_s(Y,\cdot)$ in (19) are playing the role of $\overline{F}$ and $\overline{G}$, respectively.

To prove the converse, let us define, for $t < t'$,

$$\alpha(x) = \begin{cases} 0, & \text{if } x \leq t' \\ 1, & \text{if } x > t' \end{cases}$$

and

$$\beta(x) = \begin{cases} 0, & \text{if } x \leq t \\ 1, & \text{if } x > t. \end{cases}$$

Then (19) gives

$$\frac{\int_{t'}^\infty \overline{T}_{s-1}(X,x)dx}{\int_{t'}^\infty \overline{T}_{s-1}(Y,x)dx} \leq \frac{\int_t^\infty \overline{T}_{s-1}(X,x)dx}{\int_t^\infty \overline{T}_{s-1}(Y,x)dx},$$

implying that

$$\frac{\overline{T}_s(X,x)}{\overline{T}_s(Y,x)} \text{ is decreasing in } x.$$

Hence, by (4), the result follows. □

Corollary 28: *Let α and β be as defined in the above theorem and X have distribution F and Y have G. Then*

(*a*) $X \leq_{\mathrm{hr}} Y$ *if and only if*

$$\frac{\int_0^\infty \alpha(x)dF(x)}{\int_0^\infty \beta(x)dF(x)} \leq \frac{\int_0^\infty \alpha(x)dG(x)}{\int_0^\infty \beta(x)dG(x)};$$

(*b*) $X \leq_{\mathrm{mrl}} Y$ *if and only if*

$$\frac{\int_0^\infty \alpha(x)\overline{F}(x)dx}{\int_0^\infty \beta(x)\overline{F}(x)dx} \leq \frac{\int_0^\infty \alpha(x)\overline{G}(x)dx}{\int_0^\infty \beta(x)\overline{G}(x)dx};$$

(c) $X \leq_{\text{vrl}} Y$ *if and only if*

$$\frac{\int_0^\infty \alpha(x)\overline{F}_1(x)dx}{\int_0^\infty \beta(x)\overline{F}_1(x)dx} \leq \frac{\int_0^\infty \alpha(x)\overline{G}_1(x)dx}{\int_0^\infty \beta(x)\overline{G}_1(x)dx},$$

where $\overline{F}_1$ *and* $\overline{G}_1$ *are the survival functions of the equilibrium distributions of* X *and* Y, *respectively.*

Corollary 28(a) may be found in Capéraà[7], whereas (b) is due to Joag-Dev *et al.*[18].

3.4. *Characterizations in Terms of Laplace Transform*

The characterizatons of s-FR and s-ST orderings in terms of Laplace transform can be found in Nanda[27]. Below we give Laplace transform characterizations of s-CX and s-CV orderings. Before proving the theorem we introduce the following notations.

For a nonnegative random variable X having distribution function F and survival function $\overline{F} \equiv 1 - F$ with finite moments of all orders, define the Laplace transform as

$$\phi_X(\lambda) = \int_0^\infty e^{-\lambda x} dF(x), \quad \lambda > 0$$

and denote

$$a_\lambda^X(n) = \frac{(-1)^n}{n!}\frac{d^n}{d\lambda^n}\left[\frac{1-\phi_X(\lambda)}{\lambda}\right], \quad n \in \mathcal{N},\ \lambda > 0,$$

and

$$\alpha_\lambda^X(n) = \lambda^n a_\lambda^X(n-1), \quad n \in \mathcal{N}^+,\ \lambda > 0,$$

with $\alpha_\lambda^X(0) = 1$ for $\lambda > 0$.

Similarly, for a nonnegative random variable Y with distribution function G and survival function $\overline{G} \equiv 1 - G$, define $\alpha_\lambda^Y(n)$ accordingly. It can be shown that $\alpha_\lambda^X(n)$ and $\alpha_\lambda^Y(n)$ are survival functions; denote the corresponding discrete random variables by N_λ^X and N_λ^Y. Write, for $n \in \mathcal{N}^+$ and $x \geq 0$,

$$\Gamma_\lambda g(n) = \int_0^\infty \lambda e^{-\lambda x}\frac{(\lambda x)^{n-1}}{(n-1)!} g(x)dx$$

for any function g defined on $[0, \infty)$, provided the above integral is finite. Take $\Gamma_\lambda g(0) = 1$ for all $\lambda > 0$.

If Z is a random variable with support $\mathcal{N}$, then we define

$$\overline{T}_s(Z,n) = \sum_{k=n+1}^{\infty} \frac{\overline{T}_{s-1}(Z,k)}{\mu_{s-1}(Z)}$$

for $s \in \mathcal{N}^+$ with $\overline{T}_0(Z,\cdot)$ the mass function of Z and

$$\mu_s(Z) = \sum_{k=1}^{\infty} \overline{T}_s(Z,k)$$

for $s \in \mathcal{N}^+$ with $\mu_0(Z) = 1$.

The following lemma can be proved by induction on s with the help of a result of Kebir[21] (reproduced in Nanda[25]) and Lemma 2.3 of Nanda[25]. The proof is omitted.

Lemma 29: *For $n \in \mathcal{N}$ and $s \in \mathcal{N}^+$,*
(i) $\overline{T}_s(N_\lambda^X, n) = \Gamma_\lambda \overline{T}_s(X,n)$;
(ii) $\overline{T}_s(N_\lambda^Y, n) = \Gamma_\lambda \overline{T}_s(Y,n)$.

The following definition may be obtained in Fagiuoli and Pellerey[12].

Definition 30: Let $s \in \mathcal{N}^+$. For two random variables X and Y with support $\mathcal{N}$, X is said to be smaller than Y in
(a) s-CX ordering ($X \leq_{s-\mathrm{CX}} Y$) if

$$\sum_{k=n}^{\infty} \overline{T}_s(X,k) \leq \sum_{k=n}^{\infty} \overline{T}_s(Y,k) \text{ for all } n \in \mathcal{N};$$

(b) s-CV ordering ($X \leq_{s-\mathrm{CV}} Y$) if

$$\sum_{k=0}^{n} \overline{T}_s(X,k) \leq \sum_{k=0}^{n} \overline{T}_s(Y,k) \text{ for all } n \in \mathcal{N}.$$

Suppose that a device is subject to a sequence of shocks occuring randomly as events in a Poisson process with constant intensity λ; each shock causes a random amount of damage to the system and the damage accumulates additively. The device fails as soon as the accumulated damage exceeds a random threshold. Consider such two systems with different failure thresholds. The following theorem says that if the thresholds are ordered in s-CX order, the discrete lifetime of the systems experienced by a sequence of exponentially distributed shocks are ordered in the same sense and conversely.

Theorem 31: For $s \in \mathcal{N}^+$,

$$X \leq_{s-\text{CX}} Y \Longleftrightarrow N_\lambda^X \leq_{s-\text{CX}} N_\lambda^Y, \quad \forall \lambda > 0.$$

Proof: By Lemma 29, $N_\lambda^X \leq_{s-\text{CX}} N_\lambda^Y$ if and only if, for all $n \geq 2$,

$$\mu_s(X) \int_0^\infty e^{-\lambda x}(\lambda x)^{n-2}\overline{T}_{s+1}(X,x)dx$$
$$\leq \mu_s(Y) \int_0^\infty e^{-\lambda x}(\lambda x)^{n-2}\overline{T}_{s+1}(Y,x)dx. \tag{20}$$

Necessity: Suppose $X \leq_{s-\text{CX}} Y$. Then $\overline{T}_{s+1}(X,x)\mu_s(X) \leq \overline{T}_{s+1}(Y,x)$ $\mu_s(Y)$, which gives the result.

Sufficiency: Note that (20) is equivalent to

$$\Gamma_\lambda \overline{T}_{s+1}(X, n-1)\mu_s(X) \leq \Gamma_\lambda \overline{T}_{s+1}(Y, n-1)\mu_s(Y)$$

for all $n \geq 2$. Now taking limit as $n \longrightarrow \infty$ and using Lemma 3.1 of Nanda[27], we get,

$$X \leq_{s-\text{CX}} Y \Longleftrightarrow \overline{T}_{s+1}(X,x)\mu_s(X) \leq \overline{T}_{s+1}(Y,x)\mu_s(Y),$$

which gives the required result. □

Corollary 32: *$X \leq_{\text{icx}} Y$ if and only if $N_\lambda^X \leq_{\text{icx}} N_\lambda^Y$.*[42]

On the same line as Theorem 31, one can prove the following.

Theorem 33: For $s \in \mathcal{N}^+$,

$$X \leq_{s-\text{CV}} Y \Longleftrightarrow N_\lambda^X \leq_{s-\text{CV}} N_\lambda^Y, \quad \forall \lambda > 0.$$

For $s \geq 1$, $X \leq_{s-\text{ST}} Y$ if and only if $\overline{T}_s(X,t) \leq \overline{T}_s(Y,t)$ for all $t \geq 0$. Under $E(X^k) = E(Y^k)$, $k = 1, 2, \ldots, s-1$, the above expression is equivalent to saying that

$$\overline{F}_X^s(t) \leq \overline{F}_Y^s(t) \text{ for all } t \geq 0,$$

where $\overline{F}_T^k(t) = \int_t^\infty \overline{F}_T^{k-1}(u)du$ and $\overline{F}_T^1(t) = \overline{F}_T(t)$, the survival function of the random variable T. So, by Theorem 3.3 of Denuit *et al.*[10], and Theorem 2.1 of Denuit *et al.*[11], we have the following theorem.

Theorem 34: Let X and Y be two real-valued random variables such that $E(X^k) = E(Y^k)$ for all $k = 0, 1, 2, \ldots, s-1$ and $s \in \mathcal{N}^+$. If $X \leq_{s-\text{ST}} Y$ and $E[\phi(X)] = E[\phi(Y)]$ for any function ϕ whose s^{th} derivative $\phi^{(s)}(\cdot)$ is nonnegative, then $X =_{\text{st}} Y$, where $=_{st}$ means equality in law.

Since a function ϕ is s-convex if and only if $\phi^{(s)}(t) \geq 0$ for all t, the above result is true for any s-convex function. Also, since any absolutely monotone function is s-convex, the above theorem is true for any absolutely monotone function. Theorem 34 is somewhat related to Theorem 3 of Carletti and Pellerey[8].

3.5. *Preservation under Mixtures of Distributions*

Consider a family of distribution functions $\{G_\theta, \theta \in \mathcal{X}\}$ where $\mathcal{X} \subseteq \mathcal{R}$. Let $X_\theta \sim G_\theta$ and Θ_i be a random variable with support in $\mathcal{X}$ and having distribution function F_i. Write $Y_i =_{st} X(\Theta_i)$, $i = 1, 2$. The following lemma can be proved using induction on $s \in \mathcal{N}$ and using the same technique as used in proving Lemma 14 and the proof is omitted.

Lemma 35: *For any* $s \in \mathcal{N}$,

$$(i)\ \overline{T}_s(Y_i, y) = \frac{\int_{\mathcal{X}} \Pi_{j=1}^{s-1} \mu_j(X_\theta) \overline{T}_s(X_\theta, y) dF_i(\theta)}{\int_{\mathcal{X}} \Pi_{j=1}^{s-1} \mu_j(X_\theta) dF_i(\theta)};$$

$$(ii)\ \mu_s(Y_i) = \frac{\int_{\mathcal{X}} \Pi_{j=1}^{s} \mu_j(X_\theta) dF_i(\theta)}{\int_{\mathcal{X}} \Pi_{j=1}^{s-1} \mu_j(X_\theta) dF_i(\theta)},$$

where $\Pi_{j=1}^{k} \mu_j(X_\theta) \equiv 1$ *for* $k < 1$.

The following lemma is used to prove Theorem 37. For a similar type of result one may refer to Fagiuoli and Pellerey[13].

Lemma 36: *Let* $\psi(\theta, x)$ *be any* TP_2 *(totally positive of order 2) function (and not necessarily a survival function) in* $\theta \in \mathcal{X}$ *and* $x \in \mathcal{R}$ *and* $\overline{F}_i(\theta)$ *be a survival function in* θ *and be* TP_2 *in* $i \in \{1, 2\}$ *and* $\theta \in \mathcal{X}$. *Assume that* $\psi(\theta, x)$ *is increasing in* θ *for every* x. *Then*

$$H_i(x) = \int_{\mathcal{X}} \psi(\theta, x) dF_i(\theta) \text{ is } \mathrm{TP}_2 \text{ in } x \in \mathcal{R} \text{ and } i \in \{1, 2\},$$

where $\overline{F}_i \equiv 1 - F_i$. *Conversely, if* $H_i(x)$ *is* TP_2 *in* $i \in \{1, 2\}$ *and* $x \in \mathcal{R}$ *whenever* $\overline{F}_i(\theta)$ *is* TP_2 *in* $i \in \{1, 2\}$ *and* $\theta \in \mathcal{X}$, *then* $\psi(\theta, x)$ *is* TP_2 *in* $\theta \in \mathcal{X}$ *and* $x \in \mathcal{R}$.

For an extensive study of TP_2 functions, one may refer to Karlin[20]. Below is a mixture type property of the s-FR ordering.

Theorem 37: Suppose $s \in \mathcal{N}^+$. Let

$$X(\theta) \leq_{s-\text{FR}} X(\theta') \text{ whenever } \theta \leq \theta' \tag{21}$$

and

$$\Theta_1 \leq_{\text{hr}} \Theta_2. \tag{22}$$

Then $Y_1 \leq_{s-\text{FR}} Y_2$.

Proof: Note that, by Lemma 35,

$$\frac{\overline{T}_s(Y_1, y)}{\overline{T}_s(Y_2, y)} = C \cdot \frac{\int_{\mathcal{X}} \Pi_{j=1}^{s-1} \mu_j(X_\theta) \overline{T}_s(X_\theta, y) dF_1(\theta)}{\int_{\mathcal{X}} \Pi_{j=1}^{s-1} \mu_j(X_\theta) \overline{T}_s(X_\theta, y) dF_2(\theta)}, \tag{23}$$

where $C = \left[\int_{\mathcal{X}} \Pi_{j=1}^{s-1} \mu_j(X_\theta) dF_2(\theta)\right] / \left[\int_{\mathcal{X}} \Pi_{j=1}^{s-1} \mu_j(X_\theta) dF_1(\theta)\right]$. Now (21) implies $\overline{T}_s(X_\theta, y)$ is TP_2 in $\theta \in \mathcal{X}$ and $y \in \mathcal{R}$. So, $\Pi_{j=1}^{s-1} \mu_j(X_\theta) \overline{T}_s(X_\theta, y)$ is increasing in θ for each y. (22) gives $\overline{F}_i(\theta)$ is TP_2 in $i \in \{1, 2\}$ and $\theta \in \mathcal{X}$. Hence, by Lemma 36, (23) is decreasing in y, giving the required result. □

Corollary 38: *Let* $\Theta_1 \leq_{\text{hr}} \Theta_2$.

(*a*) *If* $X(\theta) \leq_{\text{hr}} X(\theta')$ *for all* $\theta \leq \theta'$, *then* $Y_1 \leq_{\text{hr}} Y_2$.
(*b*) *If* $X(\theta) \leq_{\text{mrl}} X(\theta')$ *for all* $\theta \leq \theta'$, *then* $Y_1 \leq_{\text{mrl}} Y_2$.
(*c*) *If* $X(\theta) \leq_{\text{vrl}} X(\theta')$ *for all* $\theta \leq \theta'$, *then* $Y_1 \leq_{\text{vrl}} Y_2$.

Corollary 38(a) is given in Shaked and Wong[42], and Corollary 38(b) generalizes a result of Ahmed *et al.*[1], where they gave the result when Θ_1 and Θ_2 are likelihood ratio ordered.

4. Generalized Aging Properties

Different researchers have characterized different aging classes. Recently Nanda[25] has characterized the s-IFR, s-IFRA, s-NBU and their duals using Laplace transforms. For definitions of these aging classes, see Nanda[25] or Fagiuoli and Pellerey[12]. In this section, we characterize s-IFR (s-DFR) and s-NBUFR (s-NWUFR) classes (Definition 2), which generalizes many existing results in the literature.

4.1. *Characterizations in Terms of Residual Lives*

The following theorem characterizes s-IFR (s-DFR) distributions by s-FR ordering between residual life distributions. This says that if there are two items of the same product, of which one is new and the other is used, then the life of the new item is more than the remaining life of the used one in s-FR sense if and only if the life of the new item is s-IFR. In other words, if we have two items, one is older than the other, then the remaining life of the former is smaller than that of the later in s-FR sense if and only if the lifetime of a new item of the same product is s-IFR.

Theorem 39: Let $s \in \mathcal{N}$. Then the following conditions are equivalent:

(a) X is s-IFR (s-DFR).

(b) $X_t \geq_{s-\mathrm{FR}} (\leq_{s-\mathrm{FR}}) X_{t'}$ for all $t \leq t'$.

(c) $X \geq_{s-\mathrm{FR}} (\leq_{s-\mathrm{FR}}) X_t$ for all $t \geq 0$.

(d) $X + t \leq_{s-\mathrm{FR}} (\geq_{s-\mathrm{FR}}) X + t'$ for all $t \leq t'$.

Proof: We prove the result for $s \in \mathcal{N}^+$. The result for $s = 0$ was given in Shaked and Shanthikumar[40]. Suppose (a) holds. Now, by Lemma 14,

$$\frac{\overline{T}_s(X_t, x)}{\overline{T}_s(X_{t'}, x)} = \frac{\overline{T}_s(X, x+t)}{\overline{T}_s(X, t)} \cdot \frac{\overline{T}_s(X, t')}{\overline{T}_s(X, x+t')},$$

which is increasing in x for all $t' \geq t \geq 0$ if and only if

$$\frac{\overline{T}_s(X, x)}{\overline{T}_s(X, x+x')} \text{ is increasing in } x$$

for all $x' > 0$, which is true by (13). Hence (b) follows.

Now suppose (b) holds. Take $t = 0$ and then we get (c).

Suppose (c) holds. Then

$$\frac{\overline{T}_s(X_t, x)}{\overline{T}_s(X, x)} \text{ is decreasing in } x,$$

which, by Lemma 14, implies

$$\frac{\overline{T}_s(X, x+t)}{\overline{T}_s(X, x)} \text{ is decreasing in } x. \tag{24}$$

Note that

$$\frac{\overline{T}_s(X+t, y)}{\overline{T}_s(X+t', y)} = \frac{\overline{T}_s(X, y-t)}{\overline{T}_s(X, y-t')} = \frac{\overline{T}_s(X, x+t^*)}{\overline{T}_s(X, x)},$$

where $x = y - t'$ and $t^* = t' - t$. Hence, $\frac{\overline{T}_s(X+t,y)}{\overline{T}_s(X+t',y)}$ is decerasing in y if and only if $\frac{\overline{T}_s(X,x+t^*)}{\overline{T}_s(X,x)}$ is decreasing in x. Hence (24) gives (d).

Suppose (d) holds. Then

$$\frac{\overline{T}_s(X+t,x)}{\overline{T}_s(X+t',x)} = \frac{\overline{T}_s(X,x-t)}{\overline{T}_s(X,x-t')}$$

is decreasing in x for all $t < t'$, which implies

$$\frac{\overline{T}_s(X,y+y')}{\overline{T}_s(X,y)} \text{ is decreasing in } y$$

for all $y' > 0$, which is (a). □

Corollary 40: (a) *The following statements are equivalent:*
(i) X *is* IFR (DFR).
(ii) $X_t \geq_{\rm hr} (\leq_{\rm hr}) X_{t'}$ *for all* $t \leq t'$.
(iii) $X \geq_{\rm hr} (\leq_{\rm hr}) X_t$ *for all* $t \geq 0$.
(iv) $X + t \leq_{\rm hr} (\geq_{\rm hr}) X + t'$ *for all* $t \leq t'$.
(b) *The following statements are equivalent:*
(i) X *is* DMRL (IMRL).
(ii) $X_t \geq_{\rm mrl} (\leq_{\rm mrl}) X_{t'}$ *for all* $t \leq t'$.
(iii) $X \geq_{\rm mrl} (\leq_{\rm mrl}) X_t$ *for all* $t \geq 0$.
(iv) $X + t \leq_{\rm mrl} (\geq_{\rm mrl}) X + t'$ *for all* $t \leq t'$.
(c) *The following statements are equivalent:*
(i) X *is* DVRL (IVRL).
(ii) $X_t \geq_{\rm vrl} (\leq_{\rm vrl}) X_{t'}$ *for all* $t \leq t'$.
(iii) $X \geq_{\rm vrl} (\leq_{\rm vrl}) X_t$ *for all* $t \geq 0$.
(iv) $X + t \leq_{\rm vrl} (\geq_{\rm vrl}) X + t'$ *for all* $t \leq t'$.

Corollary 40(a) and (b) can be found in Shaked and Shanthikumar[41].

The following theorem, which characterizes s-IFR (s-DFR) aging distributions in terms of s-ST orderings, is a generalization of Theorem 1.A.13(a) of Shaked and Shanthikumar[41]. It says that, for two used items, the remaining life of the more used item is probabilistically less than that of the less used item if and only if the lifetime of a new unit of the same product has some positive aging property.

Theorem 41: For $s \in \mathcal{N}$,

$$X \text{ is } s\text{–IFR}\,(\text{s–DFR}) \Longleftrightarrow \mathrm{X_t} \geq_{\mathrm{s-ST}} (\leq_{\mathrm{s-ST}})\mathrm{X_{t'}}, \quad \forall \mathrm{t} \leq \mathrm{t'}.$$

Proof: Take $s \in \mathcal{N}^+$. By Lemma 14, $X_t \geq_{s-\mathrm{ST}} (\leq_{s-\mathrm{ST}}) X_{t'}$ if and only if

$$\frac{\overline{T}_s(X, x+t)}{\overline{T}_s(X,t)} \geq (\leq) \frac{\overline{T}_s(X, x+t')}{\overline{T}_s(X,t')}$$

for all $x \geq 0$ and $0 \leq t \leq t'$.

This can equivalently be written as

$$\frac{\overline{T}_s(X, x+t)}{\overline{T}_s(X,t)} \text{ is decreasing (increasing) in } t$$

for all $x \geq 0$, which, by (13), proves the result for $s \in \mathcal{N}^+$. For $s = 0$, the proof follows by writing the density function of X_t. □

Corollary 42: *(i) X is* ILR (DLR) $\Longleftrightarrow \mathrm{X_t} \geq_{\mathrm{wlr}} (\leq_{\mathrm{wlr}}) \mathrm{X_{t'}}$ *for all* $t \leq t'$.
(ii) X is IFR (DFR) $\Longleftrightarrow \mathrm{X_t} \geq_{\mathrm{st}} (\leq_{\mathrm{st}}) \mathrm{X_{t'}}$ *for all* $t \leq t'$.
(iii) X is DMRL (IMRL) $\Longleftrightarrow \mathrm{X_t} \geq_{\mathrm{hmrl}} (\leq_{\mathrm{hmrl}}) \mathrm{X_{t'}}$ *for all* $t \leq t'$.

The following theorem which characterizes the s-NBU (s-NWU) class is stated without proof.

Theorem 43: For $s \in \mathcal{N}$,

$$X \text{ is } s\text{–NBU (s–NWU)} \Longleftrightarrow \mathrm{X} \geq_{\mathrm{s-ST}} (\leq_{\mathrm{s-ST}}) \mathrm{X_t}, \quad \forall \mathrm{t} \geq 0.$$

Corollary 44: *X is* NBU (NWU) $\Longleftrightarrow \mathrm{X} \geq_{\mathrm{st}} (\leq_{\mathrm{st}}) \mathrm{X_t}$ *for all* $t \geq 0$.

4.2. *Characterizations in Terms of Equilibrium Distributions*

The following is a characterization of s-IFR (s-DFR) distributions in terms of s-FR order between a random variable and its equilibrium random variable.

Theorem 45: For $s \in \mathcal{N}$,

$$X \geq_{s-\mathrm{FR}} (\leq_{s-\mathrm{FR}}) A_X \Longleftrightarrow X \text{ is } (s+1)\text{–IFR}\,((\mathrm{s}+1)\text{–DFR}).$$

Proof: Take $s \in \mathcal{N}^+$. Then $X \geq_{s-\mathrm{FR}} (\leq_{s-\mathrm{FR}}) A_X$ if and only if

$$\frac{\overline{T}_s(X, x)}{\overline{T}_s(A_X, x)} \text{ is increasing (decreasing) in } x,$$

which, by Lemma 21, reduces to

$$\frac{\overline{T}_{s+1}(X,x)}{\overline{T}_s(X,x)} \text{ is decreasing (increasing) in } x.$$

This can equivalently be written as

$$\frac{\int_x^\infty \overline{T}_s(X,t)dt}{\overline{T}_s(X,x)} \text{ is decreasing (increasing) in } x,$$

which, by (2), gives the required result for $s \in \mathcal{N}^+$. For $s = 0$, the result follows by writing the density function of A_X. □

Corollary 46: *(i) X is* IFR (DFR) *if and only if* $X \geq_{\rm lr} (\leq_{\rm lr}) A_X$.[46]
(ii) X is DMRL (IMRL) *if and only if* $X \geq_{\rm hr} (\leq_{\rm hr}) A_X$.[45]
(iii) X is DVRL (IVRL) *if and only if* $X \geq_{\rm mrl} (\leq_{\rm mrl}) A_X$.

The following is a characterization of s-NBUFR property of a random variable.

Theorem 47: For $s \in \mathcal{N}$,

$$X \geq_{s-\rm ST} (\leq_{s-\rm ST}) A_X \Longleftrightarrow X \text{ is } (s+1)\text{–NBUFR} ((\text{s}+1)\text{–NWUFR}).$$

Proof: Take $s \in \mathcal{N}^+$. By Lemma 21, $X \geq_{s-\rm ST} (\leq_{s-\rm ST}) A_X$ if and only if, for all $x \geq 0$,

$$\overline{T}_s(X,x) \geq (\leq) \overline{T}_{s+1}(X,x).$$

This, by (1), can be written as

$$\overline{T}_s(X,x)\mu_s(X) \geq (\leq) \int_x^\infty \overline{T}_s(X,u)du,$$

which, by (2), reduces to

$$r_{s+1}(X,x) \geq (\leq) \frac{1}{\mu_s(X)} = r_{s+1}(X,0).$$

The case $s = 0$ follows by writing the density function of A_X. Hence the result. □

Corollary 48: *(a) X is* NBUE (NWUE) *if and only if* $X \geq_{\rm st} (\leq_{\rm st}) A_X$.[46]
(b) X is NBUFR (NWUFR) *if and only if* $X \geq_{\rm wlr} (\leq_{\rm wlr}) A_X$.

4.3. *Characterizations in Terms of Laplace Transform*

Some characterizations of generalized aging classes in terms of Laplace transforms are given below. Similar type of characterization for other generalized aging classes are given in Nanda[25].

In the following theorem, a characterization of s-NBUFR distributions is given.

Theorem 49: For $s = 2, 3, \ldots$, X is s-NBUFR if and only if

$$\Gamma_\lambda \overline{T}_s(X, n) \leq \Gamma_\lambda \overline{T}_{s-1}(X, n), \quad \forall \lambda > 0,\, n \in \mathcal{N}.$$

Proof: *Necessity*: By (11), X is s-NBUFR if and only if

$$r_s(X, x) \geq r_s(X, 0) \text{ for all } x \geq 0.$$

This can equivalently be written as

$$\frac{\overline{T}_{s-1}(X, x)}{\int_x^\infty \overline{T}_{s-1}(X, u)du} \geq \frac{\overline{T}_{s-1}(X, 0)}{\int_0^\infty \overline{T}_{s-1}(X, u)du},$$

which, on simplification, gives

$$\overline{T}_{s-1}(X, x) \geq \overline{T}_s(X, x) \text{ for } s = 2, 3, \ldots.$$

Now by multiplying both sides of the above expression by $\frac{\lambda e^{-\lambda x}(\lambda x)^{n-1}}{(n-1)!}$ and intrgrating with respect to x, we get

$$\Gamma_\lambda \overline{T}_s(X, n) \leq \Gamma_\lambda \overline{T}_{s-1}(X, n).$$

Sufficiency: Let C be the set of all continuity points of $\overline{T}_s(\cdot)$ and $\overline{T}_{s-1}(\cdot)$ both. Given that, for $\lambda > 0$, $s \in \{2, 3, \ldots\}$ and $n \in \mathcal{N}$,

$$\Gamma_\lambda \overline{T}_s(X, n) \leq \Gamma_\lambda \overline{T}_{s-1}(X, n).$$

Now taking limits as $n \to \infty$ on both sides of the above expression we have, by Lemma 2.4 of Nanda[25],

$$\overline{T}_s(X, x) \leq \overline{T}_{s-1}(X, x),$$

for all $x \in C$, which gives

$$\frac{\overline{T}_{s-1}(X, x)}{\overline{T}_s(X, x)\mu_{s-1}(X)} \geq \frac{\overline{T}_{s-1}(X, 0)}{\mu_{s-1}(X)}$$

for $s = 2, 3, \ldots$. After simplification, we get

$$r_s(X, x) \geq r_s(X, 0).$$

Hence the result follows by (11). □

Corollary 50: *X is* NBUE *if and only if*

$$\Gamma_\lambda \overline{F}_1(X,n) \le \Gamma_\lambda \overline{F}(X,n), \quad \forall\, \lambda > 0,\, n \in \mathcal{N},$$

where $\overline{F}$ is the survival function of X and $\overline{F}_1$ is the survival function of the equilibrium distribution of X.

Theorem 51: For $s \in \mathcal{N}^+$, X is s-NBUCX if and only if

$$\Gamma_\lambda \overline{T}_{s+1}(X_t,n) \;\le\; \Gamma_\lambda \overline{T}_s(X,n), \quad \forall t \ge 0,\, n \in \mathcal{N}.$$

Proof: *Necessity*: By (12), X is s-NBUCX if and only if

$$\int_t^\infty \overline{T}_s(X,x+u)du \;\le\; \overline{T}_s(X,x) \int_t^\infty \overline{T}_s(X,u)du$$

for $x, t \ge 0$. Or,

$$\frac{\overline{T}_{s+1}(X,x+t)}{\overline{T}_{s+1}(X,t)} \le \overline{T}_s(X,x)$$

for $x, t \ge 0$, which is equivalent to

$$\overline{T}_{s+1}(X_t,x) \;\le\; \overline{T}_s(X,x).$$

Now by multiplying both sides of the above expression by $\frac{\lambda e^{-\lambda x}(\lambda x)^{n-1}}{(n-1)!}$ and intrgrating with respect to x, we get

$$\Gamma_\lambda \overline{T}_{s+1}(X_t,n) \;\le\; \Gamma_\lambda \overline{T}_s(X,n).$$

Sufficiency: This can be proved on the same line as the proof of the sufficient part of Theorem 49. □

The dual results of Theorems 49 and 51 for s-NWUFR and s-NWUCX can be proved on the same line.

4.4. *Other Properties*

The following lemma will be used in proving another theorem.

Lemma 52: *For $s \in \mathcal{N}^+$,*
(*i*) $\mu_s(aX) = a\mu_s(X)$;
(*ii*) $\overline{T}_s(aX,x) = \overline{T}_s(X,x/a)$;
(*iii*) $r_s(aX,x) = a^{-1} r_s(X,x/a)$.

The following theorem generalizes a theorem of Aly and Kochar[2] and Theorems 1.C.18 and 1.D.7 of Shaked and Shanthikumar[41]. It has a nice interpretation in insurance as follows:

Suppose X denotes the risk the direct insurer faces and ϕ the corresponding reinsurance contract. One important reinsurance agreement is *quota-share treaty* defined as $\phi(X) = aX$ for $a \in [0,1]$. The following theorem states that if the risk has s-IFR distribution, then the quota-share treaty is less than the risk in the sense of s-FR order.

Theorem 53: Let X be a nonnegative s-IFR random variable. Then for any $0 < a \leq 1$ and $s \in \mathcal{N}^+$, $aX \leq_{s-\mathrm{FR}} X$.

Corollary 54: (*i*) *If* X *is* IFR, *then* $aX \leq_{\mathrm{hr}} X$ *for all* $0 < a \leq 1$.
(*ii*) *If* X *is* DMRL, *then* $aX \leq_{\mathrm{mrl}} X$ *for all* $0 < a \leq 1$.
(*iii*) *If* X *is* DVRL, *then* $aX \leq_{\mathrm{vrl}} X$ *for all* $0 < a \leq 1$.

The above theorem is not true for s=0. A counterexample may be obtained in Hu *et al.*[17].

The following interesting lemma, which is used to prove Theorem 56, is a generalization of Lemma A.1 of Pellerey *et al.*[33]. They proved the result for $s = 1$, but the proof of the general result given below follows on the same line as their proof with suitable modifications and hence omitted.

Lemma 55: *For* $s \in \mathcal{N}^+$, *if* $r_s(X,x)$ *is logconcave, then it is increasing.*

It is well known that if a random variable X is $(s-1)$-IFR, then it is s-IFR. But the converse, in general, is not true. But under certain mild condition, the converse is also true as given in the following theorem.

Theorem 56: Let $s \in \mathcal{N}^+$. If $r_s(X,x)$ is logconcave, then X is $(s-1)$-IFR.

Proof: Using (1), (2) and (3), one can write

$$\overline{T}_{s-1}(X,x) = \mu_{s-1}(X) r_s(X,x) \overline{T}_s(X,0) \exp\left\{-\int_0^x r_s(X,u)du\right\},$$

which gives

$$\ln \overline{T}_{s-1}(X,x) = \left[\ln \mu_{s-1}(X)\overline{T}_s(X,0)\right] + \ln r_s(X,x) - \int_0^x r_s(X,u)du. \quad (25)$$

By the assumption and Lemma 55, $\ln r_s(X, x)$ and $-\int_0^x r_s(X, u)du$ are both concave. Thus (25) gives $\ln \overline{T}_{s-1}(X, x)$ is concave. Hence, the result follows by (14). □

Corollary 57: (*a*) *If* X *has logconcave failure rate function, then it is* ILR.
(*b*) *If* X *has logconvex mean residual life function, then it is* IFR.

Acknowledgements

Thanks are due to Professor Moshe Shaked and Professor Franco Pellerey for providing the authors with various references and suggestions. The first author was partially supported by one grant of Chinese Academy of Sciences and the National 973 Fundamental Research Program on Financial Engineering (Grant No: G1998030418).

References

1. A. N. Ahmed, A. A. Soliman and S. E. Khider, On some partial ordering of interest in reliability, *Microelectron. Reliab.* **36**, 1337-1346 (1996).
2. E. A. A. Aly and S. C. Kochar, On hazard rate ordering of dependent variables, *Adv. Appl. Probab.* **25**, 477-482 (1993).
3. J. Averous and M. Meste, Tailweight and life distributions, *Statist. Probab. Lett.* **8**, 381-387 (1989).
4. L. Bondesson, On preservation of classes of life distributions under reliability operations: some complementary results, *Naval Res. Logist. Quart.* **30**, 443-447 (1983).
5. M. Brown and J. G. Shanthikumar, Comparing the variability of random variables and point processes, *Probab. Eng. Inform. Sci.* **12**, 425-444 (1998).
6. J. Cao and Y. Wang, The NBUC and NWUC classes of life distributions, *J. Appl. Probab.* **28**, 473-479 (1991).
7. P. Capéraà, Tail ordering and asymptotic efficiency of rank tests, *Ann. Statist.* **16**, 470-478 (1988).
8. M. Carletti and F. Pellerey, A new necessary condition for higher order stochastic dominances with applications, *Ricerche di Matematica*, **47**, 373-381 (1998).
9. J. V. Deshpande, S. C. Kochar and H. Singh, Aspects of positive ageing, *J. Appl. Probab.* **23**, 748-758 (1986).
10. M. Denuit, C. Lefevre and M. Shaked, The s-convex orders among real random variables, with applications, *Math. Inequalities Appl.* **1**, 585-613 (1998).
11. M. Denuit, C. Lefevre and M. Shaked, On the theory of high convexity stochastic orders, *Statist. Probab. Lett.* **47(3)**, 287-293 (2000).

12. E. Fagiuoli and F. Pellerey, New partial orderings and applications, *Naval Res. Logist.* **40**, 829-842 (1993).
13. E. Fagiuoli and F. Pellerey, Mean residual life and increasing convex comparison of shock models, *Statist. Probab. Lett.* **20**, 337-345 (1994).
14. P. C. Fishburn, Stochastic dominance and moments of distributions, *Math. Oper. Res.* **5**, 94-100 (1980).
15. R. C. Gupta and S. N. U. A. Kirmani, On order relations between reliability measures, *Comm. Statist. Stochastic Models*, **3**, 149-156 (1987).
16. O. Hesselager, Closure properties of some partial orderings under mixing, *Insurance Math. Econom.* **22**, 163-170 (1998).
17. T. Hu, A. K. Nanda, H. Xie and Z. Zhu, Properties of some stochastic orders: a unified study. Preprint. (2001).
18. K. Joag-Dev, S. Kochar and F. Proschan, A general composition theorem and its applications to certain partial orderings of distributions, *Statist. Probab. Lett.* **22**, 111-119 (1995).
19. R. Kaas, A. E. Van Heerwaarden and M. J. Goovaerts, *Ordering of Actuarial Risks* (Caire Education Series 1, Brussels, 1994).
20. S. Karlin, *Total Positivity*, Vol I (Stanford University Press, 1968).
21. Y. Kebir, Laplace transform characterization of probabilistic orderings, *Probab. Eng. Inform. Sci.* **8**, 69-77 (1994).
22. R. L. Launer, Inequalities for NBUE and NWUE life distributions, *Oper. Res.* **32**, 660-667 (1984).
23. J. Lynch, G. Mimmack and F. Proschan, Uniform stochastic orderings and total positivity, *Canad. J. Statist.* **15**, 63-69 (1987).
24. S. P. Mukherjee and A. Chatterjee, Stochastic dominance of higher orders and its implications, *Comm. Statist. Theory Methods* **21**, 1977-1986 (1992).
25. A. K. Nanda, Generalized ageing classes in terms of Laplace transforms, *Sankhyā*, **A62**, 258-266 (2000).
26. A. K. Nanda, On improvement and deterioration of a repairable system, *IAPQR Trans. Reliab.* **22**, 107-113 (1997).
27. A. K. Nanda, Stochastic orders in terms of Laplace transforms, *Calcutta Statist. Assoc. Bull.* **179&180**, 195-201 (1995).
28. A. K. Nanda and K. Jain, Some weighted distribution results on univariate and bivariate cases, *J. Statist. Plan. Infer.* **77**, 169-180 (1999).
29. A. K. Nanda, K. Jain and H. Singh, Properties of moments for s-order equilibrium distributions, *J. Appl. Probab.* **33**, 1108-1111 (1996).
30. A. K. Nanda, K. Jain and H. Singh, On closure of some partial orderings under mixtures, *J. Appl. Probab.* **33**, 698-706 (1996).
31. P. C. O'Brien, Stochastic dominance and moment inequalities, *Math. Oper. Res.* **9**, 475-477 (1984).
32. J. E. Pečarić, F. Proschan and Y. L. Tong, *Convex Functions, Partial Orderings, and Statistical Applications* (Academic Press, New York, 1992).
33. F. Pellerey, M. Shaked and J. Zinn, Nonhomogeneous Poisson process and logconcavity, *Probab. Eng. Inform. Sci.* **14**, 353-373 (2000).

34. T. Rolski, Order relations in the set of probability distribution functions and their applications in queueing theory, *Dissertationes Math.* **132**, 5-47 (1976).
35. S. M. Ross, *Stochastic Processes* (John Wiley & Sons, New York, 1983).
36. G. Rowell and K. Siegrist, Relative aging of distributions, *Probab. Eng. Inform. Sci.* **12**, 469-478 (1998).
37. D. Sengupta and J. V. Deshpande, Some results on the relative aging of two life distributions, *J. Appl. Probab.* **31**, 991-1003 (1994).
38. D. Sengupta and A. K. Nanda, Log-concave and concave distributions in reliability, *Naval Res. Logist.* **46**, 419-433 (1999).
39. M. Shaked, On mixtures from exponential families, *J. Roy. Statist. Soc.* **B42**, 192-198 (1980).
40. M. Shaked and J. G. Shanthikumar, Characterization of some first passage times using log-concavity and log-convexity as aging notions, *Probab. Eng. Inform. Sci.*, **1**, 279-291 (1987).
41. M. Shaked and J. G. Shanthikumar, *Stochastic Orders and Their Applications* (Academic Press, New York, 1994).
42. M. Shaked and T. Wong, Preservation of stochastic orderings under random mapping by point processes, *Probab. Eng. Inform. Sci.* **9**, 563-580 (1995).
43. H. Singh, On partial orderings of life distributions, *Naval Res. Logist.* **36**, 103-110 (1989).
44. D. Stoyan, *Comparison Methods for Queues and Other Stochastic Models* (John Wiley & Sons, New York, 1983).
45. W. Whitt, Uniform conditional variability ordering of probability distributions, *J. Appl. Probab.* **22**, 619-633 (1985).
46. W. Whitt, The renewal process stationary-excess operator, *J. Appl. Probab.* **22**, 156-167 (1985).
47. G. E. Willmot, Bound for compound distributions based on mean residual lifetimes and equilibrium distributions, *Insurance Math. Econom.* **21**, 25-42 (1997).
48. R. W. Wolff, *Stochastic Modelling and the Theory of Queues* (Prentice Hall, Englewood Cliffs, NJ, 1989).
49. R. E. Barlow and F. Proschan, *Statistical Theory of Reliability and Life Testing* (To Begin With, Silver Spring, MD, 1981; Holt, Rinehart, and Winston, New York, 1975).

CHAPTER 13

DEPENDENCE AND MULTIVARIATE AGING: THE ROLE OF LEVEL SETS OF THE SURVIVAL FUNCTION

Bruno Bassan* and Fabio Spizzichino[†] [a]

Dipartimento di Matematica, Università "La Sapienza"
P. A. Moro 5, 00185 Roma, Italy
*E-mail: *bassan@mat.uniroma1.it,*
E-mail: †fabio.spizzichino@uniroma1.it

We describe several notions of multivariate aging for exchangeable lifetimes by means of properties of the level sets of the joint survival function. These properties are characterized in terms of dependence properties of a suitably defined distribution.

1. Introduction

In many problems involving the analysis of failure and survival data it is often natural to assume that the dependent components under consideration are similar. For example, think of many identical components in a parallel system, subject to the same possibly unknown environmental influence. Hence, exchangeable laws play a relevant role. Furthermore, the analysis of the exchangeable case allows us to focus on the essential properties of aging, without undue interference from other minor nuisance aspects such as differences among individuals.

Usually, in defining multivariate notions of aging a requirement is that the 1-dimensional marginals have the aging property which is being extended. Thus, for example, some definitions of multivariate IFR which can be found in the literature are such that the one-dimensional marginals are IFR. When dealing with exchangeable laws, and more generally in the Bayesian subjective approach, it is neither necessary nor appropriate to

[a]Work partially supported by CNR and MURST.

insist on this requirement on the 1-dimensional marginals. For example, consider n exchangeable lifetimes, which are i.i.d. conditionally on a parameter θ and are such that, for every value of θ, the conditional univariate law is IFR. It is quite natural for a multivariate notion of IFR to include the n lifetimes under consideration. However, the unconditional univariate marginals need not be IFR, as it is well-known.

One can approach multivariate aging in the following way: take a univariate aging notion, and take n i.i.d. lifetimes with this aging property; find suitable properties of this n-dimensional joint law, and extend it to the exchangeable case. For example, the joint law of n i.i.d. IFR lifetimes is Schur-concave, and Schur-concavity turns out to be a good tool in multivariate aging, yielding a reasonable definition of "multivariate IFR" for exchangeable laws. See Barlow and Mendel (1992), Barlow and Spizzichino (1993) and Spizzichino (1992, 2001) for a discussion of the role of Schur-concavity in subjective multivariate aging.

A specific approach in this vein hinges on the stochastic comparison of residual lifetimes, as developed in Bassan and Spizzichino (1999, 2000). This approach is based on inequalities like

$$\mathcal{L}\left(X-x|X>x, Y>y\right) \geq_{*} \mathcal{L}\left(Y-y|X>x, Y>y\right) \quad \forall 0 \leq x \leq y, \tag{1}$$

where $\mathcal{L}\left(Z\right)$ denotes the law of the random variable Z and where $\geq_*$ can be one of the usual orderings, such as the stochastic, hazard rate or likelihood ratio ordering. This approach takes directly into consideration relevant issues of multivariate aging. For example, it has the advantage of yielding (often) an immediate intuitive explanation: the property in (1) can be interpreted as asserting that if items of different age coexist in a same system, and they are all functioning (alive), then the younger is preferred to the elder, no matter what their ages are.

We stress that these multivariate notions of aging are based on one-dimensional stochastic comparisons of residual lifetimes. Other notions of multivariate aging, on the contrary, are based on multivariate stochastic comparisons (see Shaked and Shantikumar (1994) for a review). Other notions, finally, do not involve at all comparisons of residual lifetimes (see e.g. Savits (1985)).

The aging notions that involve multivariate stochastic comparisons are typically of a dynamic type. The main feature is the fact that all surviving individuals are of the same age. As we pointed out before, in our approach we compare residual lifetimes of individuals with different ages. For further references and details about differences and similarities among these approaches, see Spizzichino (2001).

A further approach to multivariate aging, related to the one dealt with in Bassan and Spizzichino (1999,2000), hinges on the analysis of the level sets of the joint survival function. This idea goes back to Barlow and Spizzichino (1993), and is related to the analysis of Schur-concavity and Schur-convexity of the joint survival function. In fact, the joint survival $\bar{F}$ is Schur-concave if and only if the level sets

$$A_k = \{\mathbf{x}|\bar{F}(\mathbf{x}) \geq k\}$$

are Schur-concave.

In this note we resume the idea of considering the level sets of the joint survival function, and we introduce multivariate notions of aging by considering certain properties of the level sets of the joint survival function of n i.i.d. lifetimes with a given univariate aging property.

More specifically, the main idea underlying the procedure dealt with in this note can be summarized as follows:

- consider a univariate aging notion P (e.g., NBU, IFR, etc.)
- take the joint survival $\bar{F}$ of n i.i.d. lifetimes, and prove results of the following type: each lifetime has the property P if and only if the level sets of $\bar{F}$ satisfy the property $\tilde{P}$;
- define a multivariate aging notion as follows: an exchangeable law $\bar{H}$ is multivariate-P if the level sets of its joint survival function satisfy the property $\tilde{P}$.

In order to characterize the level sets of the survival function, we resort to suitable *multivariate aging functions*. Indeed, as we shall see, they will display certain dependence properties in correspondence to aging properties of the univariate laws. This approach will allow us, in particular, to introduce aging notions, such as Multivariate IFRA, which do not allow a representation in terms of Schur-properties or orderings among residual lifetimes.

The paper is organized as follows. In Section 2 we introduce the multivariate aging function and we develop the ideas sketched above. The relevant dependence properties of the multivariate aging function will be expressed in terms of *copulas* (see Joe (1997) and Nelsen (1999)). In Section 3 we consider a class of laws, the *time transformed exponentials* (TTE) models (see Barlow and Mendel (1992)). For these models, the notions of aging and dependence considered here take a simpler form. In particular, the multivariate aging functions for these models are *Archimedean copulas* (see (5) below), and this allows us to use many of the results contained in a recent paper by Averous and Dortet-Bernadet (2000). For the TTE models, we introduce the notion of multivariate IFRA. In Section 4, finally, we draw some conclusions and we show some relations among the definitions introduced here and other notions existing in the literature.

Some notation: we restrict mainly to the case of $n = 2$ lifetimes, for simplicity; many of the results can be extended immediately. The random variables X and Y are positive and exchangeable. Their joint survival function is denoted by $\bar{F}$ and the univariate marginal survival by $\bar{G}$.

2. Level Sets of $\bar{F}$ and Multivariate Aging Function

The idea of assessing the aging of a joint law by considering the level sets of the survival function and/or of a suitably defined auxiliary function has been considered in Barlow and Spizzichino (1993). They used the function

$$h = \bar{G}^{-1} \circ \bar{F}, \tag{2}$$

(i.e., obviously, $h(x, y) = \bar{G}^{-1}(\bar{F}(x, y))$; recall that $\bar{G}$ is the 1-dimensional survival). The function h has the same level curves as $\bar{F}$. Furthermore, $\bar{F}$ is Schur-concave iff h is Schur-convex iff the level sets $\tilde{A}_u = \{\mathbf{x} | h(\mathbf{x}) \leq u\}$ are Schur-concave (in the sense that their indicator functions are Schur-concave). Moreover, the joint survival function with i.i.d. IFR marginals has convex level sets.

Example 1: *(Schur-constant laws)* Let $\bar{F}(x, y) = \Phi(x + y)$, for some Φ. Then $h(x, y) = x + y$.

Example 2: *(Time Transformed Exponentials Models)* Let

$$\bar{F}(x, y) = W(R(x) + R(y)).$$

See Section 3 for details. Then $h(x, y) = R^{-1}(R(x) + R(y))$.

Example 3: *(Marshall-Olkin model)* Let

$$\bar{F}(x, y) = \exp\{-\lambda(x + y) - \lambda'(x \vee y)\}.$$

Then

$$h(x,y) = \frac{\lambda(x+y) + \lambda'(x \vee y)}{\lambda + \lambda'}$$

Example 4: *(Degenerate models)* Let $\mathbb{P}(X = Y) = 1$. Then $\bar{F}(x,y) = \psi(x \vee y)$, for some ψ, and $h(x,y) = x \vee y$.

Here, we extend the discussion about relations between multivariate notions of aging and relevant properties of the level sets of $\bar{F}$. In order to describe the level curves of $\bar{F}$, we shall consider, instead of h, the function

$$B(u,v) := \exp\left\{-\bar{G}^{-1}\left(\bar{F}\left(-\log u, -\log v\right)\right)\right\}, \qquad u,v \in [0,1]$$

We call it the *bivariate aging function.* When there is ambiguity we write $B_{\bar{F}}$. Here are some examples and preliminary remarks.

Example 5: *(Schur-constant laws)* In this case, we have $B(u,v) = uv$.

Example 6: *(Time Transformed Exponentials Models)* As we shall see later, we have

$$B(u,v) = \phi^{-1}\left(\phi(u) + \phi(v)\right),$$

with $\phi(z) = R(-\log z)$.

Example 7: *(Marshall-Olkin model)* Easy computations show that

$$B(u,v) = (uv)^{\frac{\lambda}{\lambda+\lambda'}}(u \wedge v)^{\frac{\lambda'}{\lambda+\lambda'}}$$

Example 8: *(Degenerate models)* Here, $B(u,v) = u \wedge v$.

(1) The function B can be obtained from the function h defined in (2) as follows:

$$B(u,v) = \exp\left\{-h\left(-\log u, -\log v\right)\right\}.$$

(2) B conveys information concerning the level sets of $\bar{F}$. In fact, (x,y) and (x',y') belong to the same level curve of $\bar{F}$ (and hence of h) iff (e^{-x}, e^{-y}) and $\left(e^{-x'}, e^{-y'}\right)$ belong to the same level curve of B.

(3) B has all the formal properties of a copula, except possibly for the rectangle inequality

$$B(u+a, v+b) + B(u,v) - B(u+a,v) - B(u,v+b) \geq 0$$

Within certain classes of laws, it is possible to characterize those for which B is actually a copula. Recall that a copula is a joint survival function of two random variables, each uniformly distributed on [0,1]; recall also that for every bivariate law F with marginals F_1 and F_2, there exists a copula C_F such that $F(x,y) = C_F(F_1(x), F_2(y))$. The interest of copulas is in the fact that the copula embodies the dependence structure of a multivariate law. In particular, independence for $\bar{F}$ is described by the condition $C_F(u,v) = uv$.

(4) Similarly to copulas, the function B, together with $\bar{G}$, characterizes the law:

$$\bar{F}(x,y) = \bar{G}\left(-\log B\left(e^{-x}, e^{-y}\right)\right) \tag{3}$$

The first two remarks encourage us to consider B as a "good" multivariate aging function, since they show that B is related to the level curves of $\bar{F}$.

The third remark suggests that we should consider *dependence* properties of B in order to describe *aging* properties of $\bar{F}$. For example, B is the copula corresponding to independence ($B(u,v) = uv$) if and only if $\bar{F}$ is Schur-constant.

The following procedure summarizes how we shall introduce multivariate aging notions by means of the function B.

(i) Prove results of this type:
Let X and Y be i.i.d. with one-dimensional survival function $\bar{H}$ and joint survival $\bar{F}_0(x,y) = \bar{H}(x)\bar{H}(y)$. Then the following are equivalent:

- $\bar{H}$ satisfies the univariate aging property P
- $B_{\bar{F}_0}$ satisfies the dependence property Q.

(ii) Define: an exchangeable law $\bar{F}$ is multivariate-P if $B_{\bar{F}}$ is Q

The following proposition summarizes some of the equivalences of the type described above. Its proof can be obtained by adapting results contained in Averous and Dortet-Bernadet (2000). We recall here some concepts of positive dependence, and we refer to Joe (1997) for further details.

- (X,Y) is PQD (*positive quadrant dependent*) if

$$\mathbb{P}(X \le x)\,\mathbb{P}(Y \le y) \le \mathbb{P}(X \le x, Y \le y), \quad \forall x, y$$

- Y is LTD (*left tail decreasing*) in X if

$$\mathbb{P}(Y \leq y | X \leq x') \leq \mathbb{P}(Y \leq y | X \leq x), \quad \forall x < x', \quad \forall y$$

- Y is SI (*stochastically increasing*) in X if

$$\mathbb{P}(Y \leq y | X = x') \leq \mathbb{P}(Y \leq y | X = x), \quad \forall x < x', \quad \forall y$$

As we said before, a copula is a joint distribution function (of two uniformly distributed random variables), and hence the same dependence concepts apply to copulas. For example, since given a copula C there exist two uniform random variables U, V such that $C(u, v) = \mathbb{P}(U \leq u, V \leq v)$, a copula is PQD if $C(u, v) \geq uv$ for every $0 \leq u, v \leq 1$.

Proposition 9: *Let X, Y be i.i.d. and let $\bar{F}_0$ be their joint survival function. Then*

(1) X and Y are NBU if and only if B_{F_0} is PQD (positive quadrant dependent)
(2) X and Y are IFR if and only if B_{F_0} is LTD (left tail decreasing)
(3) X and Y have log-concave density (and hence are PF_2) if and only if B_{F_0} is SI (stochastically increasing).

Proof: The proof can be obtained by adapting results contained in Averous and Dortet-Bernadet (2000). We prove the first statement, for the sake of clarity.

By independence, we can write:

$$\begin{aligned} B(u, v) &= \exp\left\{-\bar{G}^{-1}\left(\bar{F}\left(-\log u, -\log v\right)\right)\right\} \\ &= \exp\left\{-\bar{G}^{-1}\left(\bar{G}\left(-\log u\right)\bar{G}\left(-\log v\right)\right)\right\}. \end{aligned}$$

On the other hand, we notice that $\bar{G}$ is NBU if and only if

$$\bar{G}\left(-\log u\right)\bar{G}\left(-\log v\right) \geq \bar{G}\left(-\log uv\right).$$

Then, if (and only if) $\bar{G}$ is NBU, we obtain

$$B(u, v) \geq \exp\left\{-\bar{G}^{-1}\left(\bar{G}\left(-\log uv\right)\right)\right\} = uv,$$

which corresponds to the PQD property of B. □

Motivated by the above proposition, we may give the following definitions of multivariate aging notions for random vectors which are exchangeable but not necessarily independent.

Definition 10: Let (X,Y) be an exchangeable vector with bivariate aging function B.

(1) We say that (X,Y) is *multivariate-NBU* if B is PQD
(2) We say that (X,Y) is *multivariate-IFR* if B is LTD
(3) We say that (X,Y) is *multivariate-PF*$_2$ if B is SI

We shall see later how these aging notions relate to existing notions given in terms of Schur-properties or comparison of residual lifetimes. The main interest of the function B, though, is in extending those aging notions, such as IFRA, for which no representation in terms of comparison among residual lifetimes is available. We shall deal with this issue in the next Section.

3. Aging and Dependence for Time Transformed Exponential Models

A relevant class is the family of the so called *time transformed exponentials models*(TTE), namely the laws such that there exist functions W and R defined on $\mathbb{R}_+$ such that

$$\bar{F}(x,y) = W(R(x) + R(y)) \tag{4}$$

See Barlow and Mendel (1992). If $\bar{F}$ has the representation above, we write $\bar{F} \sim TTE(W,R)$. Here are some considerations.

Remark 11:

(1) Obviously, in order for (4) to yield a distribution function, certain properties should be satisfied by W and R: W is a convex survival function, which we assume strictly decreasing, and R is continuous and strictly increasing, with $R(0) = 0$ and $\lim_{x\to\infty} R(x) = \infty$.
(2) The class of TTE laws includes the independent laws (take $W(z) = \exp\{-z\}$) and the Schur-constant laws (take $R(z) = z$).
(3) If $\bar{F} \sim TTE(W,R)$ then

$$B_{\bar{F}}(u,v) = \phi^{-1}(\phi(u) + \phi(v)), \tag{5}$$

where $\phi(u) = R(-\log u)$. Observe, in particular, that B depends on R but not on W. B is a copula iff ϕ is convex. Copulas of the form (5) are called *Archimedean* copulas. See Joe (1997), Nelsen (1999) and references therein.

(4) It can be shown that the survival copula of $\bar{F}$ is also an Archimedean copula, depending on W but not on R. Recall that the survival copula $K_{\bar{F}}$ is the copula which allows us to write the joint survival function in terms of the marginal survival functions:

$$\bar{F}(x,y) = K_{\bar{F}}\left(\bar{G}\left(x\right),\bar{G}\left(y\right)\right).$$

See Joe (1997). In the TTE case, we have

$$K_{\bar{F}}\left(u,v\right) = W\left(W^{-1}(u) + W^{-1}(v)\right). \tag{6}$$

Thus we may say that W modulates dependence and R modulates multivariate aging. Their combination modulates univariate aging of the marginal $\bar{G} = W \circ R$.

(5) A characterization of the TTE class in terms of mixtures can be found in Barlow and Mendel (1992).

(6) Let Θ be a positive random variable with distribution Π, and let X, Y be i.i.d. conditionally on Θ, with

$$\mathbb{P}\left(X > x | \Theta = \theta\right) = \exp\left\{-\theta R(x)\right\}.$$

Then, conditionally on $\Theta = \theta$, (X, Y) are $TTE(W_{(\theta)}, R)$, with $W_{(\theta)}(z) = e^{-\theta z}$. Furthermore, the unconditional law of (X,Y) is $TTE(W,R)$, with $W(z) = \int_0^\infty \exp\left\{-\theta z\right\} \Pi(d\theta)$ and the same R as above.

The following proposition sheds some light on the structure of the level sets of TTE laws. We write $W_0(x) = e^{-x}$.

Proposition 12:

(1) The TTE class is the family of all joint survival functions $\bar{F}$ such that there exists a law with independent marginals which has the same level curves as $\bar{F}$.

(2) The TTE class is the family of all joint survival functions $\bar{F}$ such that there exists a law with independent marginals which has the same multivariate aging function B as $\bar{F}$.

(3) The TTE class is the family of all joint survival functions $\bar{F}$ such that there exists a Schur-constant law with the same dependence structure (i.e., same copula) as $\bar{F}$.

Proof:

(1) Let $\bar{F} \sim TTE(W, R)$. This law has the same level curves as $\bar{F}_0 \sim TTE(W_0, R)$, i.e. $\bar{F}_0(x,y) = \exp\{-[R(x)+R(y)]\}$.
Conversely, let $\tilde{F}$ be a joint law having the same level curves as the independent law $\bar{F}_0(x,y) = \exp\{-[R(x)+R(y)]\}$. Then $\tilde{F}$ depends on x and y only through $R(x)+R(y)$, i.e. there exists $\tilde{W}$ such that $\tilde{F}(x,y) = \tilde{W}(R(x)+R(y))$.

(2) Let $\bar{F} \sim TTE(W, R)$. Since B depends on R but not on W, the two functions $\bar{F}$ and $\bar{F}_0 \sim TTE(W_0, R)$ admit the same B
Conversely, let $\tilde{F}$ be a joint survival having the same B as the independent law $\bar{F}_0$. Let $\tilde{G}$ and $\bar{G}_0 = \exp\{-R(x)\}$ be their 1-dimensional marginals. Recalling (5), we obtain

$$B_{\bar{F}_0}(e^{-x}, e^{-y}) = \exp\left\{-R^{-1}(R(x)+R(y))\right\}.$$

Since $\tilde{F}$ has the same B as $\bar{F}_0$, we obtain from (3)

$$\tilde{F} = \tilde{G}(R^{-1}(R(x)+R(y))).$$

This shows that $\tilde{F} \sim TTE(\tilde{G} \circ R^{-1}, R)$.

(3) The last statement is proved similarly. □

In view of the previous proposition and remarks, we may rephrase our "strategy" for introducing multivariate aging notions in the specific case of TTE models as follows. Where appropriate, we restrict ourselves to the case of $\phi(u) = R(-\log u)$ convex, so that B is a copula.

(i) Prove results of this type:
The following are equivalent

- The (univariate) survival function $\bar{G}(x) := \exp\{-R(x)\}$ satisfies the (univariate) aging property P
- The Archimedean copula (5), namely the multivariate aging function B of $\bar{F}_0 \sim TTE(W_0, R)$ satisfies the dependence property Q.

(ii) Define: an exchangeable law $\bar{F}$ is multivariate-P if $B_{\bar{F}}$ is Q

Remark 13: Observe that we say that a $TTE(W, R)$ law is multivariate-P if a univariate law is P. The univariate law involved, though, is *not* its own marginal $W(R(x))$, but the marginal of the independent law with the same B, namely $\exp\{-R(x)\}$

Remark 14: For the conditionally i.i.d. model described in point 6 of Remark 11, we see from Remark 13 above that the following are equivalent:

(1) The survival function $\exp\{-\theta[R(x)+R(y)]\}$ has the bivariate aging property P, for every $\theta > 0$.
(2) The survival function

$$\int_0^\infty \exp\{-\theta[R(x)+R(y)]\}\,\Pi(d\theta)$$

is bivariate-P.

We can now define the notion of multivariate IFRA for TTE laws. It is in terms of the *positive K dependence* (PKD) concept defined as follows: an Archimedean copula $C(u,v) = \phi^{-1}(\phi(u)+\phi(v))$ is PKD if

$$v - \frac{\phi(v)}{\phi'(v)} \le v - v\log v, \quad \forall 0 < v \le 1$$

Proposition 15: *Let X, Y be i.i.d., with $\bar{G}(x) = \exp\{-R(x)\}$. Then X and Y are IFRA if and only if B is PKD. More precisely, the following are equivalent:*

- *$\phi = R \circ (-\log)$ satisfies*

$$v - \frac{\phi(v)}{\phi'(v)} \le v - v\log v, \tag{7}$$

  so that B, if it is a copula, is PKD
- *R is star shaped (i.e. $R(x)/x$ is increasing), so that $\exp\{-R(x)\}$ is IFRA.*

Proof: Again, the proof can be obtained by adapting results of Averous and Dortet-Bernadet (2000). We sketch some details, for the sake of clarity.

The function R is star-shaped iff $-R^{-1} = \log\phi^{-1}$ is star-shaped iff

$$0 \le \frac{d}{dt}\frac{\log\phi^{-1}(t)}{t}.$$

Evaluating this derivative at $v = \phi^{-1}(t)$, we see that it is positive iff

$$\frac{\phi(v)}{v\phi'(v)} - \log v > 0,$$

i.e. iff the condition (7) characterizing PKD copulas holds. □

Definition 16: Let (X, Y) be a $TTE(W, R)$ law. We say that (X, Y) is *multivariate IFRA* if B is PKD.

Remark 17: This notion of MIFRA applies only to TTE laws, since the dependence property (PKD) which is satisfied by the aging function B of the joint law with i.i.d. IFRA components, is defined only for Archimedean copulas.

Remark 18: TTE models exhibiting negative dependence and the multivariate IFRA property are easily constructed taking R star-shaped and W such that the corresponding copula (6) is negatively dependent.

4. Relations with Other Notions of Aging

We summarize the discussion above, collecting also some previous results. In particular, we see that the multivariate notions of NBU and IFR dealt with above admit a characterization in terms of stochastic comparison of residual lifetimes. We don't have an analogous characterization for the MIFRA and MPF_2 notions. In fact, even in the one-dimensional case the notion of IFRA does not have any immediate characterization in terms of residual lifetimes. As far as PF_2 is concerned, such a characterization exists (see Shaked and Shantikumar (1994)) in terms of likelihood ratio ordering:

$$f_T \quad \mathrm{PF}_2 \;\Leftrightarrow \mathcal{L}(\mathrm{T}-\mathrm{t}_1|\mathrm{T}>\mathrm{t}_1) \geq_{\mathrm{LR}} \mathcal{L}(\mathrm{T}-\mathrm{t}_2|\mathrm{T}>\mathrm{t}_2), \qquad \mathrm{t}_1<\mathrm{t}_2 \tag{8}$$

However, we do not expect that a multivariate analog of (8) hold for the MPF_2 notion, since this multivariate analog would involve densities, whereas MPF_2 as specified in Definition 10 involves level curves of survival functions.

We write $D(x, y) = \{X > x, Y > y\}$.

Proposition 19: *Let (X, Y) be an exchangeable vector with survival function $\bar{F}$. The following are equivalent:*

(A) *(X, Y) is multivariate-NBU (i.e., B is PQD)*
(B) $\mathcal{L}(X|D(0, y)) \geq_{\mathrm{ST}} \mathcal{L}(Y - y|D(0, y))$

If, furthermore, (X, Y) is TTE(W, R), then the following statement is equivalent to the previous two:

(C) *R is superadditive, so that* $\exp\{-R(x)\}$ *is NBU.*

Proof: The proof can be obtained by putting together results in this note and in Bassan and Spizzichino (1999). We write it in detail for the sake of clarity.

Let B be PQD. Then

$$\begin{aligned}\bar{F}(x,y) &= \bar{G}\left(-\log B(e^{-x},e^{-y})\right)\\ &\geq \bar{G}(-\log(e^{-x}e^{-y}) = \bar{G}(x+y) = \bar{F}(0,x+y).\end{aligned}$$

Hence,

$$\mathbb{P}\left(X \geq t|Y \geq y\right) = \frac{\bar{F}(t,y)}{\bar{F}(0,y)} \geq \frac{\bar{F}(0,t+y)}{\bar{F}(0,y)} = \mathbb{P}\left(Y - y \geq t|Y \geq y\right),$$

so that (B) holds. The converse implication is proved similarly.

In order to prove the equivalence of (A) and (C), observe once again that if $(X,Y) \sim TTE(W,R)$, then its law has the same B as the independent law $TTE(W_0,R)$. Thus, we may apply Proposition 9 and we get the desired conclusion. □

Proposition 20: *Let (X,Y) be an exchangeable vector with survival function $\bar{F}$. The following are equivalent:*

(i) $\mathcal{L}\left(X-x|D(x,y)\right) \geq_{\mathrm{ST}} \mathcal{L}\left(Y-y|D(x,y)\right), \quad x<y$

(ii) *$\bar{F}$ is Schur-concave*

(iii) $B\left(u,v\right) \geq B\left(u',v\frac{u}{u'}\right)$, *for* $v \leq u < u'$

If, furthermore, (X,Y) is TTE(W,R), then each of the following statements is equivalent to the previous ones:

(iv) *(X,Y) is multivariate-IFR (i.e. B is an LTD copula)*

(v) *R is convex, so that* $\exp\left\{-R(x)\right\}$ *is IFR.*

Proof: The equivalence of (i) and (ii) was proved in Spizzichino (1992). As far as (ii) and (iii) are concerned, notice that Schur-concavity of $\bar{F}$ is equivalent to the condition

$$\bar{F}\left(x+a,y-a\right) \geq \bar{F}\left(x,y\right), \mathrm{for x < y, a < y - x.}$$

By recalling (3), the latter becomes

$$\bar{G}\left(-\log B\left(e^{-x}e^{-a},e^{-y}e^{a}\right)\right) \geq \bar{G}\left(-\log B\left(e^{-x},e^{-y}\right)\right).$$

The latter is immediately seen to be equivalent to (iii), by letting

$$u = e^{-x}e^{-a}, \mathrm{v = e^{-y}e^{a}, u' = e^{-x}},$$

and by taking into account that $\bar{G}(-\log x)$ is increasing.

The proof of the equivalence of Schur-concavity and (v) for the case of TTE can be found e.g. in Barlow and Mendel (1992) or Marshall and Olkin (1979).

Finally, the equivalence of (iv) and (v) can be proved, again, by adapting results of Averous and Dortet-Bernadet (2000). □

Acknowledgements

We had very interesting discussions with Richard Barlow, and we are very grateful to him.

References

1. Averous J. and Dortet-Bernadet J.L. (2000), Dependence for Archimedean copulas and aging properties of their generating functions, mimeo.
2. Barlow R. E. and Mendel M. B. (1992), de Finetti-type representations for life distributions, *Journal of the American Statistical Association* **87**, 1116–1122.
3. Barlow R. E. and Mendel M. B. (1993), Similarity as a characteristic of Wear-out. In *Reliability and Decision Making* (R.E. Barlow, C.A. Clarotti, F. Spizzichino, Eds.), Chapman and Hall, London.
4. Barlow R. E and Spizzichino F. (1993), Schur-concave survival functions and survival analysis, *J. Comp. and Appl. Math.* **46**, 437-447.
5. Bassan B. and Spizzichino F. (1999), Stochastic comparison for residual lifetimes and Bayesian notions of multivariate aging. *Advances in Applied Probability* **31**, 1078–1094.
6. Bassan B. and Spizzichino F. (2000), On a multivariate notion of New Better than Used, *Proceedings of the Conference "Mathematical Methods of Reliability", Bordeaux*, 167–169
7. Joe H. (1997), *Multivariate Models and Dependence Concepts*, Chapman & Hall, London
8. Marshall A. W. and Olkin I. (1979), *Inequalities: Theory of Majorization and Its Applications*, Academic Press, New York.
9. Nelsen R. B. (1999), *An Introduction to Copulas*, Springer, New York
10. Savits T. H. (1985), A multivariate IFR distribution, *Journal of Applied Probability* **22**, 197–204.
11. Shaked M. and Shanthikumar J. G. (1994), *Stochastic orders and their applications*, Academic Press, London.
12. Spizzichino F.(1992), Reliability decision problems under conditions of ageing. In *Bayesian Statistics 4* (J. Bernardo, J. Berger, A.P. Dawid, A.F.M. Smith Eds.) Clarendon Press, Oxford, 803-811.
13. Spizzichino F. (2001), *Subjective Probability Models for Lifetimes*, CRC Press, Boca Raton.

CHAPTER 14

DEPENDENCE AND AGEING PROPERTIES OF BIVARIATE LOMAX DISTRIBUTION

C. D. Lai

Statistics, Massey University
Palmerston North, New Zealand
E-mail: C.Lai@massey.ac.nz

M. Xie

Industrial and Systems Engineering
National University of Singapore, Singapore
E-mail: mxie@nus.edu.sg

I. G. Bairamov

Department of Statistics
Ankara University, Ankara, Turkey
E-mail: Ismihan.Bayramov@science.ankara.edu.tr

The notions of dependence between two variables in a bivariate distributions are useful concepts in reliability theory and lifetime data analysis. Apart from the covariance and the correlation coefficient which are of particular interest as they give a measure of the strength of the dependence, other notions such as 'association' also become relevant in many situations. In this paper, the dependence notions and the correlation of the bivariate Lomax distribution are studied. The maximum and minimum of the correlation coefficient are also obtained. It is shown that the bivariate Lomax distribution has a reasonably wide admissible range that compares well with the Farlie-Gumbel-Morgenstern bivariate distribution with various marginals. Furthermore, some ageing properties of the Lomax distribution are investigated and the condition for monotonic aging are studied for under different bivariate ageing definitions.

1. Introduction

Bivariate distributions are useful in the study of two random variables that are dependent. This is for example the case when we are interested in the lifetime of two components in reliability analysis. Several bivariate and multivariate distributions have been proposed in the literature to model these situations. One such distribution is the bivariate Lomax distribution which has been studied extensively.

In the context of reliability, two exponential components (independent or dependent) in a given system are often being operated under a similar environmental. If the two exponentials are independent for a given environmental factor, then the bivariate Lomax of the type given by Lindley and Singpurwalla[14] would result. If two exponential components follow Gumbel's bivariate exponential law, then a more general form of the bivariate Lomax distribution occurs (Sankaran and Nair[18]). This joint distribution also arises through the characterisations based on some ageing properties such as hazard rates and mean residual life, see Roy[16], Fang and Joe[8], Lai[13], Wesolowski[19], Roy and Gupta[17], Ma[15] and Asadi[3].

There are a number of notions for dependence. Among these, positive quadrant dependence and association are mostly commonly cited ones. In this paper, some dependence properties are first shown for the bivariate Lomax distribution. Although the correlation property is well known for the special case of the bivariate Lomax distribution, this property has not been studied for the general model. In this paper, we fill up this gap by providing a correlation analysis which is important for any bivariate probability models. In particular, the maximum and minimum of the correlation coefficient are obtained and thus gives an admissible range which compares well with others such as the Farlie-Gumbel-Morgenstern bivariate distribution with different marginals.

This paper is organised as follows. After a short discussion on the bivariate Lomax distribution and model characteristics, the dependence notions and the correlation are studied. The maximum and minimum of the correlation coefficient are also obtained. It is further shown that the bivariate Lomax distribution has a reasonably wide admissible range that compares well with the Farlie-Gumbel-Morgenstern bivariate distribution with various marginals. Furthermore, some aging properties of this joint distribution are studied.

2. The Bivariate Lomax Distribution and Its Applications

The bivariate Lomax distributions have appeared in different forms, the one under the present discussion is of a general form with four parameters. The joint survival function is given by (see, for example, Sankaran and Nair[18]):

$$\bar{F}(x,y) = (1+ax+by+\theta xy)^{-c}, 0 \le \theta \le (c+1)ab : a, b, c > 0 \tag{1}$$

with probability density function

$$f(x,y) = \frac{c[c(b+\theta x)(a+\theta y)+ab-\theta]}{(1+ax+by+\theta xy)^{c+2}}. \tag{2}$$

It is well known that

$$E[X] = \frac{1}{a(c-1)}, E[Y] = \frac{1}{b(c-1)}, \ c > 1$$

and

$$\text{var}(X) = \frac{c}{(c-1)^2(c-2)a^2}, \ \text{var}(Y) = \frac{c}{(c-1)^2(c-2)b^2}, c > 2. \tag{3}$$

In order to have a well-defined bivariate Lomax distribution, we need to restrict ourselves to having $c > 2$ so that the second moments will exist.

The bivariate Lomax distribution (1) is a generalisation of the one considered by Lindley and Singpurwalla[14] and it may be derived from another bivariate distribution instead of from two independent univariate distributions.

(a) Begin with two independent gammas with scale parameters θ_1, θ_2 which also jointly have the Kibble's bivariate gamma distribution, see, Hutchinson and Lai[10](p272). The bivariate Lomax distribution would be obtained.

(b) Begin with Gumbel's bivariate distribution of the type

$$\bar{F}(x,y) = \exp\left(-\eta\left(\alpha x + \beta y + \lambda xy\right)\right).$$

Assuming that η has a gamma distribution with scale parameter m and shape parameter c, then (1) will arrive by letting $a = \alpha/m, b = \beta/m$ and $\theta = \lambda/m$. See, Sankaran and Nair[18].

Besides, the bivariate Lomax distribution may be derived through characterisations based on (i) the mean residual lives and hazard rates, (ii) the coefficient of variations of the residual lives, and (iii) truncated expectations. See Roy[16], Roy and Gupta[17] and Asadi[3].

The joint density function of (1) is

$$f(x,y) = \frac{c[c(b+\theta x)(a+\theta y)+ab-\theta]}{(1+ax+by+\theta xy)^{c+2}} \tag{4}$$

It is clear that a and b are marginal scale parameters whereas θ and c are dependence parameters. The special case $\theta = 0$ is studied in details by Hutchinson[9] having the distribution function given by

$$\bar{F}(x,y) = \frac{1}{(1+ax+by)^c}, c > 0$$

One can easily verify that the correlation coefficient of X and Y is given by

$$\rho = \operatorname{corr}(X,Y) = \frac{1}{c} \quad \text{for } c > 2 \tag{5}$$

One of our goals is to find the correlation coefficient ρ when $\theta \neq 0$ and its admissible range. First, some dependence properties will be studied.

3. Properties of Bivariate Dependence

Barlow and Proschan[4](pp142-146) discussed several notions of bivariate dependence and the relationships among them. The association and positive quadrant dependence are probably the widely referred ones and will be studied here.

3.1. *Positive (Negative) Quadrant Dependence*

Random variables X and Y are said to be positively (negatively) quadrant if

$$\bar{F}(x,y) \geq (\leq) \bar{F}_X(x)\bar{F}_Y(y) \quad \text{for all } x \geq 0, y \geq 0 \tag{6}$$

The following result gives the conditions for bivariate Lomax random variables to be positively or negatively quadrant dependent.

Proposition 1: *For the bivariate Lomax survival function, X and Y are positively (negatively) quadrant dependent if* $0 \leq \theta \leq ab$ $(ab < \theta \leq (c+1)ab)$.

Proof: We have, for $0 \leq \theta \leq (c+1)ab$, that

$$\begin{aligned}
\bar{F}(x,y) - \bar{F}_X(x)\bar{F}_Y(y) \\
&= \frac{1}{(1+ax+by+\theta xy)^c} - \frac{1}{\{(1+ax)(1+by)\}^c} \\
&= \frac{1}{(1+ax+by+\theta xy)^c} - \frac{1}{\{1+ax+by+abxy)\}^c}
\end{aligned}$$

Thus, the above difference is greater than or less than zero depending whether $0 \leq \theta \leq ab$ or $ab < \theta \leq (c+1)ab$.

Therefore X and Y are positively (negatively) quadrant dependent if $0 \leq \theta \leq ab$ ($ab < \theta \leq (c+1)ab$). We also note that positive (negative) quadrant dependence of (X,Y) implies positive (negative) correlation of (X,Y). Hence the conditions given in Proposition 1 also implied positive (negative) correlation. □

3.2. *Association*

Given two random variables X and Y; we say that Y is right tail increasing in X if $P(Y > y|X > x)$ is increasing in x for all y. (See p22 of Joe[11]). This is a positive dependence condition since large values of one variable tend to accompany large values of another, and similarly for small values.

Proposition 2: *Random variables X and Y following bivariate Lomax distribution are right tail increasing if $\theta \leq ab$ and right tail decreasing if $\theta \geq ab$. Furthermore, X and Y are associated if $\theta \leq ab$.*

Proof: Consider the bivariate Lomax distribution with the form given in (1).

$$\frac{\bar{F}(x,y)}{\bar{F}_X(x)} = \frac{(1+ax)^c}{(1+ax+by+\theta xy)^c}, 0 \leq \theta \leq (c+1)ab.$$

It follows that

$$\frac{\partial}{\partial x}\frac{\bar{F}(x,y)}{\bar{F}_X(x)} = \frac{cy(1+ax)^{c-1}(ab-\theta)}{(1+ax+by+\theta xy)^{c-1}}$$

which is a positive or a negative function in x for all y depending on whether $\theta \leq ab$ or $\theta \geq ab$. This shows that X and Y are right tail increasing if $\theta \leq ab$ and right tail decreasing if $\theta \geq ab$. (See also p142 of Barlow and Proschan[4]). Now from the same reference, it is known that right tail increasing implies association. Thus $X$and Y are associated if $\theta \leq ab$. □

4. Correlation Coefficients

For any useful bivariate model, the correlation structure ought to be reasonably simple having a workable admissible range for the correlation coefficient. The properties of the correlation coefficient of Lomax distribution will be studied in this section. Since a and b are marginal scale parameters we may assume without loss of generality that $a = b = 1$.

Theorem 3: The correlation coefficient for the bivariate Lomax distribution when $a = b = 1$ is given by

$$\begin{aligned} \rho &= \frac{(1-\theta)}{c(c-1)^2} \times F(1,2;c+1;(1-\theta)) \Big/ \frac{c}{(c-1)^2(c-2)} \\ &= \frac{(1-\theta)(c-2)}{c^2} F(1,2;c+1;(1-\theta)), 0 \leq \theta \leq (c+1) \end{aligned} \tag{7}$$

where $F(a,b;c;z)$ is the Gauss' hypergeometric function (see, for example, Chapter 15, Abramowitz and Stegun[1]).

Proof: The covariance can be expressed in term of the Hoeffding's formula:

$$\begin{aligned} \operatorname{cov}(X,Y) &= \iint [F(x,y) - F(x)F(y)]dxdy \\ &= \int_0^\infty \int_0^\infty \frac{dxdy}{(1+x+y+\theta xy)^c} - \frac{dxdy}{(1+x+y+xy)^c} \\ &= \int_0^\infty \frac{dy}{(c-1)(1+\theta y)(1+y)^{c-1}} - \int_0^\infty \frac{dy}{(c-1)(1+y)^c} \end{aligned} \tag{8}$$

This is simplified to

$$\begin{aligned} &\operatorname{cov}(X,Y) \\ &= \frac{1}{(c-1)} \int_0^\infty \frac{ydy}{(1+\theta y)(1+y)(1+y)^{c-1}} \\ &= \frac{1}{(c-1)} \int_0^1 t(1-t)^{c-2} \left(1-(1-\theta)t\right)^{-1} dt, t = \frac{y}{1+y} \\ &= \frac{(1-\theta)}{c(c-1)^2} \times F(1,2;c+1;(1-\theta)) \end{aligned} \tag{9}$$

by utilising the integral representation of the Gauss hypergeometric function. (See p558 of Abramowitz and Stegun[1]). It now follows from (3) that the correlation coefficient is as given by (7). This completes the proof of the theorem. □

Hence, it follows immediately from (8) that $\text{cov}(X,Y) \geq 0$ for $0 < \theta \leq 1$ and $\text{cov}(X,Y) < 0$ for $1 < \theta \leq c+1$.

Corollary 4: *For $a \neq 1, b \neq 1$, the correlation is*

$$\rho = \frac{(ab-\theta)(c-2)}{c^2} F\left(1,2;c+1;(1-\theta/ab)\right), 0 \leq \theta \leq (c+1)ab \qquad (10)$$

Proof: Define $X^* = aX, Y^* = bY$ and $\theta^* = \theta/(ab)$. Then

$$\text{corr}(X,Y) = \text{corr}(X^*,Y^*).$$

The latter is given by (7) with θ being replaced by θ^*. Equation (10) now follows.

□

It is now obvious that for $a,b > 0$, the covariance is positive for $0 \leq \theta \leq ab$ and negative for $ab < \theta \leq (c+1)ab$. This is not surprising as we have already verified that X and Y are positively (negatively) quadrant dependent if $0 \leq \theta \leq ab$ $(ab < \theta \leq (c+1)ab)$.

It is interesting to look at the correlation for integer value of c designated by n. Directly from (6) together with $a = b = 1$ or $ab = 1$, we have from (7) that

$$\begin{aligned}
\text{cov}\,(X,Y) &= \int_0^\infty \frac{dy}{(n-1)(1+\theta y)(1+y)^{n-1}} - \int_0^\infty \frac{dy}{(n-1)(1+y)^n} \\
&= \int_0^\infty \frac{1}{(n-1)} \left(\sum_{i=1}^{n-1} \frac{-\theta^{n-1-i}}{(\theta-1)^{n-i}(1+y)^i} + \frac{\theta^{n-1}}{(\theta-1)^{n-1}(1+\theta y)} \right) dy - \int_0^\infty \frac{dy}{(n-1)(1+y)^n} \\
&= \frac{\theta^{n-2}}{(n-1)(\theta-1)^{n-1}} \log\left(\frac{1+\theta y}{1+y} \right)\Bigg|_0^\infty - \sum_{i=2}^{n-1} \frac{-\theta^{n-1-i}}{(n-1)(i-1)(\theta-1)^{n-i}} - \frac{1}{(n-1)^2} \\
&= \frac{\theta^{n-2}}{(n-1)(\theta-1)^{n-1}} \log\theta - \sum_{i=2}^{n-1} \frac{-\theta^{n-1-i}}{(n-1)(i-1)(\theta-1)^{n-i}} - \frac{1}{(n-1)^2}
\end{aligned} \qquad (11)$$

and thus

$$\begin{aligned}
&\text{corr}\,(X,Y) \\
&= \left(\frac{\theta^{n-2}}{(n-1)(\theta-1)^{n-1}} \log\theta - \sum_{i=2}^{n-1} \frac{\theta^{n-1-i}}{n(i-1)(\theta-1)^{n-i}} - \frac{1}{(n-1)^2} \right) \Big/ \frac{n}{(n-1)^2(n-2)} \\
&= \left(\frac{\theta^{n-2}}{(\theta-1)^{n-1}} \log\theta - \sum_{i=2}^{n-1} \frac{\theta^{n-1-i}}{(i-1)(\theta-1)^{n-i}} - \frac{1}{n-1} \right) \Big/ \frac{n}{(n-1)(n-2)}, n \geq 3
\end{aligned} \qquad (12)$$

4.1. *Some Special Cases*

As an example, we now look at some special, but interesting cases. Note that in the first case, the correlation does not exist because of equation (3). However, the following direct calculation shows that the covariance still exists.

(i) $c = n = 2$.
It follows from (8) that

$$\begin{aligned}\operatorname{cov}(X,Y) &= \int_0^\infty \frac{dy}{(c-1)(1+\theta y)(1+y)^1} - \int_0^\infty \frac{dy}{(c-1)(1+y)^2}\\ &= \int_0^\infty \frac{dy}{(\theta-1)}\left[\frac{\theta}{1+\theta y} - \frac{1}{1+y}\right] - \int_0^\infty \frac{dy}{(1+y)^2}\\ &= \frac{\log\theta}{(\theta-1)} - 1.\end{aligned}$$

Table 1. Correlation coefficients ρ for various values of θ and c

	c					
θ	4.000	5.000	6.000	7.000	8.000	9.000
0.1	0.196	0.165	0.141	0.123	0.108	0.097
0.2	0.157	0.137	0.119	0.105	0.093	0.084
0.3	0.127	0.137	0.119	0.105	0.093	0.084
0.4	0.100	0.092	0.082	0.073	0.065	0.059
0.5	0.079	0.073	0.065	0.059	0.053	0.048
0.6	0.060	0.056	0.050	0.046	0.041	0.038
0.7	0.043	0.040	0.037	0.033	0.030	0.028
0.8	0.027	0.026	0.024	0.022	0.020	0.018
0.9	0.013	0.012	0.011	0.010	0.010	0.009
1.0	0.000	0.000	0.000	0.000	0.000	0.000
1.5	-0.052	-0.052	-0.049	-0.046	-0.042	-0.039
2.0	-0.091	-0.092	-0.088	-0.083	-0.078	-0.073
2.5	-0.121	-0.124	-0.120	-0.114	-0.108	-0.102
3.0	-0.146	-0.151	-0.147	-0.141	-0.134	-0.127
3.5	-0.167	-0.174	-0.171	-0.164	-0.157	-0.149
4.0	-0.184	-0.193	-0.191	-0.185	-0.178	-0.170
4.5	-0.200	-0.211	-0.210	-0.204	-0.197	-0.188
5.0	-0.213	-0.226	-0.226	-0.221	-0.214	-0.205
6.0		-0.253	-0.255	-0.251	-0.244	-0.235
6.5			-0.268	-0.264	-0.257	-0.249
7.0			-0.279	-0.276	-0.269	-0.261
7.5				-0.287	-0.281	-0.273
8.0				-0.297	-0.291	-0.284

Thus, the covariance exist for $c = 2$ even though the correlation does not exist since the marginal variance does not exist for this value of c, see equation (3).

(ii) $c > 2$

Both the covariance and the correlation will exist. Consider $c = n = 3$. It follows from (11) that

$$\begin{aligned}\rho = \operatorname{corr}(X, Y) &= \left[\tfrac{1}{2(\theta-1)^2}\theta\log\theta - \tfrac{1}{2(\theta-1)} - \tfrac{1}{4}\right] / \tfrac{3}{4}\\ &= \left[\tfrac{2}{3(\theta-1)^2}\theta\log\theta - \tfrac{2}{3(\theta-1)} - \tfrac{1}{3}\right]\end{aligned}$$

For the sake of some comparison, Table 1 provides an indicative trend of the values of the correlation for $c = 4, ..., 9$.

The values in Table 1 indicates that for a given c, the correlation ρ decreases as θ increases. However, it does not decrease uniformly over c, for example, $\rho = -0.213(c{=}4, \theta = 5)$ and $\rho = -0.205(c{=}8, \theta = 5)$.

4.2. *The Admissible Range of the Correlation Coefficient*

As observed from the paragraph that precedes Corollary 4, the covariance is negative for $1 < \theta \leq c + 1$ and positive for $0 \leq \theta \leq 1$ assuming $ab = 1$. It follows immediately from (8) that for given c, the correlation ρ is a monotonely decreasing function of $\theta \in [0, c{+}1]$. It would be of much interest to establish the admissible range of the correlation coefficient ρ.

Theorem 5: For given c, the value of ρ lies in the following interval

$$-\frac{(c-2)}{c}F(1, 2; c+1; -c) \leq \rho \leq 1/c. \tag{13}$$

Proof: Since ρ is a decreasing in θ, it is now obvious that the maximum of ρ occurs when $\theta = 0$. Equation (5) gives

$$\rho_{\max} = 1/c, \quad c > 0 \tag{14}$$

The minimum would occur when $\theta \to c + 1$. That is, for a given c, the minimal correlation is

$$\rho_{\min} = -\frac{(c-2)}{c}F(1, 2; c+1; -c). \tag{15}$$

□

To find the admissible interval for ρ we need to evaluate the supremum and the infimum of ρ over all possible values of c. It follows from (13) above that $\sup_{c>2}\rho=0.5$ so that $\rho < 0.5$. This value is also the least upper bound of ρ.

Since the function $\frac{(c-2)}{c}F(1,2;c+1;-c)$ is an increasing function of c, the maximal negative correlation is found by letting c tend to infinity, i.e.,

$$\lim_{c\to\infty} -\frac{(c-2)}{c}F(1,2;c+1;-c) \to -0.403$$

obtained by numerical computation via (11). Thus the admissible range for ρ is $(-0.403, 0.5)$.

This reasonably wide admissible range compares well with the well known Farlie-Gumbel-Morgenstern bivariate distribution having the ranges of correlation given by (i) $-\frac{1}{3}$ to $\frac{1}{3}$ for uniform marginals, (ii) $-\frac{1}{4}$ to $\frac{1}{4}$ for exponential marginals and (iii) $-\frac{1}{\pi}$ to $\frac{1}{\pi}$, see, for example, Hutchinson and Lai[10](Section 5.2.5).

5. Some Ageing Properties

Several reliability properties are given in Sankaran and Nair[18]. In this section, we investigate some additional properties pertaining to reliability analysis. Several bivariate aging definitions are available and the condition for monotonicity is studied.

5.1. *Bivariate Failure Rate of Basu*

There are several versions of a bivariate failure rate. One such version was defined by Basu[5]:

$$r(x,y) = \frac{f(x,y)}{\bar{F}(x,y)} \tag{16}$$

Thus, the bivariate failure rate for the bivariate Lomax in the sense of Basu is:

$$r(x,y) = \frac{f(x,y)}{\bar{F}(x,y)} = \frac{c[c(b+\theta x)(a+\theta y)+ab-\theta]}{(1+ax+by+\theta xy)^2}, 0 \le \theta \le (c+1)ab \tag{17}$$

If $\theta = ab$, $r(x,y)$ becomes the product of two marginal failure rate functions.

We say that F is BIFR (BDFR) if $r(x, y)$ defined by (16) is increasing (decreasing) in both x and y. Note that since $\frac{\partial r(x,y)}{\partial x} = \frac{\partial \log r(x,y)}{\partial x} \times r(x, y)$, it follows that

$$\frac{\partial \log r(x, y)}{\partial x} \geq 0 \Rightarrow \frac{\partial r(x, y)}{\partial x} \geq 0.$$

Proposition 6: *The bivariate Lomax distribution of (1) is BDFR for* $0 \leq \theta \leq ab$.

Proof: For the bivariate Lomax distribution under consideration,

$$\log r(x, y) = \log c + \log[c(b + \theta x)(a + \theta y) + ab - \theta] - 2\log(1 + ax + by + \theta xy)$$

$$\frac{\partial \log r(x, y)}{\partial x} = \frac{c(a + \theta y)\theta}{[c(b + \theta x)(a + \theta y) + ab - \theta]} - 2 \times \frac{(a + \theta y)}{(1 + ax + by + \theta xy)}.$$

For $\theta \leq ab$, $[c(b + \theta x)(a + \theta y) + ab - \theta] \geq [c(b + \theta x)(a + \theta y)]$. Hence

$$\begin{aligned} &\frac{c(a+\theta y)\theta}{[c(b+\theta x)(a+\theta y)+ab-\theta]} - 2 \times \frac{(a+\theta y)}{(1+ax+by+\theta xy)} \\ &< \frac{(a+\theta y)\theta}{[(b+\theta x)(a+\theta y)]} - 2 \times \frac{(a+\theta y)}{(1+ax+by+\theta xy)} \\ &= \frac{\theta}{(b+\theta x)} - 2 \times \frac{(a+\theta y)}{(1+ax+by+\theta xy)} \\ &\leq \frac{\theta}{(b+\theta x)} - 2 \times \frac{(a+\theta y)}{(1+ax)(1+by)} \end{aligned}$$

For $\theta = 0$, $\frac{\theta}{(b+\theta x)} - 2 \times \frac{(a+\theta y)}{(1+ax)(1+by)} < 0$

For $\theta = ab$, $\frac{\theta}{(b+\theta x)} - 2 \times \frac{(a+\theta y)}{(1+ax)(1+by)} = \frac{a}{1+ax} - \frac{2a}{1+ax} < 0$.

We also note that $\frac{\theta}{(b+\theta x)} - 2 \times \frac{(a+\theta y)}{(1+ax)(1+by)}$ is a decreasing function of θ. Therefore

$\frac{\theta}{(b+\theta x)} - 2 \times \frac{(a+\theta y)}{(1+ax)(1+by)} < 0$ for all $0 \leq \theta \leq ab$.

Hence, the proof is completed. □

5.2. *Bivariate Ageing Property According to Johnson and Kotz*

According Johnson and Kotz[12], a bivariate hazard (failure) rate is defined by the vector:

$$\mathbf{h}(x, y) = (h(x, y)_1, h_2(x, y))$$

where

$h_1(x,y) = -\frac{\partial \log \bar{F}(x,y)}{\partial x}$ and $h_2(x,y) = -\frac{\partial \log \bar{F}(x,y)}{\partial y}$.

A distribution is called a bivariate IHR (DHR) if $h_1(x,y)$ and $h_2(x,y)$ are increasing (decreasing) function of x and y, respectively.

Proposition 7: *The bivariate Lomax is a bivariate DHR in the sense of Johnson and Kotz*[12].

Proof: It is easy to verify that

$-\frac{\partial \log \bar{F}(x,y)}{\partial x} = \frac{c(a+\theta y)}{1+ax+by+\theta xy}$ and $-\frac{\partial \log \bar{F}(x,y)}{\partial y} = \frac{c(b+\theta x)}{1+ax+by+\theta xy}$ are both decreasing function with respect to x and y. Thus the bivariate Lomax is a bivariate DHR. □

Note that for univariate distributions, IFR (DFR) is the same as IHR (DHR) although the former is more frequently used in the literature

5.3. *Bivariate Ageing Property According to Esary and Marshall*

Another definition of biviate ageing is given by Esary and Marshall[6]. A bivariate distribution is said to have a BIFRA (bivariate increasing failure rate in average) if $\bar{F}^{\alpha}(x,y) \geq \bar{F}(\alpha x, \alpha y), 0 < \alpha \leq 1$.

Proposition 8: *For $\theta = 0$, (X,Y) is BIFRA. In general, $0 \leq \theta < \theta_0$, F is BIFRA and for $\theta_0 < \theta \leq ab$,$F$ is BDFRA, where $\theta_0 \in (0, ab)$.*

Proof: Consider the case when $\theta = 0$, i.e.,$\bar{F}^{\alpha}(x,y) = (1+ax+by)^{-c\alpha}, 0 < \alpha \leq 1$. By taking logarithm on both sides and noting that $\log(1+\alpha u) > \alpha \log(1+u)$, for $0 < \alpha \leq 1$ and $u \geq 0$, we have that

$$\bar{F}^{\alpha}(x,y) = (1+ax+by)^{-c\alpha} < (1+\alpha ax+\alpha by)^{-c} = \bar{F}(\alpha x, \alpha y),$$

if $0 < \alpha \leq 1$ for all $x \geq 0, y \geq 0$.

This implies that when $\theta = 0$, (X,Y) is BIFRA (bivariate increasing failure rate in average) given by Esary and Marshall (1979).

On the other hand, when $\theta = ab$

$$\bar{F}^{\alpha}(x,y) = (1+ax+by+abxy)^{-c\alpha}, 0 < \alpha \leq 1$$
$$= (1+ax)^{-c\alpha}(1+by)^{-c\alpha} \geq (1+\alpha ax)^{-c}(1+\alpha by)^{-c} = \bar{F}(\alpha x, \alpha y)$$

showing that (X, Y) is BDFRA This is not surprising since in this case, X and Y are independent and the marginals are both Lomax which is DFRA.

In general, for all $0 \leq \theta \leq ab$, we can find a value of θ, say $\theta = \theta_0 \in (0, ab)$ such that for $\theta < \theta_0$, F is BIFRA and for $\theta_0 < \theta \leq ab$,$F$ is BDFRA. This is because both functions $\bar{F}^\alpha$ and $\bar{F}$ are monotonic in θ and hence θ_0 is the solution to the functional equation:

$$\begin{aligned}\bar{F}^\alpha(x, y) &= (1 + ax + by + \theta xy)^{-c\alpha} \\ &= \bar{F}(\alpha x, \alpha y) \\ &= (1 + \alpha ax + \alpha by + \alpha^2\theta xy)^{-c}, \quad 0 < \alpha \leq 1.\end{aligned}$$

Or equivalently, $\theta = \theta_0$ is the value for which the relation holds:

$$\frac{1 + \alpha ax + \alpha by + \alpha^2\theta xy}{(1 + ax + by + \theta xy)^\alpha} = 1 \quad . \qquad \square$$

6. Discussions

In this paper, some further properties of bivariate Lomax distribution is studied with respect to the dependence, correlation coefficient and bivariate aging. The maximum and minimum of the correlation coefficient obtained can be used as an indicator when choosing a suitable bivariate model. In fact, the bivariate Lomax distribution has a reasonably wide admissible range compared with the Farlie-Gumbel-Morgenstern bivariate distribution. We also note that apart from the correlation coefficient, there are some other measures of dependence. For example, Fan *et al.*[7] proposed two nonparametric summary measures of dependence. Hence, this warrants further study and investigation of this model in data analysis.

References

1. M. Abramowitz and I .A. Stegun, *Handbook of Mathematical Functions* (Dover, New York, 1964).
2. M. Asadi, Multivariate distributions characterized by a relationship between mean residual life and hazard rate, *Metrika* **49**, 121-126 (1999).
3. M. Asadi, Some general characterizations of the bivariate Gumbel distribution and the bivariate Lomax distribution based on truncated expectations, *Journal of Multivariate Analysis* **67**, 190-202 (1998).
4. R.E. Barlow and F. Proschan, *Statistical Theory of Reliability and Life Testing: Probability Models* (To Begin with, Silver Spring, MD, 1981).

5. A.P. Basu, Bivariate failure rate, *Journal of the American Statistical Association* **66**, 103-104 (1971).
6. J.D. Esary and A.W. Marshall, Multivariate distributions with increasing hazard rate average, *Annals of Probability* **7**, 359-370 (1979).
7. J.J. Fan, R.L. Prentice and L. Hsu, A class of weighted dependence measures for bivariate failure time data, *Journal of the Royal Statistical Society, Series B* **62**(1), 181-190 (2000).
8. Z. Fang and H. Joe, Further developments on some dependence orderings for continuous bivariate distributions, *Annals of the Institute of statistical Mathematics* **44**(3), 501-517 (1992).
9. T.P. Hutchinson, Compound gamma bivariate distributions, *Metrika* **28**, 263-271 (1981).
10. T.P. Hutchinson and C.D. Lai, *Continuous Bivariate Distributions, Emphasizing Applications* (Rumsby Scientific, 1990).
11. H. Joe, *Multivariate Models and Dependence Concepts* (Chapman and Hall, London, 1997).
12. N.L. Johnson and S. Kotz, A vector multivariate hazard rate, *Journal of Multivariate Analysis* **5**, 53-66 (1975).
13. C.D. Lai, Tests of univariate and bivariate stochastic aging, *IEEE Transactions on Reliability* **R43**(2), 233-241 (1994).
14. D.V. Lindley and N.D. Singpurwalla, Multivariate distribution for the life lengths of components of a system sharing a common environment, *Journal of Applied Probability* **23**, 418-431 (1986).
15. C. Ma, Multivariate survival functions characterized by the constant product of mean remaining lives and hazard rates, *Metrika* **44**, 71-83 (1996).
16. D. Roy, A Characterization of Gumball's bivariate exponential and Lindley and Singpurwalla's bivariate Lomax distributions, *Journal if Applied Probability* **27**, 886-891 (1989); Correction, **28**, 736.
17. D. Roy and R.P. Gupta, Bivariate extension of Lomax and finite range distributions through characterization approach, *Journal of Multivariate Analysis* **59**, 22-33 (1996).
18. P.G. Sankaran and N.U. Nair, A bivariate Pareto model and its applications to reliability, *Naval Research Logistics* **40**, 1013-1020 (1993).
19. J. Wesolowski, Bivariate distributions via a Pareto conditional distribution and a regression function, *Annals of the Institute of Statistical Mathematics* **47**(1), 177-183 (1995).

CHAPTER 15

PHYSICAL FOUNDATIONS FOR LIFETIME DISTRIBUTIONS

John F. Shortle

Dept. of Systems Engineering, George Mason University
4400 University Dr., MS 4A6, Fairfax, VA 22030, U.S.A.
E-mail: jshortle@gmu.edu

Max B. Mendel

Boston Fund Services
40 Summer Street, Salem, MA 01970, U.S.A.
E-mail: max@gacf.com

This paper lays a foundation for deriving physically meaningful lifetime distributions. We start with the physical structure of the underlying lifetime space and argue that Euclidean space is not a correct representation. We propose an alternative physical structure for this space. This structure is important since it determines which quantities defined on the space are physically meaningful and which are not. We illustrate by giving several familiar quantities associated with lifetimes and showing that some are physical, but others are not. Using the physical quantities, we derive a no-aging characterization for lifetime distributions based on the idea of a usage policy.

1. Introduction

The purpose of this paper is to lay a foundation for deriving physically meaningful lifetime distributions. It is generally better to *derive* distributions from physical principles rather than simply assuming them. A well known example is the exponential distribution, which can be derived from the property that an item does not age (for example, Barlow and Proschan[2]).

Many other examples of derived lifetime distributions exist. For example, the Marshall-Olkin[5] bivariate exponential distribution can be derived from a bivariate version of the no-aging property (e.g., Barlow and Proschan[2]). Barlow and Mendel[1] have given a different generalization of the no-aging property for a finite set of N similar items (items which are exchangeable, but not necessarily independent). From this, they derive a finite version of the exponential distribution. Similar families of distributions have been derived for aging items (e.g., Barlow and Mendel[1] and Spizzichino[7]).

In order to lay a foundation for deriving physically meaningful lifetime distributions, we need to start with the correct representation of the underlying sample space. Typically, the space of lifetimes is represented as Euclidean space - that is, a vector space of real numbers with the Euclidean metric. Sec. 2 argues that this representation is physically incorrect and that another differential manifold is more appropriate.

The distinction is subtle, but important, since the physical structure of the underlying sample space directly implies which objects defined on the space are physically meaningful and which are not. In particular, Marshall's hazard gradient[4] is not physically meaningful. Sec. 3 gives examples of other familiar lifetime quantities which *are* physically meaningful. We refer to these objects as *physical invariants.*

Physical invariants are the set of mathematical objects from which to derive meaningful lifetime distributions. Sec. 4 illustrates concepts by deriving lifetime distributions based on these quantities. In particular, we give a new characterization of the no-aging property based on the idea of a usage policy. From this, we derive a finite version of the Weibull distribution.

2. Physical Characterizations of Lifetime Spaces

This section argues from a physical perspective that Euclidean space is an incorrect representation for the space of lifetimes and that another differential manifold is more appropriate.

To motivate doing this, we first consider what it means to make a physical statement about lifetimes. Intuitively, two people who measure lifetimes in different units should be able to make the same physical statements and agree (even though the statements are expressed in different units). The following example shows how this can fail.

Example: Marshall's hazard gradient[4] $\left(-\nabla \log \bar{F}(\mathbf{x})\right)$ is a multivariate generalization of the classical hazard (or failure) rate (e.g., Barlow and Proschan[2]). Intuitively, the hazard gradient points in the direction of most likely failure. However, as this example shows, this direction has no physical basis.

Let $\mathbf{X}$ be the lifetimes of N items and let $\mathbf{Y}$ be the lifetimes in different units, where $Y_i = \sqrt{X_i}$. For instance, X_i could be the *time* when an item fails and Y_i could be the amount of *wear* on an item when it fails. Suppose the lifetimes are i.i.d. exponentials - that is, the survival distribution is $\bar{F}_{\mathbf{X}}(\mathbf{x}) = \exp(-\sum x_i)$. Then, $\bar{F}_{\mathbf{Y}}(\mathbf{y}) = \exp(-\sum y_i^2)$. The hazard gradients in both sets of coordinates are:

$$-\nabla \log \bar{F}_{\mathbf{X}}(\mathbf{x}) = (1, 1, \cdots, 1), \quad -\nabla \log \bar{F}_{\mathbf{Y}}(\mathbf{y}) = (2y_1, 2y_2, \cdots, 2y_N).$$

Suppose that two items have survived to time $\mathbf{x} = (4, 0)$, which corresponds to $\mathbf{y} = (2, 0)$. Then,

$$-\nabla \log \bar{F}_{\mathbf{X}}(4, 0) = (1, 1), \quad -\nabla \log \bar{F}_{\mathbf{Y}}(2, 0) = (4, 0).$$

A person measuring lifetime in $\mathbf{X}$ coordinates would say that using both items equally accumulates hazard most quickly. A person measuring lifetime in $\mathbf{Y}$ coordinates would say that using only the first item (and not the second) accumulates hazard most quickly. *These are two physically inconsistent statements regarding accumulation of hazard.*

The basic problem is that the hazard gradient is based on direction and distance - ideas which exist in Euclidean space, but not in the physical space of lifetimes, as we will show. Euclidean space $\mathbb{E}^N$ is the vector-space of real numbers $\mathbb{R}^N$ combined with the Euclidean inner product $\langle \cdot, \cdot \rangle$. The inner product gives notions of distance and angle: $\|\mathbf{x}\| = \langle \mathbf{x}, \mathbf{x} \rangle$ and $\cos \theta = \langle \mathbf{x}, \mathbf{y} \rangle / (\|\mathbf{x}\| \|\mathbf{y}\|)$.

Let $\mathcal{L}^N$ be the space of possible lifetimes for N items. We argue that $\mathbb{E}^N$ is not a good representation for $\mathcal{L}^N$ for two reasons:

(1) $\mathcal{L}^N$ has a preferred orientation for its axes.
(2) $\mathcal{L}^N$ has no natural notion of distance.

Regarding the first point, observe that Euclidean space is invariant under rotations, since rotations preserve the value of the inner product. Another way of saying this is that there is no preferred orientation for the axes. When drawing a topological map, for example, you can pick any direction to be the x-axis of the map and still draw a physically valid map.

$\mathcal{L}^N$, on the other hand, cannot be rotated in the same way. Consider the two-item lifetime space $\mathcal{L}^2$, and rotate it by 45 degrees. The x-axis of the rotated space measures units of:

$$\sqrt{2}/2 \text{ lifetime units of item one minus } \sqrt{2}/2 \text{ lifetime units of item two.}$$

Although we can do this rotation in a technical sense, there is no *physical* reason to do so.

The second point is that $\mathcal{L}^N$ has no natural notion of distance. First, consider two items which fail after 3 and 4 hours. The "length" of this point is $\sqrt{3^2+4^2} = 5$ hours. This has little physical meaning: 5 hours is not the time it takes for both items to fail; nor is it the time it takes for the last item to fail. The only physical basis for "5 hours" comes from drawing the point (3,4) on a piece of paper (a natural Euclidean space), measuring the length of the point, and converting back to units of time.

Furthermore, "distance" in $\mathcal{L}^N$ does not behave correctly under a change of units. For example, consider two points in $\mathcal{L}^2$: $(9,9)$ and $(1,16)$. Now consider the change of units $Y_i = \sqrt{X_i}$. Observe that the Euclidean metric does not preserve orderings of distance under transformations between these units. That is,

$$\|(9,9)\| < \|(1,16)\|, \quad \text{but} \quad \|(\sqrt{9},\sqrt{9})\| > \|(\sqrt{1},\sqrt{16})\|.$$

The fact that lifetime spaces are not physical Euclidean spaces is an important point, because it follows immediately that the hazard gradient is not a meaningful quantity. A vector gradient gives the direction of greatest increase, but direction cannot be defined without the Euclidean metric.

In summary, we can characterize the physical structure of a space by the transformations that leave the space invariant. For $\mathbb{E}^N$, these are translations and rotations. For $\mathcal{L}^N$, these are changes of units of the individual items. This is because physical properties about lifetimes should not depend on the units used to measure lifetimes. Specifically, invariant transformations are smooth, increasing mappings ψ_i from one lifetime to another: $Y_i = \psi_i(X_i)$.

Space	Invariant Transformations
$\mathbb{E}^N$	Rotations, translations
$\mathcal{L}^N$	Item-wise change of units

In the language of differential geometry, the correct representation for the space of lifetimes is a collection of fiber bundles (for example, see Schutz[6]). Intuitively, this means that for each value x_1 of the lifetime of the first item, there is a line of values for the lifetime of the second item; for each pair of lifetimes (x_1, x_2), there is a line of values for the lifetime of the third item; and so on. The next section describes how we can define physically meaningful quantities in $\mathcal{L}^N$ given this structure.

3. Physical Invariants for Lifetime Distributions

This section gives several examples of physical invariants in the space of lifetimes. These objects are important since they are the mathematical set of objects from which to derive physically meaningful distributions.

Physical invariants are objects which do not change when invariant transformations are applied to the underlying space. For $\mathcal{L}^N$, these transformations are changes of units for individual lifetimes. Intuitively, two people who measure lifetimes in different units should be able to make the same physical statements about lifetimes and agree.

Survival Distribution and Hazard Potential: Two important invariant quantities are the survival distribution $\bar{F} : \mathcal{L}^N \to \mathbb{R}$ and the hazard potential $H : \mathcal{L}^N \to \mathbb{R}$.

$$\bar{F}_{\mathbf{X}}(\mathbf{x}) = P(X_1 \geq x_1, \cdots, X_N \geq x_n),$$

$$H_{\mathbf{X}}(\mathbf{x}) = -\log \bar{F}(\mathbf{x}).$$

The subscript $\mathbf{X}$ denotes the choice of units. Where obvious, we drop the subscript and write, $\bar{F}(\mathbf{x})$ for $\bar{F}_{\mathbf{X}}(\mathbf{x})$.

$\bar{F}$ is a physical invariant because an item-wise change of variables $Y_i = \psi_i(X_i)$ leaves the value of $\bar{F}$ unchanged. That is,

$$\bar{F}_{\mathbf{X}}(\mathbf{x}) = P(X_1 \geq x_1, \cdots, X_N \geq x_n) =$$

$$P(\psi_1(X_1) \geq \psi_1(x_1), \cdots, \psi_N(X_N) \geq \psi_N(x_n)) = \bar{F}_{\mathbf{Y}}(\mathbf{y}).$$

Similarly, H is an invariant since it is defined directly from $\bar{F}$.

Note that $\bar{F}$ and H are *not* invariants in $\mathbb{E}^N$, since $\bar{F}$ changes values when the space is rotated. That is, if $\mathcal{R}$ is a rotation matrix and $\mathbf{Y} = \mathcal{R}\mathbf{X}$

is the rotated vector of lifetimes, then $\bar{F}_{\mathbf{Y}}(\mathcal{R}\mathbf{x}) \neq \bar{F}_{\mathbf{X}}(\mathbf{x})$.[a]

Usage Policy: Here, we introduce a physical invariant called a usage policy. Intuitively, a usage policy specifies how items are used over time. An example is an accelerated life test in which an item is "used" faster than clock time. The following definition extends this idea to multiple items.

Specifically, a usage policy is an increasing function $\mu : \mathbb{R} \to \mathcal{L}^N$. In coordinates, the usage policy is written:

$$\mu_{\mathbf{X}}(t) = (x_1(t), x_2(t), \cdots, x_N(t)), \tag{1}$$

where $\mathbf{X}$ indicates the choice of lifetime units. For example, if the lifetimes are in units of hours, then $\mu_{\mathbf{X}}(4) = (1, 3)$ means that after four time units, item 1 has been used for one hour and item 2 has been used for three hours. Choosing $x_i(t) = t$ corresponds to using an item at the same rate as clock time. Geometrically, μ is a path in $\mathcal{L}^N$. Note that μ is an invariant since a change of coordinates $Y_i = \psi_i(X_i)$ on $\mathcal{L}^N$ gives

$$\mu_{\psi(\mathbf{X})}(t) = \mu_{\mathbf{Y}}(t) = (\psi_1(x_1(t)), \cdots, \psi_N(x_N(t))) = \psi(\mu_{\mathbf{X}}(t)).$$

Hazard Rate: Together, H and μ act as a function from $\mathbb{R}$ to $\mathbb{R}$ mapping time t to the hazard accumulated at that time:

$$H \circ \mu : \mathbb{R} \to \mathcal{L}^N \to \mathbb{R}.$$

The hazard rate is the rate at which total hazard is accumulated:

$$\dot{H} \equiv \frac{d(H \circ \mu)}{dt}.$$

In coordinates,

$$\dot{H}(t) = -\frac{1}{\bar{F}(\mathbf{x})} \sum_{i=1}^{N} \frac{\partial \bar{F}(\mathbf{x})}{\partial x_i} \dot{x}_i(t),$$

where $x_i(t)$ is the usage policy for item i (Eq. 1), and $\mathbf{x}$ is evaluated at $\mathbf{x} = \mu(t)$. It can be shown that $\dot{H}(t)$ is independent of the choice of coordinates for $\mathbf{X}$. The hazard rate for item i is

$$h_i(t) \equiv -\frac{1}{\bar{F}(\mathbf{x})} \frac{\partial \bar{F}(\mathbf{x})}{\partial x_i} \dot{x}_i(t). \tag{2}$$

[a]That is, $F_{\mathbf{X}}(\mathbf{x})$ measures the probability of the rectangle $(X_1 \geq x_1, \cdots, X_N \geq x_N)$. $F_{\mathbf{Y}}(\mathbf{y})$ measures the probability of the rectangle $(Y_1 \geq y_1, \cdots, Y_N \geq y_N)$. These two rectangles are not the same set of events, since the first rectangle in $\mathbf{X}$-space is a tilted rectangle in $\mathbf{Y}$-space - not aligned horizontally.

The hazard rate defined here should not be confused with the classical failure rate (Barlow and Proschan[2]), which is sometimes also referred to as the hazard rate. Here, h_i depends on the lifetimes of *all* items, not just item i. Only when lifetimes are independent and measured in clock time ($x_i(t) = t$) does the hazard rate in Eq. 2 match the classical failure rate.

Lemma: $h_i(t) = f(x_i)/\bar{F}(x_i)$ *if and only if* $x_i(t) = t$ *and the lifetimes* X_i *are independent and identically distributed.*

The hazard rate is also closely related to the hazard gradient:

$$-\nabla \log \bar{F}(\mathbf{x}) = \left(-\frac{1}{\bar{F}(\mathbf{x})}\frac{\partial \bar{F}(\mathbf{x})}{\partial x_1}, \cdots, -\frac{1}{\bar{F}(\mathbf{x})}\frac{\partial \bar{F}(\mathbf{x})}{\partial x_N}\right).$$

Note that the hazard rate components h_i (Eq. 2) are coordinate invariant, while the components of the hazard gradient are not. The difference is the factor of $\dot{x}_i(t)$ in Eq. 2. In a change of variables, the Jacobian of this term exactly cancels the Jacobian of $\partial \bar{F}(\mathbf{x})/\partial x_i$.

4. Models for No-Aging

To illustrate ideas, this section derives a class of lifetime distributions using the invariant quantities given in the previous section.

The basic idea we use is to apply a multivariate version of the no-aging property to usage policies. Recall that usage policies can be used to describe accelerated life tests. The theorem characterizes items which do not age in the accelerated time frame, but do age with respect to usual clock time. In other words, if time is measured with respect to the use of each item, then the items do not age. This theorem generalizes a no-aging theorem by Chick and Mendel[3]. The two are equivalent when $\mu(t) = t$.

Theorem 1: Let $\mu(t) : \mathbb{R} \to \mathbb{R}$ be an increasing function (a usage policy for a single item). Let $(X_1, \cdots, X_N)$ be the lifetimes (measured in clock time t) of N similar (exchangeable) items. The following are equivalent:

(1) $P(X_i \geq x_i + t_i | \mathbf{X} \geq \mathbf{x}) = P(X_j \geq x_j + t_j | \mathbf{X} \geq \mathbf{x})$ if

$$\mu(x_i + t_i) - \mu(x_i) = \mu(x_j + t_j) - \mu(x_j).$$

(2) $\bar{F}_{\mathbf{X}}$ is a function of λ_N, where

$$\lambda_N = \frac{N}{\sum_{i=1}^{N} \mu(x_i)}.$$

(3) Let $\bar{F}_{\mathbf{X}_n}(\mathbf{x}_n)$ be the survival function for a subset n of the items, for $0 < n < N$. $\bar{F}_{\mathbf{X}_n}(\mathbf{x}_n)$ is a $\mu(\cdot)$-isotropic survival function:

$$\bar{F}_{\mathbf{X}_n}(\mathbf{x}_n) = \int \left(1 - \frac{\lambda_N}{N}\sum_{i=1}^{n}\mu(x_i)\right)^{N-1} P(d\lambda_N). \tag{3}$$

Remarks:

(1) This is a generalization of the memoryless property. Two events which increase the usage of two different items by the same amount are equally likely, regardless of the current usage of the items.
(2) The survival function depends on individual lifetimes only through the total (or average) usage of all items. This is the no-aging property with respect to usage.
(3) This is a de Finetti-type representation for the family of $\mu(\cdot)$-isotropic survival functions (distributions for which lifetimes $\mathbf{x}_N$ with the same total usage $\sum \mu(x_i)$ are equally likely). Conditioning on the average usage λ_N of N items, the representation shows the survival function of a subset n of these items.

Example: (Also given in Barlow and Mendel[1].) If $\mu(t) = t^\alpha$, then Theorem 1 characterizes a finite version of the Weibull distribution. That is, as $N \to \infty$, $\bar{F}_{\mathbf{X}_n}$ (characterization 3) goes to a mixture of i.i.d. Weibulls, with $\lambda = \lim_{N\to\infty} \lambda_N$:

$$\bar{F}_{\mathbf{X}_n}(\mathbf{x}_n) = \int \prod_{i=1}^{n} \exp\left(-\lambda\, x_i^\alpha\right) P(d\lambda).$$

Thus, one physical interpretation is that the finite Weibull distribution describes items which do not age, but are used at a polynomial rate.

Proof of Theorem 1. Let $Y_i = \mu(X_i)$. The X_i's are the lifetimes measured in clock time; the Y_i's are the lifetimes measured in usage. To prove the theorem, we show that the above characterizations for $\mathbf{X}$ are equivalent to ℓ^1-isotropic characterizations for $\mathbf{Y}$ given in Chick and Mendel[3]. Their theorem completes the proof.

Since $\bar{F}$ is an invariant in $\mathcal{L}^N$, $\bar{F}_{\mathbf{X}}(\mathbf{x}) = \bar{F}_{\mathbf{Y}}(\mathbf{y})$. Rewriting characterization (1) in $\mathbf{Y}$ coordinates:

$$P(Y_i \geq \mu(x_i + t_i)|\mathbf{Y} \geq \mathbf{y}) = P(Y_j \geq \mu(x_j + t_j)|\mathbf{Y} \geq \mathbf{y}),$$

if $\mu(x_i+t_i)-y_i = \mu(x_j+t_j)-y_j$. Pick $h_i > 0$ such that $\mu(x_i+t_i) = y_i+h_i$ (and similarly for h_j). Then,

$$P(Y_i \geq y_i + h_i|\mathbf{Y} \geq \mathbf{y}) = P(Y_j \geq y_j + h_j|\mathbf{Y} \geq \mathbf{y}) \text{ if } h_i = h_j.$$

This is characterization 1 given in Chick and Mendel. Similarly, for characterization (2), $\bar{F}_{\mathbf{Y}_N}$ is a function of $\lambda'_N = N/\sum y_i$ since $\lambda_N = \lambda'_N$ and

$$\bar{F}_{\mathbf{Y}}(\mathbf{y}) = \bar{F}_{\mathbf{X}}(\mathbf{x}) = f(\lambda_N),$$

for some function f. This is characterization 3 given in Chick and Mendel. Finally, rewriting characterization (3):

$$\bar{F}_{\mathbf{Y}_n}(\mathbf{y}_n) = \bar{F}_{\mathbf{X}_n}(\mathbf{x}_n) = \int \left(1 - \frac{\lambda'_N}{N}\sum_{i=1}^{n} y_i\right)^{N-1} P(d\lambda'_N),$$

where we substituted $\lambda'_N = \lambda_N$ and $y_i = \mu(x_i)$ in Eq. 3. This is characterization 5 given in Chick and Mendel. □

5. Conclusions

This paper gave a foundation for deriving lifetime distributions from physical principles. Specifically, we argued that the set of mathematical objects used to derive distributions should be invariant. In other words, the objects should remain physically consistent under changes of units for individual items. Another way of saying this is that the objects should have a physical meaning independent of the units chosen. While many familiar quantities are invariant, such as the survival function and hazard function, others are not - for example, the hazard gradient. We introduced the concept of a usage policy and showed that the hazard function in conjunction with the usage policy gave a hazard rate - which is an invariant form of the hazard gradient. We used the usage policy to derive a new characterization for "no-aging" distributions and showed how this could motivate a finite version of the Weibull distribution.

References

1. R. E. Barlow and M. B. Mendel, De Finetti-type Representations for Life Distributions, *Journal of the American Statistical Association.* **87**, 420, pp. 1116-1122 (1992).
2. R. E. Barlow and F. Proschan, *Statistical Theory of Reliability and Life Testing*, Holt, Rinehart and Winston, Inc., New York, 1975.

3. S. Chick and M. B. Mendel, New Characterizations of the No-Aging Property and the ℓ_1-Isotropic Models, *Journal of Applied Probability.* **35**, 4, pp. 903-910 (1998).
4. A. W. Marshall, Some Comments on the Hazard Gradient, *Stochastic Processes and Their Applications.* **3**, 4, pp. 293-300 (1975).
5. A. W. Marshall and I. Olkin, Multivariate Exponential Distributions, *JASA.* **62**, pp. 30-44 (1967).
6. B. Schutz, *Geometrical Methods of Mathematical Physics*, Cambridge University Press, New York (1980).
7. F. Spizzichino, Reliability Decision Problems under Conditions of Aging, in *Bayesian Statistics, 4*, Oxford University Press, pp. 803-811 (1992).

PART 3

BAYESIAN ANALYSIS

CHAPTER 16

ON THE PRACTICAL IMPLEMENTATION OF THE BAYESIAN PARADIGM IN RELIABILITY AND RISK ANALYSIS

Terje Aven

University of Stavanger

P.O. Box 2557, N-4091 Stavanger, Norway

E-mail: terje.aven@tn.his.no

The Bayesian paradigm comprises a unified and consistent framework for analysing and expressing reliability and risk. Yet, we see rather few examples of applications of major reliability and risk analyses where the full Bayesian setting has been adopted with specifications of priors of unknown parameters. In this paper we discuss some of the practical challenges of implementing Bayesian thinking and methods in reliability and risk analysis, emphasizing the introduction of probability models and parameters and associated uncertainty assessments. We conclude that there is a need for a pragmatic view in order to "successfully" apply the Bayesian approach when performing reliability and risk analyses of complex systems, so that we can do the assignments of some of the probabilities without introducing parameters and adopting the somewhat sophisticated procedure of specifying prior distributions of parameters. Furthermore, if parameters are introduced they should have a physical meaning, they should represent states of the world. A simple reliability analysis example is presented to illustrate the ideas.

1. Introduction

The purpose of engineering reliability and risk analyses is to provide decision support for design and operation. In this presentation we are particularly interested in the planning of complex man-machine systems, for which operational and maintenance aspects strongly affect the performance of the system.

In the following we focus on reliability analysis, but the discussion also applies to risk analysis.

The traditional approach to reliability analysis is based on the classical statistical thinking, in the way that reliability is assumed to be a property of the system under consideration, and its value is given through the relative frequency interpretation of probability. Thus the probability of an event A, $P(A)$, represents the relative fraction of times the event occurs if the situation analyzed were hypothetically repeated an infinite number of times. This true, underlying reliability is unknown, and the purpose of the reliability analysis is to estimate its value. This is done by developing models, like reliability block diagrams, and estimating the parameters of the models. Only "hard" historical data are used to estimate the model parameters. Uncertainty is related to the accuracy and precision of the estimators compared to the true reliability.

This approach has the disadvantage that the estimators could in practice be extremely poor; in many cases a sufficient amount of high quality data is not available, which means large uncertainties in the estimators. And as a consequence, the analyses do not give a message as clear as desired.

Bayesian methods are often presented as an alternative to the classical approach. But, what is the Bayesian alternative in a reliability analysis context? In practice and in the literature we often see a mixture of classical and Bayesian analyses, cf. Aven[1,2]. The starting point is classical in the sense that it is assumed that there exists an underlying, true reliability. This reliability is unknown, and subjective probability distributions are used to express confidence related to where the true value is. Starting with specifying probability distributions on the model parameter level, procedures are developed to propagate these distributions through the model to the reliability of the system. Updating schemes for incorporating new information are presented using Bayes formula. The approach is also referred to as the "probability of frequency" framework, see Apostolakis and Wu[3] and Kaplan[4]. In this framework the concept probability is used for the subjective probability and frequency for the "objective", relative frequency based probability.

This approach to reliability analysis introduces two levels of uncertainty; the value of the observable quantities such as the number of failures of a system, the downtime, etc., and what the correct value of the reliability is. The result is that both the analysis and the results of the analysis are considered uncertain. This does not provide a good basis for communication and decision making.

Now, how does this way of thinking relate to the Bayesian approach as it is presented in the literature, see *e.g.* Barlow[5], Bernardo and Smith[6], Lindley[7], Singpurwalla[8] and Singpurwalla and Wilson[9]? As we will see from the brief summary below, the Bayesian thinking is in fact not that different from the mixture of the classical and Bayesian analysis described above.

2. The Bayesian Paradigm

The Bayesian paradigm can be summarized as follows: All probabilities are subjective probabilities, based on judgements, reflecting our uncertainty about something. Thus the terms "true reliability" and "real reliability" are meaningless. Probabilities are always conditioned on the background information, K say. To specify the probabilities related to a random quantity X, a direct assignment could be used, based on everything we know. Since this knowledge is often complex, of high dimension, and much in K may be irrelevant to X, this approach is often replaced by the use of probability (reliability) models, which is a way of abridging K so that it is manageable. Such probability models play a key role in the Bayesian approach. A probability model, $p(x|\theta)$, expresses the probability distribution of the unknown quantity X, given a parameter θ. This parameter θ is unknown, it is a random quantity and our uncertainty related to its value is specified through a prior distribution $P(\theta)$. According to the law of total probability,

$$P(X \leq x) = \int p(x|\theta) dP(\theta). \tag{1}$$

More precisely, showing the dependence on the background information K,

$$P(X \leq x|K) = \int P(X \leq x|\theta, K) dP(\theta|K). \tag{2}$$

Suppose, that were we to know θ, we would judge X independent of K, so that for all θ, $P(X \leq x|\theta, K) = P(X \leq x|\theta)$, then Equation (2) is equal to (1). Thus the uncertainty distribution of X is expressed via two probability distributions, $p(x|\theta)$ and $P(\theta|K)$. The latter distribution is the prior distribution of θ. The two distributions reflect what is commonly referred to aleatory (stochastic) uncertainty and epistemic (state of knowledge) uncertainty. To simplify notation, we will in the following normally omit the K.

Bayesian statistical inference has such a probability model as a starting point and assumes that, in addition to K, we have at our disposal the realizations of n random quantities $X_1, X_2, ..., X_n$ that are judged exchangeable with X. Random quantities $X_1, X_2, ..., X_n$ are judged exchangeable if their joint distribution is invariant under permutations of coordinates, *i.e.*,

$$F(x_1, x_2, ..., x_n) = F(x_{r_1}, x_{r_2}, ..., x_{r_n}),$$

where F is a generic joint cumulative distribution for $X_1, X_2, ..., X_n$ and equality holds for all permutation vectors $(r_1, r_2, ..., r_n)$.

Suppose the random quantities are binary, *i.e.* take either the value 0 or 1, and we consider an infinite number of quantities, then it is a well-known result from Bayesian theory that the probability that k-out-of-n are 1 is necessarily of the form

$$P(\sum_{i=1}^{n} X_i = k) = \binom{n}{k} \int_0^1 \theta^k (1-\theta)^{n-k} dP(\theta), \tag{3}$$

for some distribution P. A similar result can be obtained in other settings, see Bernardo and Smith[6]. Thus, we can think of the uncertainties (beliefs) about observable quantities as being constructed from a parametric model, where the random quantities can be viewed as independent, together with a prior distribution for the parameter. The parameters have interpretation as strong limits of (appropriate) functions of observable quantities. In the binary example, the parameter θ is interpreted as the long run frequency of 1's.

Bayesian statistics is mainly concerned about inference about θ. Starting from the prior distribution P, this distribution is updated to a posterior distribution using Bayes theorem. In our set-up this posterior distribution can be written

$$P(\theta | X_1 = x_1, X_2 = x_2, ..., X_n = x_n) \propto \prod_{i=1}^{n} p(x_i|\theta) P(\theta), \tag{4}$$

where the constant of proportionality ensures that the distributions are legitimate distributions, *i.e.*, the integral of the distribution over θ is equal to 1. From this distribution we derive the predictive distribution of X,

$$P(X = x) = \int p(x|\theta) dP(\theta | X_1 = x_1, X_2 = x_2, ..., X_n = x_n).$$

Note that the Bayesian approach, as presented here, allows for fictional parameters, based on thought experiments. Such parameters are introduced and uncertainty of these assessed. Thus, from a practical point of view, an analyst would probably not see much difference between this framework and the combined classical and Bayesian (probability of frequency) approach referred to above. Of course, Bayesians would not speak about true, objective reliability and probabilities, and the predictive form is seen as the most important one. However, in practice, Bayesian parametric analysis is often seen as an end-product of a statistical analysis. The use of and understanding of probability models gives focus on limiting values of quantities constructed through a thought experiment, which are very close to the mental constructions of probability and reliability used in the classical relative frequency approach.

In our view, applying the standard Bayesian procedures, gives too much focus on fictional parameters, established through thought experiments. Focus should be on observable quantities. We believe that there is a need for a rethinking on how to present the Bayesian way of thinking, to obtain a successful implementation in a practical setting. In a reliability analysis comprising a large number of observable quantities, a pragmatic view to the Bayesian approach is required, in order to be able to conduct the analysis. Direct probability assignments should be seen as a useful supplement to establishing probability models where it is needed to specify prior distributions of parameters. A Bayesian updating procedure may be used for expressing uncertainty related to observable quantities, but its applicability is in many cases rather limited. In most real-life cases we would not perform a formal Bayesian updating to incorporate new observations—rethinking of the whole information basis and approach to modeling is required when we conduct the analysis at a particular point in time, for example in the prestudy or concept specification phases of a project. Furthermore, we should make a sharp distinction between probability and utility. In our view it is unfortunate that these two concepts are seen as inseparable as is often done in the Bayesian literature.

These issues are further discussed in the following, using a simple example to illustrate ideas.

3. An Illustrative Example

Suppose a reliability analysis is to be conducted for a system in the design phase. Alternative system configurations are to be analyzed and evaluated to support decision making. We would like to look closer into the analysis of one of these alternative configurations.

Focus is on the system performance in the operational phase. The first issue to be discussed is how to measure this performance. We restrict here attention to reliability indices, and let us say that attention is placed on a time interval $[0, T]$, where T is one year.

To conduct the analysis, a reliability analysis team is established, comprising a reliability analyst and an engineer who is familiar with the system being analyzed. We refer to the latter person as an expert. Suppose that a brainstorming process within the team gives the following list of indices for further evaluation;

- X, representing the number of failures of the system in the time interval $[0, T]$,
- a prediction of X,
- λ defined as the expected (mean) number of failures, $\lambda = EX$,
- An uncertainty distribution $P(\lambda') = P(\lambda \leq \lambda')$ of λ,
- A distribution of X, $P(X \leq x)$.

Now, what indices should we use? The reliability analyst and the expert also discuss the need for modeling, for example using fault trees and reliability block diagrams. Would more detailed modeling give more accurate results?

The reliability analyst has briefly studied the Bayesian approach to reliability analysis and would like to apply the principles and methods of this framework. He then immediately thinks of establishing a probability model, and in this case a Poisson model is the natural first choice since we study a failure process in a time interval. So, X is treated as a random quantity with a Poisson distribution, $p(x|\lambda)$, given the parameter value λ. The value of λ is unknown, and the challenge is then to establish an uncertainty distribution over λ reflecting available knowledge. The reliability analyst establishes this framework in a rather mechanical way. He has some problem in explaining the meaning of the probability model and the parameter to the expert, and he is not sure how to determine the uncertainty distribution of λ? He remembers the Bayesian theory referring to parameters as limits of an infinite sequence of exchangeable random quantities, but this abstraction is difficult to appreciate in this case; his concern is the number of failures for the system he is analyzing, not statistical quantities representing the average performance of a fictional population.

If the reliability analyst seeks advice on how to perform this analysis, how should we guide him? How should we approach this problem?

3.1. *Analysis*

Let us start from scratch, and introduce quantities and tools as they are required. We would like to use subjective probabilities and seek a practical way of analyzing the system. Then we will see later how this would compare to the standard Bayesian paradigm.

We distinguish between two cases:

(1) Failure data from systems "similar" to the one analyzed is available, and let us assume for the sake of simplicity that these are of the form $x_1, x_2, ..., x_n$, where x_i is the number of failures observed in $[0, T]$ for system i. These data are considered relevant for the system being studied.
(2) We have no relevant data available for this type of system.

First, let us consider case 1.

The interesting quantity is X, the number of failures. We would like to predict this number. How should we do this? The data allow a prediction simply by using the mean $\overline{x}$ of the observations $x_1, x_2, ..., x_n$. But what about uncertainty in this prediction? How should we express uncertainty related to X and the prediction of X? Suppose the observations $x_1, x_2, ..., x_n$ are 1,1,2,0,1, so that $n = 5$ and the observed mean is equal to 1. In this case we have rather strong background information, and we suggest to use the Poisson distribution with mean 1 as our uncertainty distribution of X. How can this uncertainty distribution be "justified"? Well, if this distribution reflects our uncertainty about X, it *is* justified, and there is nothing more to say. This is a subjective probability distribution and there is no need for further justification. But is a Poisson distribution with mean 1 "reasonable", given the background information? We note that this distribution has a variance not larger than 1? By using this distribution, 99% of the mass is on values less than 4.

Adopting the standard Bayesian thinking, as outlined above, using the Poisson distribution with mean 1, means that we have no uncertainty about the parameter λ, which is interpreted as the long run average number of failures when considering an infinite number of exchangeable random quantities, representing similar systems as the one being analyzed. According to the Bayesian theory, ignoring the uncertainty about λ, gives misleading overprecise inference statements about X, *cf. e.g.* Bernardo and Smith[7], p. 483. This reasoning is of course valid if we work within the standard Bayesian setting, considering an infinite number of exchangeable random quantities. In our case, however, we just have one X, so what do we gain by making a reference to limiting quantities of a sequence of similar hypothetical Xs? The point is that given the observations $x_1, x_2, ..., x_5$, the choice of the Poisson distribution with mean 1, is in fact reasonable. Consider the following argumentation. Suppose that we divide the year $[0, T]$ into time periods of length T/k, where k is for example 1000. Then we may ignore the possibility of having two failures occurring in one time period, and we assign a failure probability of $1/k$ for the first time period, as we predict 1 failure in the whole interval $[0, T]$. Suppose that we have observations related to $i - 1$ time periods. Then for the next time period we should take these observations into account—using independence means ignoring available information. A possible way of balancing the prior information and the observations is to assign a failure probability of

$$c_i \frac{1}{k} + (1 - c_i) \frac{d_i}{i - 1},$$

for the ith time period, where $c_i = 1 - c(i - 1)/k$ and c is a constant such that $0 \leq c \leq 1$, and d_i is equal to the total number of events that occurred in $[0, T(i - 1)/k]$. We see that $c_1 = 1$ so that all weight is put on the initial assignment $1/k$, and $c_k = 1 - c(k - 1)/k$ meaning that the initial probability is given a weight approximately equal to $1 - c$ and the observations a weight of c. It turns out that this assignment process gives an approximate Poisson distribution for X, for suitable c values, typically in the range $c < 0.5$. This can be shown for example by using Monte Carlo simulation. The Poisson distribution is justified as long as the background information dominates the uncertainty assessment of the number of events occurring in a time period. Since in this case we have observations for five years, it is reasonable to put $c = 1/6$, giving the year being studied the same weight as the 5 years observed. Thus from a practical point of view, there is no problem in using the Poisson distribution with mean 1.

Alternatively, the Poisson approximation follows by studying the predictive distribution of X in a full Bayesian analysis, assuming that $x_1, x_2, ..., x_5$ are observations coming from a Poisson distribution, given the mean λ and using a suitable (for example a non-informative) prior distribution on λ. Restricting attention to observable quantities only, a procedure as specified in Barlow[5], Chapter 3, can be used. This procedure, in which the multinomial distribution is used to establish the Poisson distribution, is based on exact calculation of the conditional probability distribution of the number of events in subintervals, given the observed number of events for the whole interval.

Note that for the direct assignment procedure using the k time periods, the observations $x_1, x_2, ..., x_5$ are considered a part of the background information, meaning that this procedure does not involve any modeling of these data. In contrast, the more standard Bayesian approach requires that we model $x_1, x_2, ..., x_5$ as observations coming from a Poisson distribution, given the mean λ.

We conclude that a Poisson distribution with mean 1 can be used to describe the analyst's uncertainty with respect to X in this case. The background information is sufficiently strong.

Now consider the case 2 with no historical data. Then we will probably find the direct use of the Poisson distribution as described above to have too small variance. The natural approach is then to implement a full parametric Bayesian procedure as described above. But how should we interpret the various elements of the set-up? Should we speak about λ having a true value (the limit of an infinite sequence of exchangeable random quantities), introduce an uncertainty distribution over λ, and refer to the Poisson distribution as a model?

No, as long as λ is fictional, not a state of the world (the nature) and thus not observable, a true value of λ does not exist and the Poisson distribution is not a representation of the world. Instead we suggest the following interpretation.

The Poisson probability distribution $p(x|\lambda)$ is a candidate for our subjective probability for the event $X = x$, and $P(\lambda)$ is a confidence measure, reflecting for a given value of λ, the confidence we have in $p(x|\lambda)$ for being able to predict X. If we have several X_is, similar to X, and λ is our choice, we believe that about $p(x|\lambda) \cdot 100\%$ of the X_is will take a value equal to x, and $P(\lambda)$ reflects for a given value of λ, the confidence we have in $p(x|\lambda)$ for being able to predict the number of X_is taking the value x. Following this interpretation, note that $p(x|\lambda)$ is not a model, and $P(\lambda)$ is not an uncertainty measure. We refer to this as the confidence interpretation.

If a suitable infinite (or large) population of "similar units" can be defined, in which X and the X_is belong, then the above standard Bayesian framework applies as the parameter λ represents a state of the world, an observable quantity. Then $P(\lambda)$ is a measure of uncertainty and $p(x|\lambda)$ is truly is a model—a representation of the portion of units in the population having the property that the number of failures is equal to x. We may refer to the variation in this population, modeled by $p(x|\lambda)$, as aleatory uncertainty, but still the uncertainty related to the values of the X_is is seen as a result of lack of knowledge, *i.e.*, the uncertainty is epistemic. This nomenclature is in line with the basic thinking of *e.g.* Winkler[10], but not with that commonly used in the standard Bayesian framework; see the introduction above.

The above analysis is a tool for predicting X and assessing associated uncertainties. When we have little data available, modeling is required to get insight into the uncertainty related to X and hopefully reduce the uncertainty. The modeling also makes it possible to see the effects of changes in the system and to identify risk and unreliability contributors. In the following we outline how one could carry out the modeling for the system being analyzed, using a decomposition approach.

3.2. *Modeling Using Decomposition*

The main step of the modeling is to define the system as being composed of components (or subsystems), and link the performance of the components and the system by a structure function. We apply the theory of monotone (coherent) systems, as presented in Barlow and Proschan[11] and Aven and Jensen[12].

Let $X_t(i)$ be a binary stochastic process with right-continuous sample paths representing the state of component i, $i = 1, 2, \ldots, n$; $X_t(i) = 1$ if component i is functioning at time t and $X_t(i) = 0$ if component i is not functioning at time t. We assume that all components are functioning at time 0, *i.e.*, $X_0(i) = 1$. Let T_{im}, $m = 1, 2, \ldots$, represent the positive length of the mth operation period of component i, and let R_{im}, $m = 1, 2, \ldots$, represent the positive length of the mth repair time for component i, see Fig. 1. We have

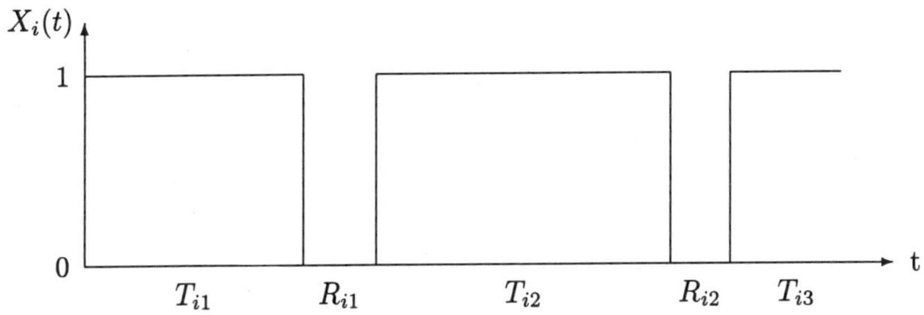

Fig. 1. Time evolution of a failure and repair process for component i starting at time $t = 0$ in the operating state

$$X_t(i) = I(T_{i1} > t) + \sum_{k=1}^{\infty} I(S^{\circ}_{ik} \leq t, S^{\circ}_{ik} + T_{i(k+1)} > t),$$

where

$$S^{\circ}_{ik} = \sum_{m=1}^{k} (T_{im} + R_{im}), \quad k \in \{1, 2, ...\}$$

and I is the indicator function which is equal to 1 if the argument is true and 0 otherwise.

Let $\Phi : \{0,1\}^n \to \{0,1\}$ be the structure function of the system. We assume that this function is monotone, *i.e.*, $\Phi(\mathbf{x})$, where $\mathbf{x} = (x_1, x_2, \ldots, x_n)$, is a non-decreasing function in each argument x_i, and $\Phi(\mathbf{1}) = 1$ and $\Phi(\mathbf{0}) = 0$ where $\mathbf{1} = (1, 1, \ldots, 1)$ and $\mathbf{0} = (0, 0, \ldots, 0)$. At time t the states of the components are given by

$$\mathbf{X}_t = (X_t(1), X_t(2), \ldots, X_t(n)).$$

Focus is on the performance measure N_t, representing the number of system failures in [0,t]. We see that

$$N_t = \sum_t \max\{0, \Phi(\mathbf{X}_{t-}) - \Phi(\mathbf{X}_t)\}. \tag{5}$$

The modeling gives insight into the performance of the system and the uncertainties. Given the model (5), the remaining uncertainties are related to the values of the component lifetimes and repair times. The quantities T_{im} and R_{im} are unknown and we express our uncertainty related to what will be the true values by probability distributions.

The question is now how to assess these uncertainties and specify the probability distributions. Ideally, a simultaneous distribution for all lifetimes and repair times should be provided, but this is not feasible in practice. So we need to simplify. Fix time t. Suppose we have strong background information concerning the component lifetimes and the repair times. Then as a simplification of the uncertainty assessments, we could judge all T_{im} and R_{im} to be independent and use the same distribution F_i for all lifetimes and the same distribution G_i for all repair times of component i. This is of course a rather strong simplification; we ignore learning when observing some of the lifetimes and repair times. But as discussed above, in some cases the background information is such that we could justify the use of independence. Suppose for example that we use exponentially distributed lifetimes and fixed repair times. Then we can argue along the same lines as above for the Poisson example, that the Poisson process is reasonable to use when considering operational time (we ignore the downtimes), with the parameter λ, the expected number of failures per unit of time, given by the observed mean. Of course in the general case we would use a full Bayesian analysis. But let us first see how we would proceed when using independence.

Applying the above modeling and the uncertainty distributions for the lifetimes and repair times, an associated uncertainty distribution for N_t can be computed, see *e.g.* Aven and Jensen[12]. In most cases, approximations need to be used. As a predictor of N_t, it is common to use $\lambda_\Phi t$, where λ_Φ is the steady state system failure rate given by (see *e.g.* Aven and Jensen[12] and Barlow and Proschan[11]):

$$\lambda_\Phi = \sum_{i=1}^{n} \frac{h(1_i, \mathbf{A}) - h(0_i, \mathbf{A})}{\mu_{F_i} + \mu_{G_i}},$$

where

$$\mu_{F_i} = ET_{im} \qquad \mu_{G_i} = ER_{im},$$

$h(x_i, \mathbf{p})$ is the reliability function when it is given that $X_i = x_i$, and $\mathbf{A}$ is the vector of component steady state availabilities, given by

$$A_i = \frac{\mu_{F_i}}{\mu_{F_i} + \mu_{G_i}}.$$

To express uncertainties related to N_t we normally utilize approximations to the Poisson and Normal distributions, see Aven and Jensen[12]. The approximation to the Normal distribution gives excellent results if the expected number of system failures is quite high, typically greater than 5. If that many failures are not expected to occur, *i.e.*, the system is "highly available", the Poisson distribution gives good approximation for the distribution of N_t.

In practice, Monte Carlo simulation is often used to obtain the uncertainty distribution of N_t, given the uncertainty distributions on the component level. When performing a Monte Carlo study we generate a sequence of independent, identically distributed random variables, say $N_t^{(1)}, N_t^{(2)}, \ldots, N_t^{(k)}$ based on the same uncertainty distributions on the component level and the model (5). The simulations are performed in a classical statistical setting where we have repeated experiments and the starting point is a probability that we wish to determine. By using the sample $N_t^{(1)}, N_t^{(2)}, \ldots, N_t^{(k)}$ we can determine the uncertainty distribution of N_t, and its mean and variance.

Now, let us look more closely into the problem of specifying the uncertainty distributions on the component level, $F_i(t)$ and $G_i(t)$. One approach is the following: The company has "standardized its opinion", meaning that the reliability expertise in the company have agreed on which uncertainty distributions to be used on the component level for this type of application. The experts have looked at relevant historical data, they have analyzed the possible causes of failures and the associated time needed for repair. Based on this they have concluded what would give the best predictions and what reflects best their view on what would be the value of the lifetimes and repair times. We do not speak about a true underlying probability distributions and associated parameters. To make this clear, suppose the lifetime distribution of a component is exponential, *i.e.*, $F(t) = 1 - e^{-\lambda t}$, where λ is the failure rate. We may then adopt the following interpretation: The lifetime T is an observable quantity, but now it is unknown. Our uncertainty related to the value of T is described through the lifetime distribution $F(t)$. In the exponential case we have to choose a specific λ. There exists no true value that we need to estimate, but λ is simply a parameter in a mathematical class of functions, which we use to describe our uncertainty related to T. When specifying the distribution function $F(t)$ and the value of λ, we may think of these quantities as representing the average long run portion of lifetimes less than or equal to t and the number of failures per unit of operational time, respectively. This way of thinking does not mean that we define $F(t)$ and λ by the asymptotic limits; it is just a cognitive tool for facilitating the specification of $F(t)$ and λ, and it is therefore not in conflict with the above interpretation: the distribution we choose is a measure of uncertainty related to the value of the finite number of lifetimes in the time interval of interest, and there is no correct or true value of this distribution nor of the parameter λ.

If we can define an infinite or large population of "similar units", then we can of course also see the distribution as a model of the portion of lifetimes having values equal to or less than t.

If the background information is such that we cannot justify the use of independence, we would apply a full Bayesian analysis, with the specification of a distribution class $p(x|\theta)$, with parameter θ, and a prior (posterior) distribution $P(\theta)$. The interpretation of these elements, is analogous to the one given for the Poisson case above. Clearly, to run a full Bayesian analysis in a case like this, would be challenging since we need to specify a high number of prior (posterior) distributions.

4. Conclusions

The alternative to the classical approach to reliability and risk analysis is the Bayesian approach, where the concept of probability is used as the analyst's measure of uncertainty or degree of belief. This alternative approach has, however, not been commonly accepted; there is still a lot of scepticism among many reliability and risk analysts when speaking about subjective probabilities. However, most reliability and risk analysts do in fact use some subjective methods when carrying out reliability and risk analyses. For example, subjective probabilities are commonly developed for the branches of the event trees. But a total adoption of the Bayesian theory is surprisingly not often seen among reliability and risk analysts. Perhaps one reason for this is lack of practical implementation guidelines. When studying the Bayesian paradigm, it is not clear how we should implement the theory in practice. We find the Bayesian literature very technical and theoretical. The literature is to a large extent concerned about mathematical and statistical aspects of the Bayesian paradigm. The more practical challenges of adopting the Bayesian approach are seldomly addressed.

As stated in the introduction, we see the need for a rethinking how to present the Bayesian paradigm in a practical setting. The aim of the discussion in this paper has been to give a basis for such a thinking.

The presentation of the Bayesian paradigm should in our view has a clear focus and an understanding of what can be considered as technicalities. A possible way of structuring the various elements of the analysis is shown in Fig. 2, which highlights the way the reliability (risk) analyst uses the model and probability calculus. The figure is read as follows: A reliability (risk) analyst (or an analyst team) conducts a reliability (risk) analysis. Focus is on the world, and in particular some future observable quantities reflecting the world; Y and $\mathbf{X} = (X_1, X_2, ..., X_n)$. Based on the analyst's understanding of the world the analyst develops a model (several models), that relates the overall system performance measure Y to $\mathbf{X}$, which is a vector of quantities on a more detailed level. The analyst assesses uncertainties of $\mathbf{X}$, and that could mean the need for simplifications in the assessments, for example using independence between the quantities X_i as discussed in the previous section. Using probability calculus, the uncertainty assessments of $\mathbf{X}$ together with the model g, gives the results of the analysis, *i.e.*, the probability distribution of Y and a prediction of Y. The uncertainties are a result of lack of knowledge, *i.e.*, it is epistemic.

This way of presenting the subjectivistic approach to reliability and risk analysis is sometimes referred to as the predictive, epistemic approach, cf. Aven[1,2,13] and Apeland et al. [14]. The essential steps of the analysis can be summarized as follows:

(1) Identify the overall system performance measures (observable quantities on a high level). These are typically associated with the objectives of the system performance.
(2) Develop a deterministic model of the system, linking the system performance measures and observable quantities on a more detailed level.
(3) Collect and systematize information about these low level observable quantities.
(4) Use probabilities to express uncertainty of these observable quantities.
(5) Calculate the uncertainty distributions of the performance measures and determine suitable predictions from these distributions.

Sometimes a model is not developed as the analysis is just a transformation from historical data to a probability distribution and predictions related to a performance measure. Often the predictions are derived directly from the historical data without using the probability distributions.

The reliability or risk description needs then to be evaluated and related

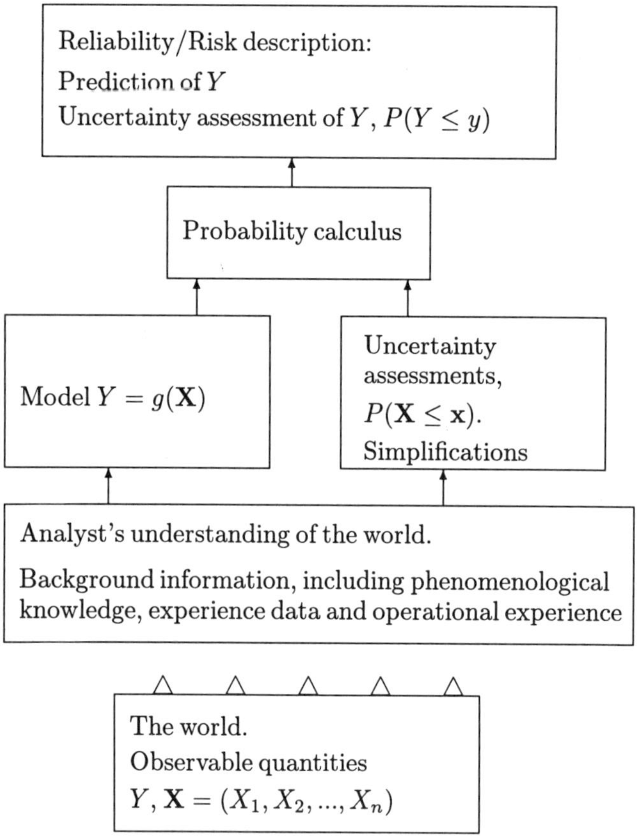

Fig. 2. Basic elements of a reliability or risk analysis

to costs, and other aspects, to support decision making. A utility-based analysis could be carried out as a tool for identifying a "good" decision alternative. We refer to Aven[13] for a discussion on when to use such an analysis.

The above way of thinking, emphasizing observable quantities and using the reliability or risk analysis as a tool for prediction, is in line with the modern, predictive Bayesian theory, as described in *e.g.* Bernardo and Smith[6], Barlow[5] and Barlow and Clarotti[15]. The objective of this presentation, which is partly based on Aven[13], has been to add some perspectives to this thinking, in order to strengthen the practical applicability of the subjectivistic, Bayesian paradigm.

More research is needed on how to implement the subjectivistic, Bayesian ideas in practice. In particular, more work should be done on establishing procedures for quantifying uncertainty and guiding the analyst in developing suitable models, including probability models. We also see a strong need for research on how to link reliability (risk) analysis and formal decision analysis.

References

1. T. Aven, In Proceedings of European Safety and Reliability Conference (ESREL), Eds. Cottam, M.P. et al. (Balkema, Rotterdam, 2000) pp. 21-28.
2. T. Aven, In *Recent Advances in Reliability Theory: Methodology, Practice and Inference*, Eds. N. Liminios and M. Nikulin (Birkhauser, Boston, 2000) pp. 23-28.
3. G. Apostolakis, and J.S. Wu, In *Reliability and Decision Making*, Eds. R.E. Barlow, C.A. Clarotti (Chapman & Hill, London, 1993) pp. 311-322 (1993).
4. S. Kaplan, *Nuclear Technology*, **102**, 137-142 (1992).
5. R.E. Barlow, *Engineering Reliability* (SIAM, Philadephia 1998).
6. J.M. Bernardo, and A. Smith, *Bayesian Theory* (Wiley & Sons., New York, 1994).
7. D.V. Lindley, *The Statistician*, **49**, 293-337 (2000).
8. N.D. Singpurwalla, *SIAM Review*, **30**, 264-281 (1988).
9. N.D. Singpurwalla and S.P. Wilson, *Statistical Methods in Software Engineering* (Springer Verlag, New York 1999).
10. R.L. Winkler, *Reliability Engineering and System Safety* **54**, 127-132 (1996).
11. R.E. Barlow and F. Proschan, *Statistical Theory of Reliability and Life Testing* (Holt, Rinehart and Winston, New York, 1975).
12. T. Aven and U. Jensen, *Stochastic Models in Reliability* (Springer Verlag, New York, 1999).
13. T. Aven, *How to Approach Risk and Uncertainty to Support Decision Making* (Wiley, New York, 2002) to appear.
14. S. Apeland, T. Aven and T. Nilsen, *Reliability Engineering and System Safety*, Quantifying uncertainty under a predictive, epistemic approach to risk analysis. To appear, 2001.
15. R.E. Barlow and C.A. Clarotti, *Reliability and Decision Making*, Preface (Chapman & Hill, London, 1993).

CHAPTER 17

A WEIBULL WEAROUT TEST: FULL BAYESIAN APPROACH

Telba Zalkind Irony

Food and Drug Administration
1350 Piccard Dr., HFZ-542, Rockville, MD 20850, U.S.A.
E-mail: tzi@cdrh.fda.gov

Marcelo Lauretto* and Carlos Alberto de Bragança Pereira†

Av. Afrânio Peixoto 188, CEP 05507-000 Butantan
São Paulo, SP - Brasil
*E-mail: *lauretto@supremum.com,*
E-mail: †cpereira@ime.usp.br

Julio Michael Stern

Department of Mathematics and Statistics, SUNY - Binghamton
Binghamton NY 13902-6000, U.S.A.
E-mail: jstern@ime.usp.br

The Full Bayesian Significance Test (FBST) for precise hypotheses is presented, with some applications relevant to reliability theory. The FBST is an alternative to significance tests or, equivalently, to *p-values*. In the FBST we compute the evidence of the precise hypothesis. This evidence is the probability of the complement of a credible set "tangent" to the sub-manifold (of the parameter space) that defines the null hypothesis. We use the FBST in an application requiring a quality control of used components, based on remaining life statistics.

1. Introduction

The Full Bayesian Significance Test (FBST) is presented in Pereira and Stern (1999b) as a coherent Bayesian significance test. The FBST is intuitive and has a geometric characterization. It can be easily implemented using modern numerical optimization and integration techniques. The method is "Full" Bayesian and is based on the analysis of credible sets. By Full we mean that we need only the knowledge of the parameter space represented by its posterior distribution. The FBST needs no additional assumption, like a positive probability for the precise hypothesis, that generates the Lindley's paradox effect. The FBST regards likelihoods as the proper means for representing statistical information, a principle stated by Royall (1997) to simplify and unify statistical analysis. Another important aspect of the FBST is its consistency with the "benefit of the doubt" juridical principle. These remarks will be understood in the sequel.

Significance tests are regarded as procedures for measuring the consistency of data with a null hypothesis, Cox (1977) and Kempthorne and Folks (1971). *p-values* are a tail area under the null hypothesis, calculated in the sample space, not in the parameter space where the hypothesis is formulated.

Bayesian significance tests defined in the literature, like Bayes Factor or the posterior probability of the null hypothesis, consider the *p-value* as a measure of evidence of the null hypothesis and present alternative Bayesian measures of evidence, Aitkin (1991), Berger and Delampady (1987), Berger *et al.* (1997), Irony and Pereira (1986, 1995), Pereira and Wechsler (1993), Sellke *et al.* (1999). As pointed out in Cox (1977), the first difficulty to define the *p-value* is the way the sample space is ordered under the null hypothesis. Pereira and Wechsler (1993) suggests a *p-value* that always regards the alternative hypothesis. One can find a great deal of objections agaist each of these measures of evidence. The most important argument against Bayesian tests for precise hypothesis is presented by Lindley (1957). The literature is full of objections to the classical *p-value*. The book by Royall (1997) and its review by Vieland *et al.* (1998) presents interesting and relevant arguments motivating statisticians to start thinking about new methods of measuring evidence. In a more philosophical terms, Carnap (1962), de Finetti (1989), Good (1983) and Popper (1989) discuss, in a great detail, the concept of evidence.

2. Motivation

In order to illustrate the FBST we discus a well known problem. Given a sample from a normal distribution with unknown parameters, we want to test if the standard deviation is equal to a constant. The hypothesis

$\sigma = c$ is a straight line. We have a precise hypothesis since it is defined by a manifold (surface) of dimension (one) strictly smaller than the dimension of the parameter space (two).

It can be shown that the conjugate family for the Normal Distribution is a family of bivariate distributions, where the conditional distribution of the mean, μ, for a fixed precision, $\rho = 1/\sigma^2$, is normal, and the marginal distribution of the precision, ρ, is gamma, DeGroot (1970), Lindley (1978). We use the standard improper priors, uniform on $]-\infty, +\infty[$ for μ, and $1/\rho$ on $]0, +\infty[$ for ρ, in order to get a fair comparison with p-values, DeGroot (1970). Hence we have the parameter space, hypothesis and posterior joint distribution:

$$\Theta = \{(\mu, \rho) \in R \times R_+\} \ , \ \ \Theta_0 = \{(\mu, \rho) \in \Theta \,|\, \rho = c\}$$

$$f(\mu, \rho \,|\, x) \propto \sqrt{\rho}\; exp(-n\rho(\mu - m)^2/2) exp(-b\rho)\rho^{a-1}$$

$$x = [x_1 \dots x_n] \ , \ a = \frac{n-1}{2} \ , \ m = \frac{1}{n}\sum_{i=1}^{n} x_i \ , \ b = \frac{n}{2}\sum_{i=1}^{n}(x_i - m)^2$$

Figure 1 shows the plot of some level curves of the posterior density function, including the level curve tangent to the hypothesis manifold. At the tangency point, θ^*, the posterior density attains its maximum, f^*, on the hypothesis. The interior of the tangent level curve, T^*, includes all points with posterior density greater than f^*, i.e. it is the highest probability density set tangent to the hypothesis.

The posterior probability of T^*, κ^*, gives an indication of inconsistency between the posterior and the hypothesis: Small values of κ^* indicate that the hypothesis traverses high density regions, favoring the hypothesis. Therefore we define $Ev(H) = 1 - \kappa^*$ as the measure of evidence (for the precise hypothesis).

In Figure 1 we test $c = 1$ with $n = 16$ observations of mean $m = 10$ and standard deviation $s = 1.02$, 1.1, and 1.5. We present the FBST evidence, Ev, and the standard χ^2-test, $chi2$.

It is clear that this example is only an illustration: there is no need of new methods to test the standard deviation of a normal distribution. However, efficient numerical optimization and integration computer programs, make it straightforward to extend the FBST to more complex structures. In sections 6 and 7 we present an important application involving the Weibull

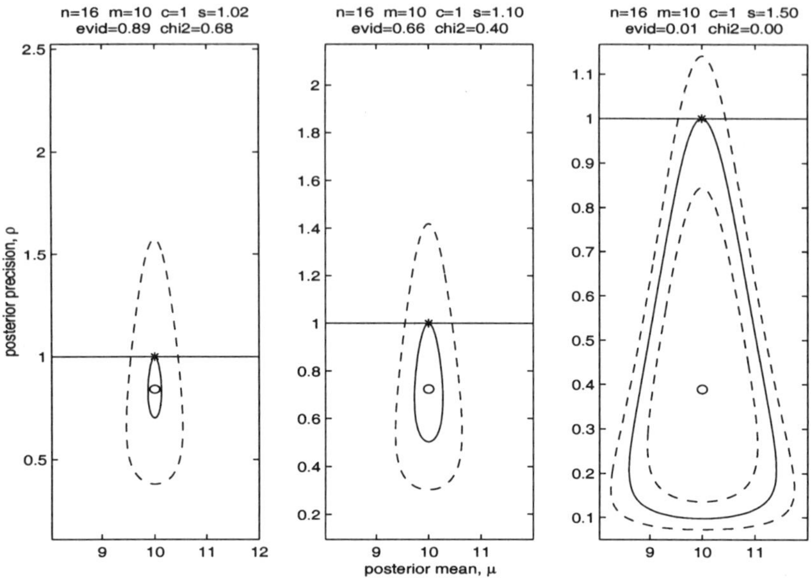

Fig. 1. Tangent and other Highest Probability Density Sets

distribution, requiring a quality control test for used components, based on remaining life data. This problem appears in engineering as well as biological and pharmacological applications. The FBST is exact and performs well even for small samples and low frequencies. In the next section we give a more formal definition of the FBST.

3. The Evidence Calculus

Consider the random variable D that, when observed, produces the data d. The statistical space is represented by the triplet (Ξ, Δ, Θ) where Ξ is the sample space, the set of possible values of d, Δ is the family of measurable subsets of Ξ and Θ is the parameter space. We define now a prior model (Θ, B, π_d), which is a probability space defined over Θ. Note that in this model $Pr\{A \mid \theta\}$ has to be Θ measurable. As usual, after observing data d, we obtain the posterior probability model (Θ, B, π_d), where π_d is the conditional probability measure on $\boldsymbol{B}$ given the observed sample point, d. In this paper we restrict ourselves to the case where the functions π_d has a probability density function f.

To define our procedure we should concentrate only on the posterior probability space (Θ, B, π_d). First, we define T_φ as the subset of the parameter space where the posterior density is greater than φ.

$$T_\varphi = \{\theta \in \Theta \,|\, f(\theta) \geq \varphi\}$$

The credibility of T_φ is its posterior probability,

$$\kappa = \pi_d(T_\varphi) = \int_{T_\varphi} f(\theta \,|\, d) d\theta = \int_\Theta f_\varphi(\theta \,|\, d) d\theta$$

where $f_\varphi(x) = f(x)$ if $f(x) \geq \varphi$ and zero otherwise.

Now, we define f^* as the maximum of the posterior density over the null hypothesis, attained at the argument θ^*,

$$\theta^* \in arg \max_{\theta \in \Theta_0} f(\theta) \ , \quad f^* = f(\theta^*)$$

and define $T^* = T_{f^*}$ as the set "tangent" to the null hypothesis, H, whose credibility is κ^*.

The measure of evidence we propose in this article is the complement of the probability of the set T^*. That is, the evidence of the null hypothesis is

$$Ev(H) = 1 - \kappa^* \ \text{ or } \ 1 - \pi_d(T^*)$$

If the probability of the set T^* is "large", it means that the null set is in a region of low probability and the evidence in the data is against the null hypothesis. On the other hand, if the probability of T^* is "small", then the null hypothesis is in a region of high probability and the evidence in the data is in its favor. In the next section we give an operational construction of the FBST.

4. Numerical Optimization and Integration

We restrict the parameter space, Θ, to be always a subset of R^n, and the hypothesis is defined as a further restricted subset $\Theta_0 \subset \Theta \subseteq R^n$. Usually, Θ_0 is defined by vector valued inequality and equality constraints:

$$\Theta_0 = \{\theta \in \Theta \,|\, g(\theta) \leq 0 \wedge h(\theta) = 0\}.$$

Since we are working with precise hypotheses, we have at least one equality constraint, hence $dim(\Theta_0) < dim(\Theta)$. Let $f(\theta)$ be the posterior probability density function, as defined in the last section.

The computation of the evidence measure defined in the last section is performed in two steps, a numerical optimization step, and a numerical integration step. The numerical optimization step consists of finding an argument θ^* that maximizes the posterior density $f(\theta)$ under the null hypothesis. The numerical integration step consists of integrating the posterior density over the region where it is greater than $f(\theta^*)$. That is,

- Numerical Optimization step:

$$\theta^* \in arg \max_{\theta \in \Theta_0} f(\theta) \ , \quad \varphi = f^* = f(\theta^*)$$

- Numerical Integration step:

$$\kappa^* = \int_{\Theta} f_{\varphi}(\theta \,|\, d) d\theta$$

where $f_{\varphi}(x) = f(x)$ if $f(x) \geq \varphi$ and zero otherwise.

Efficient computational algorithms are available, for local and global optimization as well as for numerical integration, Bazaraa *et al.* (1993), Horst *et al.* (1995), Luenberger (1984), Nocedal and Wright (1999), Pinter (1996), Krommer and Ueberhuber (1998), and Sloan and Joe (1994). Computer codes for several such algorithms can be found at software libraries as ACM, GSL and NAG, or at internet sites as *www.ornl.gov* and *www-rocq.inria.fr*.

We notice that the method used to obtain T^* and to calculate κ^* can be used under general conditions. Our purpose, however, is to discuss precise hypothesis testing, i.e. $dim(\Theta_0) < dim(\Theta)$, under absolute continuity of the posterior probability model, the case for which most solutions presented in the literature are controversial.

5. Weibull Distribution

The two parameter Weibull probability density, reliability (or survival probability) and hazard functions, for a failure time $t \geq 0$, given the shape, and characteristic life (or scale) parameters, $\beta > 0$, and $\gamma > 0$, are:

$$\begin{aligned} w(t \,|\, \beta, \gamma) &= (\beta\, t^{\beta-1} / \gamma^{\beta})\, exp(-(t/\gamma)^{\beta}) \\ r(t \,|\, \beta, \gamma) &= exp(-(t/\gamma)^{\beta}) \\ z(t \,|\, \beta, \gamma) &\equiv w(\ )/r(\ ) = \beta\, t^{\beta-1} / \gamma^{\beta} \end{aligned}$$

The mean and variance of a Weibull variate are given by:

$$\mu = \gamma\,\Gamma(1 + 1/\beta)$$
$$\sigma^2 = \gamma^2(\Gamma(1 + 2/\beta) + \Gamma^2(1 + 1/\beta))$$

By altering the parameter, β, $W(t \mid \beta, \gamma)$ takes a variety of shapes, Dodson(1994). Some values of shape parameter are important special cases: for $\beta = 1$, W is the exponential distribution; for $\beta = 2$, W is the Rayleigh distribution; for $\beta = 2.5$, W approximates the lognormal distribution; for $\beta = 3.6$, W approximates the normal distribution; and for $\beta = 5.0$, W approximates the peaked normal distribution. The flexibility of the Weibull distribution makes it very useful for empirical modeling, specially in quality control and reliability. The regions $\beta < 1$, $\beta = 1$, and $\beta > 1$ correspond to decreasing, constant and increasing hazard rates. These three regions are also known as infant mortality, memoryless, and wearout failures. γ is approximately the 63rd percentile of the life time, regardless of the shape parameter.

The Weibull also has important theoretical properties. If n i.i.d. random variables have Weibull distribution, $X_i \sim w(t \mid \beta, \gamma)$, then the first failure is a Weibull variate with characteristic life $\gamma/n^{1/\beta}$, i.e. $X_{[1,n]} \sim w(t \mid \beta, \gamma/n^{1/\beta})$. This kind of property allows a characterization of the Weibull as a limiting life distribution in the context of extreme value theory, Barlow and Prochan (1975).

The affine transformation $t = t' + \alpha$ leads to the three parameter truncated Weibull distribution. A location (or threshold) parameter, $\alpha > 0$ represents beginning observation of a (truncated) Weibull variate at $t = 0$, after it has already survived the period $[-\alpha, 0[$. The three parameter truncated Weibull is given by:

$$w(t \mid \alpha, \beta, \gamma) = (\beta\,(t + \alpha)^{\beta-1}/\gamma^\beta)\,exp(-((t + \alpha)/\gamma)^\beta)/r(\alpha \mid \beta, \gamma)$$
$$r(t \mid \alpha, \beta, \gamma) = exp(-((t + \alpha)/\gamma)^\beta)/r(\alpha \mid \beta, \gamma)$$

6. Display Panels

We were faced with the problem of testing the wearout of a lot of used display panels. A panel displays 12 to 18 characters. Each character is displayed as a 5×8 matrix of pixels, and each pixel is made of 2 (RG) or 3

(RGB) individual color elements, (like a light emitting diode or gas plasma device). A panel fails when the first individual color element fails. The construction characteristics of a display panel makes the Weibull distribution specially well suited to model its life time. The color elements are "burned in" at the production process, so we assume they are not at the infant mortality region, i.e. we assume the Weibull's shape parameter to be greater than one, with wearout or increasing hazard rates.

The panels in question were purchased as used components, taken from surplus machines. The dealer informed the machines had been operated for a given time, and also informed the mean life of the panels at those machines. Only working panels were acquired. The acquired panels were installed as components on machines of a different type. The use intensity of the panels at each type of machine corresponds to a different time scale, so mean lifes are not directly comparable. The shape parameter however is an intrinsic characteristic of the panel. The used time over mean life ratio, $\rho = \alpha/\mu$, is adimensional, and can therefore be used as an intrinsic measure of wearout. We have recorded the time to failure, or times of withdrawal with no failure, of the panels at the new machines, and want to use this data to corroborate (or not) the wearout information provided by the surplus equipment dealer.

7. The Model

The problem described at the preceding sections can be tested using the FBST, with parameter space, hypothesis and posterior joint density:

$$\Theta = \{(\alpha, \beta, \gamma) \in \,]0, \infty] \times [1, \infty] \times [0, \infty[\,\}$$
$$\Theta_0 = \{(\alpha, \beta, \gamma) \in \Theta \,|\, \alpha = \rho\mu(\beta, \gamma)\,\}$$
$$f(\alpha, \beta, \gamma \,|\, D) \propto \prod_{i=1}^{n} w(t_i \,|\, \alpha, \beta, \gamma) \prod_{j=1}^{m} r(t_j \,|\, \alpha, \beta, \gamma)$$

where the data D are all the recorded failure times, $t_i > 0$, and the times of withdrawal with no failure, $t_j > 0$.

At the optimization step it is better, for numerical stability, to maximize the log-likelihood, $fl(\;)$. Given a sample with n recorded failures and m withdrawals,

$$wl_i = \log(\beta) + (\beta - 1)\log(t_i + \alpha) - \beta\log(\gamma) - ((t_i + \alpha)/\gamma)^{\beta} + (\alpha/\gamma)^{\beta}$$

$$rl_j = -((t_j + \alpha)/\gamma)^\beta + (\alpha/\gamma)^\beta$$

$$fl = \sum_{i=1}^{n} wl_i + \sum_{j=1}^{m} rl_j$$

the hypothesis being represented by the constraint

$$h(\alpha, \beta, \gamma) = \rho\gamma\Gamma(1 + 1/\beta) - \alpha = 0$$

The gradients of $fl(\,)$ and $h(\,)$ analytical expressions, to be given to the optimizer, are:

$dwl =$

$$[\,(\beta - 1)/(t + \alpha) - ((t + \alpha)/\gamma)^\beta \beta/(t + \alpha) + (\alpha/\gamma)^\beta \beta/\alpha\,,$$
$$1/\beta + \log(t + \alpha) - \log(\gamma) - ((t + \alpha)/\gamma)^\beta \log((t + \alpha)/\gamma) + (\alpha/\gamma)^\beta \log(\alpha/\gamma)\,,$$
$$-\beta/\gamma + ((t + \alpha)/\gamma)^\beta \beta/\gamma - (\alpha/\gamma)^\beta \beta/\gamma\,]$$

$drl =$

$$[\,-((t + \alpha)/\gamma)^\beta \beta/(t + \alpha) + (\alpha/\gamma)^\beta \beta/\alpha\,,$$
$$-((t + \alpha)/\gamma)^\beta \log((t + \alpha)/\gamma) + (\alpha/\gamma)^\beta \log(\alpha/\gamma)\,,$$
$$((t + \alpha)/\gamma)^\beta \beta/\gamma\,,\; -(\alpha/\gamma)^\beta \beta/\gamma\,]$$

$dh =$

$$[\,-1\,,\; -\rho\gamma\Gamma'(1 + 1/\beta)\Gamma(1 + 1/\beta)/\beta^2\,,\; \rho\Gamma(1 + 1/\beta)\,]$$

For gamma and digamma functions efficient algorithms see Spanier and Oldham (1987).

8. Numerical Example

Table 1 displays 45 failure times (in years), plus 5 withdrawals, for a small lot of 50 panels, in a 3.5 years long experiment. The panels have supposedly been used, prior to acquisition, for 30% of its mean life, i.e. we want to test $\rho = 0.3$. In general, some prior distribution of the shape parameter is needed to stabilize the model. Knowing color elements' life time to be approximately normal, we consider $\beta \in [3.0, 4.0]$. Table 2 displays the evidence of some values of ρ. The maximum likelihood estimates of the Weilbull's parameters are $\alpha = 1.25$, $\beta = 3.28$ and $\gamma = 3.54$; so the estimates $\mu = 3.17$ and $\rho = 0.39$. The FBST corroborates the hypothesis $\rho = 0.3$ with an evidence of 98%.

Table 1. Failure times and withdrawals in years, $n = 45, m = 5$

0.01	0.19	0.51	0.57	0.70	0.73	0.75	0.75	1.11	1.16
1.21	1.22	1.24	1.48	1.54	1.59	1.61	1.61	1.62	1.62
1.71	1.75	1.77	1.79	1.88	1.90	1.93	2.01	2.16	2.18
2.30	2.30	2.41	2.44	2.57	2.61	2.62	2.72	2.76	2.84
2.96	2.98	3.19	3.25	3.31	+1.19	+3.50	+3.50	+3.50	+3.50

Table 2. Evidence for some values of ρ

ρ	0.05	0.10	0.20	0.30	0.40	0.50	0.60	0.70	0.80	0.90
Evid	0.04	0.14	0.46	0.98	1.00	0.98	0.84	0.47	0.21	0.01

9. Final Remarks

The theory presented in this paper, grew out of the necessity of the authors' activities in the role of audit, control or certification agents, Pereira and Stern (1999a). These activities made the authors (sometimes painfully) aware of the benefit of the doubt juridical principle, or safe harbor liability rule. This kind of principle establishes that there is no liability as long as there is a reasonable basis for belief, effectively placing the burden of proof on the plaintiff, who, in a lawsuit, must prove false a defendant's misstatement. Such a rule also prevents the plaintiff from making any assumption not explicitly stated by the defendant, or tacitly implied by existing law or regulation. The use of an a priori point mass on the null hypothesis, as on standard Bayesian tests, can be regarded as such an ad hoc assumption.

As audit, control or certification agents, the authors had to check compliance with given requirements and specifications, formulated as precise hypotheses on contingency tables. In Pereira *et al.* (1999b) we describe several applications based on contingency tables, comparing the use of FBST with standard Bayesian and Classical tests. The applications presented in this paper are very similar in spirit, but we are not aware of any standard exact test in the literature. The implementation of FBST is immediate and trivial, as long as good numerical optimization and integration programs are at hand. In the applications in this paper, as well in those in Pereira *et al.* (1999b), it is desirable or necessary to use a test with the following characteristics:

- Be formulated directly in the original parameter space.
- Take into account the full geometry of the null hypothesis as a manifold (surface) imbedded in the whole parameter space.

- Have an intrinsically geometric definition, independent of any non-geometric aspect, like the particular parameterization of the (manifold representing the) null hypothesis being used.
- Be consistent with the benefit of the doubt juridical principle (or safe harbor liability rule), i.e. consider in the "most favorable way" the claim stated by the hypothesis.
- Consider only the observed sample, allowing no ad hoc artifice (that could lead to judicial contention), like a positive prior probability distribution on the precise hypothesis.
- Consider the alternative hypothesis in equal standing with the null hypothesis, in the sense that increasing sample size should make the test converge to the right (accept/reject) decision.
- Give an intuitive and simple measure of significance for the null hypothesis, ideally, a probability in the parameter space.

FBST has all these theoretical characteristics, and straightforward (computational) implementation. Moreover, as shown in Madruga *et al.* (2001), the FBST is also in perfect harmony with the Bayesian decision theory of Rubin (1987), in the sense that there are specific loss functions which render the FBST.

We remark that the evidence calculus defining the FBST takes place entirely in the parameter space where the prior was assessed by the scientist, Lindley (1983). We call it the "original" parameter space, although acknowledging that the parameterization choice for the statistical model semantics is somewhat arbitrary. We also acknowledge that the FBST is not invariant under general change of parameterization.

The FBST is in sharp contrast with the traditional schemes for dimensional reduction, like the elimination of so called "nuisance" parameters. In these "reduced" models the hypothesis is projected into a single point, greatly simplifying several procedures. Problems with the traditional approach are presented in Pereira and Lindley (1987). The traditional reduction or projection schemes are also incompatible with the benefit of doubt principle, as stated earlier. In fact, preserving the original parameter space, in its full dimension, is the key for the intrinsic regularization mechanism of the FBST, when it is used in the context of model selection, Pereira and Stern (2000,2001).

Of course, there is a price to be paid for working with the original parameter space, in its full dimension: A considerable computational work load. But computational difficulties can be overcome with the used of efficient continuous optimization and numerical integration algorithms. Large problems can also benefit from program vectorization and parallelization techniques. Dedicated vectorized or parallel machines may be expensive and not always available, but most of the algorithms needed can benefit from asynchronous and coarse grain parallelism, a resource easily available, although rarely used, on any PC or workstation network through MPI, Portable Parallel Programming Message-Passing Interface, or similar distributed processing environments, Wilson and Lu (1996).

Finally, we notice that statements like "increase sample size to reject (accept) the hypothesis" made by many users of frequentist (standard Bayesian) tests, do not hold for the FBST. Increasing the sample size makes the FBST converge to the Boolean truth indicator of hypothesis being tested. In this sense, the FBST has good acceptance/rejection symmetry, even if the safe harbor rule prevents this symmetry from being perfect, introducing an offset for small samples. We believe that the existence of a precise hypothesis test with the FBST's symmetry properties has important consequences in knowledge theory, given the role played by the completely asymmetric standard statistical tests in some epistemological systems, Carnap (1962), Popper (1989).

References

1. Aitkin, M. (1991). Posterior Bayes Factors. *J R Statist Soc B* **53**(1), 111–142.
2. Barlow, R.E. and Prochan, F. (1988) *Statistical Theory of Reliability and Life Testing.* NY: Holt, Rinehart and Winston Inc.
3. Barlow, RE and Prochan, F (1988) Life Distributions Models and Incomplete Data. In: *Handbook of Statistics 7: Quality Control and Reliability.* (Krishnaiah, PR and Rao, CR eds.) Amsterdam: North-Holand pp 225-250.
4. Bazaraa, M.S., Sherali, H.D., and Shetty, C.M. (1993) *Nonliear Programming: Theory and Algorithms.* NY: Wiley.
5. Berger, J.O. and M Delampady, M. (1987). Testing precise hypothesis. *Statistical Science* **3**, 315–352.
6. Berger, J.O., Boukai, B. and Wang, Y. (1997). Unified frequentist and Bayesian testing of a precise hypothesis. *Statistical Science* **3**, 315–352.
7. Bratley, P., Fox B.L. and Schrage, L. *A Guide to Simulation.* Springer-Verlag, 1987.
8. Dodson, B. (1994). *Weibull Analysis.* Milwaukee: ASQC Quality Press.
9. Carnap, R. (1962). *Logical Foundations of Probability.* Univ. of Chicago

Press.
10. Cox, D.R. (1977). The role of significance tests. *Scand J Statist* **4**, 49–70
11. DeGroot, M.H. (1970). *Optimal Statistical Decisions.* NY: McGraw-Hill.
12. Evans, M. (1997). Bayesian Inference Procedures Derived via the Concept of Relative Surprise. *Communications in Statistics* **26**, 1125–1143.
13. Galassi, M., Davies J., Theiler, J., Gough, B., Priedhorsky, R., Jungman, G. and Booth, M. (1999). *GSL - GNU Scientific Library Reference Manual V-0.5.* WWW: lanl.gov.
14. Gomez, C. (1999). *Engineering and Scientific Computing with Scilab.* Berlin: Birkhäuser.
15. Good, I.J. (1983). *Good thinking: The foundations of probability and its applications.* University of Minnesota Press.
16. Finnetti, B. de (1989). *Decisão.* in Enciclopédia EINAUDI. Romano, R. (edtr.). O Porto: Imprensa Nacional.
17. Horst, R., Pardalos P.M. and Thoai N.V. (1995). *Introduction to Global Optimization.* Boston: Kluwer.
18. Irony, T.Z. and Pereira, C.A.B. (1986). Exact test for equality of two proportions: Fisher×Bayes. *J Statist Comp & Simulation* **25**, 93–114.
19. Irony, T.Z. and Pereira C.A.B. (1995). Bayesian Hypothesis Test: Using surface integrals to distribute prior information among hypotheses. *Resenhas* **2**, 27–46.
20. Kempthorne, O. and Folks, L. (1971). *Probability, Statistics, and Data Analysis.* Ames: Iowa State U.Press.
21. Knuth, D.E. *The Art of Computer Programming, vol 2 - Seminumerical Algorithms.* Addison Wesley, 1996.
23. Krommer, A.R. and Ueberhuber C.W. (1998). *Computational Integration.* Philadelphia: SIAM.
24. L'Ecuyer, P. Efficient and Portable Combined Pseudo-Random Number Generators. *Commun. ACM*, 1988.
25. Lepage, G.P (1978). VEGAS: An Adaptive Multidimensional Integration Program. *J. Comput. Phys.* **27**, 192–205.
26. Lehmann, E.L. (1986). *Testing Statistical Hypothesis.* NY: Wiley.
27. Lindley, D.V. (1957). A Statistical Paradox. *Biometrika* **44**, 187–192.
28. Lindley, D.V. (1978). The Bayesian Approach. *Scand J Statist* **5**, 1-26.
29. Luenberger, D.G. (1984). *Linear and Nonlinear Programming.* Reading: Addison-Wesley.
30. Madruga, M.R. Esteves, L.G. and Wechsler, S. (2001). On the Bayesianity of Pereira-Stern Tests. To appear in *TEST.*
31. Nocedal, J. and Wright, S. (1999). *Numerical Optimization.* NY: Springer.
32. Pereira,C.A.B. and Lindley,D.V. (1987). Examples Questioning the Use of Partial Likelihood. *The Statistician* **36**, 15–20.
33. Pereira,C.A.B. and Wechsler,S. (1993). On the Concept of *p-value. Braz J Prob Statist* **7**, 159–177.
34. Pereira,C.A.B. Stern,J.M. (1999a). A Dynamic Software Certification and

Verification Procedure. *Proc. ISAS-99 - International Conference on Information Systems Analysis and Synthesis* II, 426–435.
35. Pereira,C.A.B. Stern,J.M. (1999b). Evidence and Credibility: Full Bayesian Significance Test for Precise Hypotheses. *Entropy* **1**, 69–80.
36. Pereira,C.A.B. Stern,J.M. (2000). *Intrinsic Regularization in Model Selection using the Full Bayesian Significance Test.* Technical Report RT-MAC-2000-6, Dept. of Computer Science, University of Sao Paulo.
37. Pereira,C.A.B. Stern,J.M. (2001). Model Selection: Full Bayesian Approach. To appear in *Environmetrics.*
38. Pintér, J.D. (1996). *Global Optimization in Action. Continuous and Lipschitz Optimization: Algorithms, Implementations and Applications.* Boston: Kluwer.
39. Popper, K.R. (1989). *Conjectures and Refutations: The Growth of Scientific Knowledge.* London: Routledge.
40. Royall, R. (1997). *Statistical Evidence: A Likelihood Paradigm.* London: Chapman & Hall.
41. Rubin, H. (1987). A Weak System of Axioms for "Rational" Behavior and the Non-Separability of Utility from Prior. *Statistics and Decisions* **5**, 47–58.
42. Sellke, T., Bayarri, M.J. and Berger, J. (1999). Calibration of p-values for Testing Precise Null Hypotheses. *ISDS Discussion Paper 99-13.*
43. Sloan, I.R. and Joe, S. (1994). *Lattice Methods for Multiple Integration.* Oxford: Oxford University Press.
44. Spanier, J. and Oldham, K.B. (1987). *An Atlas of Functions.* NY: Hemisphere Publishing.
45. Vieland, V.J. and Hodge, S.E. (1998). Book Reviews: *Statistical Evidence* by R Royall (1997). *Am J Hum Genet* **63**, 283–289.
46. Wichmann, B.A. and Hill, I.D. (1982). An Efficient and Portable Pseudo-Random Number Generator. *Appl. Stat.*, **31**, 188-190.
47. Wilson,G.V. and Lu,P. (1996). *Parallel Programming Using C++.* Cambridge: MIT Press.

CHAPTER 18

BAYESIAN NONPARAMETRIC ESTIMATION OF A MONOTONE HAZARD RATE

Man-wai Ho* and A. Y. Lo†[a]

Department of Information and Systems Management
University of Science and Technology
Clear Water Bay, Hong Kong
E-mail: **waiho@ust.hk,*
E-mail: *†imaylo@ust.hk*

The Bayesian nonparametric estimation of a monotone hazard rate for a reliability model is considered. In general a posterior mean of a mixture hazard rate with respect to a gamma-type process prior could be written as an average over all possible partitions of the first K integers, where K is the number of the complete data. Numerical approximations to posterior quantities based on Gibbs samplers which have stationary distributions on partitions are studied. Examples are given.

1. Introduction

The Bayesian nonparametric estimation of a monotone hazard rate in the reliability context was considered by Dykstra and Laud[3] based on gamma process priors. Lo and Weng[16] model a hazard rate as a mixture of parametric kernel functions, and show that the monotone hazard rate case corresponds to a uniform kernel. They also propose approximations to posterior quantities using the Chinese restaurant process of sampling partitions.[1,12] However, the Monte Carlo sample of partitions is essentially taken from the prior distribution and it ignores the importance weights contributed by the data. As a result, the Chinese restaurant process does not yield useful Monte Carlo approximations, and Lo, Brunner and Chan[15] propose

[a]Research supported in part by Hong Kong RGC Competitive Earmarked Research Grant 6189/98E.

an importance sampler, called a weighted Chinese restaurant process (iid-WCR), for numerical evaluations of posterior quantities in the Bayesian mixture models, which seems to work better. Hayakawa, Paul and Vignaux implement the iidWCR to an IFR testing problem.[8] Recently James[11] extends work of Lo and Weng[16] to allow for additional regression parameters, and this extension provides the flexibility needed for various applications including the Cox proportional hazard model. It should be clear that the method developed by James could also be applied to proportional hazard models with time-dependent covariate functions[2]. Ishwaran and James discuss, among other schemes, extensions of the iidWCR to fit semiparametric hazard rate models.[10]

The above-mentioned works concentrate on iid Monte Carlo methods for sampling partitions. Monte Carlo methods that do not rely on sampling random partitions have been widely discussed; see for example Laud, Smith and Damien[13], Wolpert and Ickstadt[19], Gasbarra and Karia[4]. A survey on related work not depending on random partitions is given by Gelfand[5]. For a frequentist approach to this problem, see Huang and Wellner[9].

In this paper, a Gibbs sampler method[7,6] for sampling partitions[17] is developed to the evaluation of posterior quantities for mixture hazard rate models. The emphasis is on a monotone hazard rate model. Sec. 2 discusses the prior to posterior analysis of a monotone hazard rate model based on a gamma-type process prior. Posterior quantities are expressed as averages over partitions. A Gibbs sampler on partitions which yields Gibbs averages for approximations to posterior quantities is discussed in Sec. 3. For comparison purposes, another Gibbs chain on partitions, which takes slightly less importance weight into account in sampling within the Gibbs cycle, is discussed in Sec. 4. The latter Gibbs chain yields weighted Gibbs averages as approximations to posterior quantities. Sec. 5 gives numerical examples on the Gibbs approximations to a posterior mean of the decreasing hazard rate. The examples seem to suggest that care must be taken in using the weighted Gibbs average as an approximation.

2. Bayes Methods for a Decreasing Hazard Rate Model

This section is concerned with the Bayes estimation of a decreasing hazard function on the half line. The model can be described by a life-testing experiment of N light bulbs. The bulbs are assumed to have independent and identically distributed life times with density $f(t)$, $t \geq 0$. Then

$$r(t) = \frac{f(t)}{1 - F(t)} \tag{1}$$

is the hazard rate of the model. Suppose the experiment terminates at time τ, and let $T_1, \ldots, T_K$ be the K complete failure times and $T_{K+1} = \cdots = T_N = \tau$ be $N - K$ censored times. The likelihood function of the hazard rate is proportional to

$$\left[\prod_{i=1}^{K} r(T_i)\right] \exp\left[-\int_0^{\tau} Y(s)\, r(s)\, ds\right],$$

where $Y(s)$ is a left continuous integer-valued function which counts the number of bulbs not yet "failed" just before age s, given by

$$Y(t) = \begin{cases} N & t \leq T_1 \\ N - 1 & T_1 < t \leq T_2 \\ \vdots & \\ N - K & T_K < t \leq \tau. \end{cases}$$

$Y(.)$ is called the total time transform (TTT). For brevity, the following notation will be used.

$$T_0 = 0, T_{K+1} = \tau, \text{ and } T_{K+2} = \infty.$$

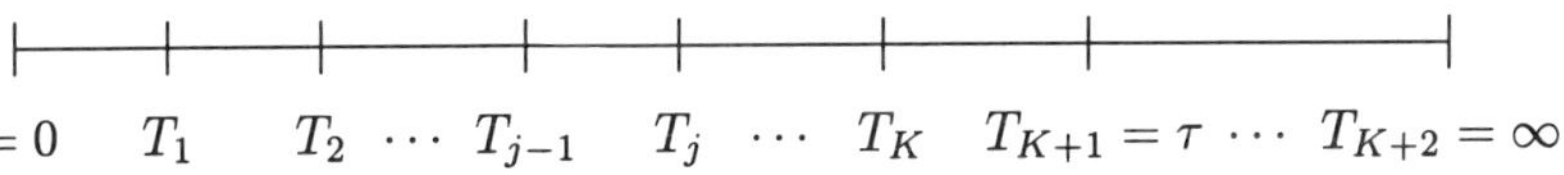

For a Bayesian decreasing hazard rate model, the hazard rate (1) can be represented by

$$r(t\,|\mu) = \int_R I_{\{0 \leq t < u\}}\, \mu(du), \quad \text{for } t \in [0, \tau], \tag{2}$$

a mixture of indicators $I_{\{0 \leq t < u\}}$ by a freely varied (locally) finite measure $\mu(du)$.[3] In this case, $\mu \in \Theta$ is the parameter. The likelihood function of $\mu \in \Theta$ becomes

$$L(\mu\,|\mathbf{T}) \propto \left[\prod_{i=1}^{K} \int_R I_{\{0 \leq T_i < u\}}\, \mu(du)\right] \times \exp\left\{-\int_0^{\tau} Y(s)\left[\int_R I_{\{0 \leq s < u\}}\, \mu(du)\right] ds\right\}. \tag{3}$$

The likelihood function of μ looks like a gamma density in μ. The idea of conjugate priors indicates that a gamma-like prior on the mixing measure μ facilitates computations of posterior quantities. One possible choice of such a prior distribution on μ is a weighted gamma process, denoted by $\mathcal{G}(d\mu|\alpha,\beta)$, with shape measure α and scale measure β (See Remark 2). Given μ, let $\mathbf{T}=(T_1,\ldots,T_K,\ldots,T_N)$ be i.i.d. observations from a reliability model with the decreasing hazard rate $r(t|\mu)$. It is known that if μ has a weighted gamma process prior distribution, the posterior distribution of μ is a mixture of weighted gamma processes.[16]

Lo, Brunner and Chan[15] discuss the structure of the posterior distribution of μ in the general mixture hazard rate models precisely using a three-step simulation experiment. Their description, specializing to the monotone hazard rate model, goes as follows: let $\mathbf{p}=\{C_i, i=1,2,\ldots,n(\mathbf{p})\}$ be a partition of $\{1,2,\ldots,K\}$, and e_i is the number of elements in cell C_i, respectively,

(i) $\mathbf{p}=\left\{C_1,\ldots,C_{n(\mathbf{p})}\right\}$ has a discrete distribution $W^*(\mathbf{p})$;
(ii) given $\mathbf{p}, u_1,\ldots,u_{n(\mathbf{p})}$ are independent and the density of u_i is

$$\alpha(du|C_i)\propto[\beta^*(u)]^{e_i}\, I_{\{0\le T_{C_i}<u\}}\alpha(du), \quad i=1,\ldots,n(\mathbf{p});$$

(iii) given $\left(\mathbf{p},u_1,\ldots,u_{n(\mathbf{p})}\right)$, μ has a $\mathcal{G}\left(d\mu\,\middle|\,\alpha+\sum_{i=1}^{n(\mathbf{p})}e_i\delta_{u_i},\beta^*\right)$ distribution,

where $T_{C_i}=\max_{j\in C_i}T_j$,

$$W^*(\mathbf{p})=\frac{\varphi^*(\mathbf{p})}{\sum_{\mathbf{p}}\varphi^*(\mathbf{p})}, \tag{4}$$

$$\varphi^*(\mathbf{p})=\prod_{1\le i\le n(\mathbf{p})}\left\{(e_i-1)!\int[\beta^*(u)]^{e_i}\, I_{\{0\le T_{C_i}<u\}}\alpha(du)\right\},$$

and

$$\beta^*(u)=\frac{\beta(u)}{1+\beta(u)\int_0^\tau Y(s)\,I_{\{0\le t<s\}}ds}.$$

The posterior mean serves as a point estimator of μ. According to (i), a posterior mean of the monotone hazard rate is a $W^*(\mathbf{p})$ – mixture of basic hazard rate $m_i^*(t|\mathbf{p})$'s: For each $t\in[0,\tau]$,

$$E[r(t|\mu)|T]=m_0^*(t)+\sum_{\mathbf{p}}W^*(\mathbf{p})\sum_{i=1}^{n(\mathbf{p})}e_i\,m_i^*(t|\mathbf{p}), \tag{5}$$

where $\sum_{\mathbf{p}}$ is summing up over all partitions of $\{1, 2, \ldots, K\}$,

$$m_0^*(t) = \int \beta^*(u) I_{\{0 \le t < u\}} \alpha(du), \tag{6}$$

and

$$m_i^*(t \,|\mathbf{p}) = \frac{\int [\beta^*(u)]^{e_i+1} I_{\{0 \le \max(t, T_{C_i}) < u\}} \alpha(du)}{\int [\beta^*(u)]^{e_i} I_{\{0 \le T_{C_i} < u\}} \alpha(du)}.$$

Remark 1: One could view $\alpha(du\,|C_i)$ as a "posterior" distribution of u_i given table C_i where $\alpha(du)$ is the "prior". In this sense,

$$m_i^*(t\,|\mathbf{p}) = \int \beta^*(u) I_{\{0 \le t < u\}} \alpha(du\,|C_i),$$

and can be called a "predictive" hazard rate given C_i.

Remark 2: The weighted gamma process (random measure) is defined in Lo[14] as follows. ν is a gamma process with a (locally) finite shape $\alpha(.)$ measure if

(i) $\nu(.)$ is an "independent increment" process,
(ii) $\nu(A)$ is a *gamma* $(\alpha(A); 1)$ random variable.

μ is a weighted gamma process with shape $\alpha(.)$ and weight $\beta(.)$ if

$$\mu(A) = \int I_{\{s \in A\}} \beta(s) \nu(ds).$$

3. A Gibbs Sampler for $W^*(\mathbf{p})$: Gibbs Averages

In the present setting, a posterior quantity can often be written as an average over partitions

$$\eta = \sum_{\mathbf{p}} \phi(t, \mathbf{p}) W^*(\mathbf{p}). \tag{7}$$

In particular,

$$\phi(t, \mathbf{p}) = m_0^*(t) + \sum_{i=1}^{n(\mathbf{p})} e_i m_i^*(t\,|\mathbf{p}), \tag{8}$$

in (5). The exact evaluation of (7) for moderate or large sample sizes K posts a formidable problem as the number of summands increases roughly as the factorial of K. One way to evaluate (7) is to use the Monte Carlo method to sample a random partition $\mathbf{p} \sim W^*(\mathbf{p})$. The exact simulation[18] of $\mathbf{p} \sim W^*(\mathbf{p})$ is not yet known, one could recourse to the Gibbs sampler which has $W^*(\mathbf{p})$ as a stationary distribution.

The key ingredient of a Gibbs sampler for a random partition $\mathbf{p} \sim W^*(\mathbf{p})$ consists of a Gibbs cycle which randomly generates a partition $\mathbf{p}$ of $\{1,\ldots,n\}$ from a given partition $\tilde{\mathbf{p}}$ of $\{1,\ldots,n\} - \{r\}$, for $r = 1,\ldots,n$. The selection is achieved by randomly seating the integer r to tables in $\tilde{\mathbf{p}} = \left\{\tilde{C}_1,\ldots,\tilde{C}_{n(\tilde{\mathbf{p}})}\right\}$, forming a $\mathbf{p}$. The required seating probability, i.e., the conditional distribution of $\mathbf{p}\,|\tilde{\mathbf{p}}$, $Pr\{\mathbf{p}\,|\tilde{\mathbf{p}}\}$, turns out to be a predictive probability, which can be determined by the distribution of $\mathbf{p}$. Since $\tilde{\mathbf{p}}$ is a function of the resulting $\mathbf{p}$ (i.e., deleting r from this $\mathbf{p}$ will get $\tilde{\mathbf{p}}$), $W^*(\mathbf{p})$ induces a distribution on $\{\tilde{\mathbf{p}}\}$. It turns out that[15]

$$Pr\{\mathbf{p}\,|\tilde{\mathbf{p}}\} \propto \tilde{e}_s \times m_s^*(x_r\,|\tilde{\mathbf{p}}), \quad s = 0,\ldots,n(\tilde{\mathbf{p}}), \tag{9}$$

where

$$m_0^*(x_r\,|\tilde{\mathbf{p}}) \equiv m_0^*(x_r) \qquad \text{and} \qquad \tilde{e}_0 \equiv 1.$$

The seating probability (9) defines a weighted Chinese restaurant process (WCR). [See Remark 3 below]. A Gibbs cycle for the corresponding Gibbs sampler is defined in two steps:

(I) creating the delete$-r$ partition $\tilde{\mathbf{p}}$ from $\mathbf{p}$, and reseating the integer r to $\tilde{\mathbf{p}}$ with

$$P\left\{r \text{ sits at table } \tilde{C}_s\,|\tilde{\mathbf{p}}\right\} \propto \tilde{e}_s \times m_s^*(x_r\,|\tilde{\mathbf{p}}), \quad s = 0,\ldots,n(\tilde{\mathbf{p}}),$$

(II) repeating (I) for $r = 1,\ldots,n$.

Steps (I) and (II) move an initial state $\mathbf{p} = \mathbf{p}_0$ to a new state $\mathbf{p} = \mathbf{p}_1$, and complete a Gibbs cycle. Repeat to get a Markov chain $\mathbf{p}_0, \mathbf{p}_1, \ldots, \mathbf{p}_M$, where $\mathbf{p}_0$ is an initial partition. The Gibbs average

$$\frac{1}{M}\sum_{m=1}^{M} \phi(t, \mathbf{p}_m) \tag{10}$$

approximates η in (7).

Remark 3: The Chinese Restaurant process (CR) with parameter $c > 0$ goes as follows: Customers $1,\ldots,n$ enters an initially empty restaurant in the order written, and customer 1 is seated at an empty table. Suppose the first $r-1$ customers have been seated, customer r is seated at an empty table with probability proportional to c; otherwise, he/she is seated at an occupied table with probability proportional to the number of customers at that table. The resulting random partition after seating $1, \ldots, n$ is Chinese Restaurant process with parameter $c > 0$. Note that, for the CR, the seating probability is given by

$$Pr\{\mathbf{p}\,|\tilde{\mathbf{p}}\} \propto \tilde{e}_s, \quad s = 0, \ldots, n(\tilde{\mathbf{p}})\,. \tag{11}$$

As (9) reweighs the seating probability of the CR (11), it is the seating probability of a weighted Chinese restaurant process.

4. An Alternative Gibbs Sampler: Weighted Gibbs Averages

Alternative Gibbs samplers with easily computable seating probabilities can be obtained via a change of measure technique. Lo, Brunner and Chan[15] propose the use of weighted averages of the form (14) for posterior approximation if the shape measure $\alpha(du)$ has a form which leads to efficient computation of the seating probabilities (9). This technique could be very useful for various applications involving, for example complex $\alpha(du)$'s. Let us illustrate this technique by assuming that $\beta^*(.)$ is a rather complicated function (i.e., not a step function), so that the seating probabilities (9) are not computable. Suppose $\mathbf{p}_0, \mathbf{p}_1, \ldots, \mathbf{p}_M, \ldots,$ is a Markov chain sequence of partitions with a stationary distribution

$$W^{**}(\mathbf{p}) = \frac{\varphi(\mathbf{p})}{\sum_{\mathbf{p}} \varphi(\mathbf{p})}, \tag{12}$$

and

$$\varphi(\mathbf{p}) = \prod_{1 \le i \le n(\mathbf{p})} \left[(e_i - 1)! \int I_{\{0 \le T_{C_i} < u\}} \alpha(du) \right].$$

The seating probability of this Gibbs sampler, $Pr\{\mathbf{p}\,|\tilde{\mathbf{p}}\}$, is (9) with $m_s^*(t\,|\tilde{\mathbf{p}})$ there replaced by

$$m_s^{**}(t\,|\tilde{\mathbf{p}}) = \frac{\int I_{\{0 \le \max(t, T_{\tilde{C}_s}) < u\}} \alpha(du)}{\int I_{\{0 \le T_{\tilde{C}_s} < u\}} \alpha(du)}, \quad s = 0, \ldots, n(\tilde{\mathbf{p}}), \tag{13}$$

where

$$m_0^{**}(t\,|\tilde{\mathbf{p}}) \equiv m_0^{**}(t) = \int I_{\{0 \le t < u\}} \alpha(du)\,.$$

Note that the Gibbs sampler derived from $W^{**}(\mathbf{p})$ does not involve the $\beta^*(.)$ part which is assumed to be troublesome. The alternative Gibbs sampler approximation, gWCR(a), to the posterior mean (5) is then given by a weighted Gibbs average: for any $t \in [0, \tau]$,

$$\frac{\sum_{m=1}^{M}\left\{\psi(t, \mathbf{p}_m)\left[\prod_{b=1}^{n(\mathbf{p}_m)} \beta^*\left(u_b^{(m)}\right)^{e_b^{(m)}}\right]\right\}}{\sum_{m=1}^{M}\left\{\prod_{b=1}^{n(\mathbf{p}_m)}\left[\beta^*\left(u_b^{(m)}\right)\right]^{e_b^{(m)}}\right\}}, \tag{14}$$

where for $m = 1, 2, \ldots, M$, $\mathbf{p}_m = \left\{C_1^{(m)}, \ldots, C_{n(\mathbf{p}_m)}^{(m)}\right\}$ with table sizes $e_1^{(m)}, \ldots, e_{n(\mathbf{p}_m)}^{(m)}$, respectively, $u_0^{(m)}$ is simulated from $\alpha(du)/\alpha(R)$, $u_i^{(m)}$ is simulated from the density proportional to $I_{\left\{0 \le T_{C_i^{(m)}} < u\right\}} \alpha(du)$, for $i = 1, \ldots, n(\mathbf{p}_m)$, and

$$\psi(t, \mathbf{p}_m) = \alpha(R) \cdot \beta^*\left(u_0^{(m)}\right) I_{\left\{0 \le t < u_0^{(m)}\right\}} + \sum_{i=1}^{n(\mathbf{p}_m)} e_i^{(m)} \beta^*\left(u_i^{(m)}\right) I_{\left\{0 \le t < u_i^{(m)}\right\}}. \tag{15}$$

5. Numerical Results Based on a Uniform Shape Probability

This section discusses decreasing hazard rate estimation using a uniform shape probability, $\alpha(du)/\alpha(R)$, from 0 to $d(\ge \tau)$. This choice of shape simplifies the analytic expression for the posterior mean; and it results in computationally efficient Monte Carlo algorithms. See the Appendix.

5.1. *Simulation Study*

Empirical data from certain parametric models with smooth decreasing hazard rates are selected to test the resolution of the gWCRs. Samples from the *Weibull* model and the piecewise constant hazard rate model are taken. In the numerical results, $\alpha(R) = 1.0$, the Monte Carlo sample size is $M = 1000$, and the pre-determined initial partition $\mathbf{p}_0$ is K singletons of tables.

Figures 1 and 2 give gWCR approximations to $E[r(t|\mu)|T]$ in (5), based on nested samples of sizes, $N = 100$, 500, and 1000, from *Weibull* and piecewise constant hazard rate distributions. The sample size of the complete observation (K) for different whole sample size (N) cases varies due to simulation, but the censoring rate is more or less constant.

Example 4: [The *Weibull* model] A *Weibull* $(0.5, 1)$ model has a hazard rate

$$h(t) = 0.5t^{-0.5}, \quad t \geq 0.$$

The lifetime of a subject, T, has the probability density function and the survivorship function

$$f(t) = 0.5t^{-0.5} \exp\left(-t^{0.5}\right), \quad t \geq 0,$$

and

$$S(t) = \exp\left(-t^{0.5}\right), \quad t \geq 0,$$

respectively. Numerical results, with a termination time $\tau = 7$ and a uniform shape probability from 0 to 14 (i.e., $d = 14$), are presented in Fig. 1.

Example 5: [The piecewise constant hazard rate model] Here we consider a piecewise constant hazard rate model, in which

$$h(t) = \begin{cases} 1 & 0 \leq t < 1 \\ 0.5 & t \geq 1. \end{cases}$$

The lifetime of a subject, T, has the probability density function and the survivorship function

$$f(t) = \begin{cases} \exp(-t) & 0 \leq t < 1 \\ 0.5 \exp(-0.5 - 0.5t) & t \geq 1, \end{cases}$$

and

$$S(t) = \begin{cases} \exp(-t) & 0 \leq t < 1 \\ \exp(-0.5 - 0.5t) & t \geq 1, \end{cases}$$

respectively. Numerical results, with a termination time $\tau = 3$ and a uniform shape probability from 0 to 6 (i.e., $d = 6$), are presented in Fig. 2.

These examples seem to suggest that care must be taken when implementing the weighted Gibbs averages as Monte Carlo estimators. It seems that the alternative Gibbs sampler gWCR(a) ignores the contribution of $\beta^*(.)$ in the sampling scheme and this appears to retard its Monte Carlo

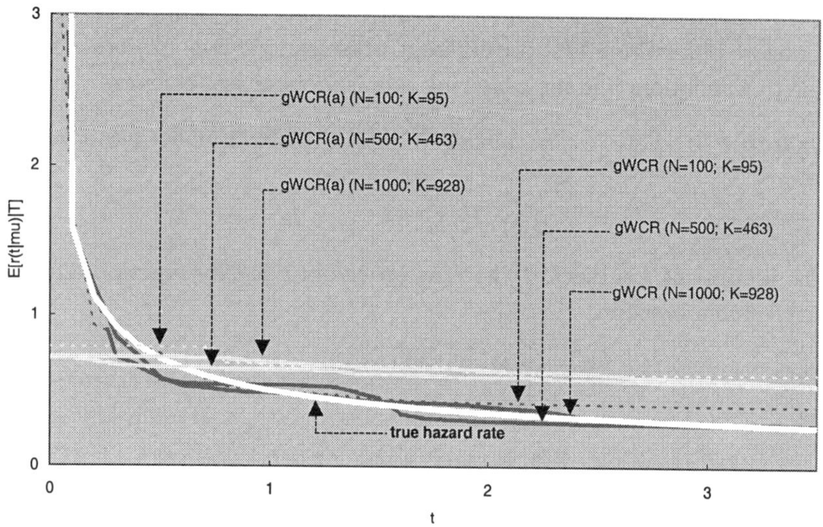

Fig. 1. The *weibull* (0.5, 1) model.

convergence. A close scrutiny at the computer printouts on the partitions reveals that the gWCR chain eventually settles down to partitions of the form $\{C_1, C_2\}$. Furthermore, T_{C_1} is close to 1, i.e., the jump of the true hazard rate curve, and T_{C_2} is close to 3 which is the time the experiment is terminated (the ratio e_1/e_2 settles down to 1/3). In this sense, gWCR correctly identifies the true population hazard rate. On the other hand, the gWCR(a) settles down to the partition consisting of one table (with 3 being the maximum observation corresponding to that table). Further work is needed to study the performance of the Gibbs weighted averages.

Appendix

For simplicity, we assume $\beta(u) = 1$, for all u, and $N > K$, then $\beta^*(u)$ has an explicit expression

$$\begin{aligned}\beta^*(u) &= \frac{1}{1 + \int_0^\tau Y(t)\, I_{\{0 \le t < u\}} dt} \\ &= \begin{cases} \dfrac{1}{1 + \sum_{i=1}^{j-1} T_i + (N - j + 1)\, u} & \text{if } T_{j-1} < u \le T_j, j = 1, \dots, K+1 \\ \dfrac{1}{1 + \sum_{i=1}^{K} T_i + (N - K)\, \tau} & \text{if } T_{K+1} = \tau < u \le \infty = T_{K+2}. \end{cases}\end{aligned}$$

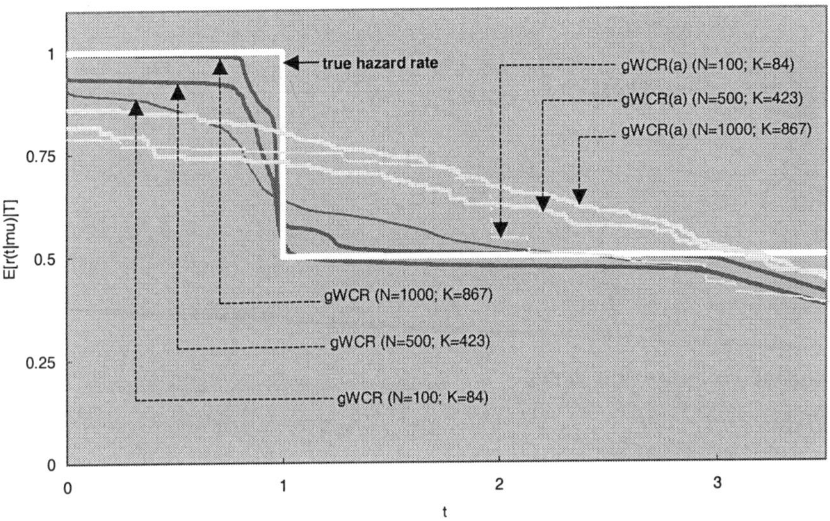

Fig. 2. The piecewise constant hazard rate model.

Taking into account that $\beta^*(u)$ is a step function, the $m_0^*(t)$ can be simplified as follows:

$$m_0^*(t) = \frac{1}{d}\int_0^\infty \beta^*(u) I_{\{t<u<d\}} du$$

$$= \begin{cases} \frac{1}{d}\left\{ \frac{1}{N-a+1} \log\left[1 + \frac{(N-a+1)(T_a - t)}{1+\sum_{i=1}^{a-1} T_i + (N-a+1)t}\right] \right. \\ \quad + \sum_{j=a+1}^{K+1} \frac{1}{N-j+1} \log\left[1 + \frac{(N-j+1)(T_j - T_{j-1})}{1+\sum_{i=1}^{j-1} T_i + (N-j+1)T_{j-1}}\right] \\ \quad \left. + \frac{d-\tau}{1+\sum_{i=1}^{K} T_i + (N-K)\tau} \right\} & \text{if } T_{a-1} < t \le T_a, \\ & \quad a = 1, \ldots, K \\ \frac{1}{d}\left\{ \frac{1}{N-K} \log\left[1 + \frac{(N-K)(T_{K+1} - t)}{1+\sum_{i=1}^{K} T_i + (N-K)t}\right] \right. \\ \quad \left. + \frac{d-\tau}{1+\sum_{i=1}^{K} T_i + (N-K)\tau} \right\} & \text{if } t \in (T_K, T_{K+1}] \\ \frac{d-t}{d\left[1+\sum_{i=1}^{K} T_i + (N-K)\tau\right]} & \text{if } T_{K+1} = \tau \\ & \quad < t \le d < \infty. \end{cases}$$

Similarly, one obtains $m_i^*(t\,|\mathbf{p})$:

$$m_i^*(t\,|\mathbf{p}) = \frac{\int_0^\infty [\beta^*(u)]^{e_i+1} I_{\{\max(t,T_{C_i})<u<d\}} du}{\int_0^\infty [\beta^*(u)]^{e_i} I_{\{T_{C_i}<u<d\}} du} = \frac{\pi_{e_i+1}(\max(t, T_{C_i}))}{\pi_{e_i}(T_{C_i})},$$

where

$$\pi_e(t) = \int_0^\infty [\beta^*(u)]^e \cdot I_{\{t<u<d\}} du, \qquad e = 0, 1, 2, \ldots.$$

Routine computation yields

(i) if $t \in (T_{a-1}, T_a]$, i.e., $T_{a-1} < t \le T_a, a = 1, 2, \ldots, K$:

$$\begin{aligned}
\pi_e(t) &= \int_t^{T_a} [\beta^*(u)]^e du + \sum_{j=a+1}^{K+1} \int_{T_{j-1}}^{T_j} [\beta^*(u)]^e du + \int_{T_K}^d [\beta^*(u)]^e du \\
&= \frac{1}{(e-1)(N-a+1)} \left\{ \left[1 + \sum_{i=1}^{a-1} T_i + (N-a+1)t \right]^{-e+1} \right. \\
&\quad \left. - \left[1 + \sum_{i=1}^{a-1} T_i + (N-a+1)T_a \right]^{-e+1} \right\} \\
&\quad + \frac{1}{e-1} \left\{ \sum_{j=a+1}^{K+1} \frac{1}{N-j+1} \left\{ \left[1 + \sum_{i=1}^{a-1} T_i + (N-j+1)T_{j-1} \right]^{-e+1} \right. \right. \\
&\quad \left. \left. - \left[1 + \sum_{i=1}^{a-1} T_i + (N-j+1)T_j \right]^{-e+1} \right\} \right\} \\
&\quad + \frac{d-\tau}{\left[1 + \sum_{i=1}^{K} T_i + (N-K)\tau \right]^e},
\end{aligned}$$

(ii) if $t \in (T_K, T_{K+1}]$:

$$\begin{aligned}
\pi_e(t) &= \int_t^{T_{K+1}} [\beta^*(u)]^e du + \int_{T_{K+1}}^d [\beta^*(u)]^e du \\
&= \frac{1}{(e-1)(N-K)} \left\{ \left[1 + \sum_{i=1}^{K} T_i + (N-K)t \right]^{-e+1} \right.
\end{aligned}$$

$$- \left[1 + \sum_{i=1}^{K} T_i + (N-K)\, T_{K+1} \right]^{-e+1} \Bigg\}$$

$$+ \frac{d - \tau}{\left[1 + \sum_{i=1}^{K} T_i + (N-K)\, \tau \right]^{e}},$$

(iii) if $t \in (\tau, d]$, i.e., $T_{K+1} = \tau < t \le d < \infty$:

$$\pi_e(t) = \frac{1}{d} \int_t^d \left[\beta^*(u)\right]^e du$$

$$= \frac{d - t}{\left[1 + \sum_{i=1}^{K} T_i + (N-K)\, \tau \right]^{e}}.$$

References

1. D. J. Aldous, Exchangeability and Related Topics, *Lecture Notes in Mathematics*, **1117** (Springer, Berlin-New York, 1985).
2. V. B. Bagdonavicius and M. S. Nikulin, Generalized proportional hazards model based on modified partial likelihood. *Lifetime Data Anal.*, **5**, 329-350 (1999).
3. R. L. Dykstra and P. Laud, A Bayesian nonparametric approach to reliability. *The Annals of Statistics*, **9**, 356-367 (1981).
4. D. Gasbarra and S. R. Karia, Analysis of competing risks by using Bayesian smoothing. *Sand. J. Statist.*, **27**, 605-617 (2000).
5. A. E. Gelfand, Approaches for semiparametric Bayesian regression. *Asymptotics, Nonparametric, and Time Series: A tribute to Madan Lal Puri* (edited by Ghosh, S.), 615-638. (Marcel Dekker, Inc. New York, 1999).
6. A. E. Gelfand and A. F. M. Smith, Sampling-based approaches to calculating marginal densities. *Journal of the American Statistical Association*, **85**, 398-409 (1990).
7. A. Geman and D. Geman, Stochastic relaxation, Gibbs distributions, and Bayesian restoration of images. *IEEE Transactions on Pattern Analysis and Machine Intelligence*, **6**, 721-741 (1984).
8. Y. Hayakawa, S. Paul and G. A. Vignaux, Testing failure data for evidence of aging via a Bayesian nonparametric method. *Proceedings of Second International Conference on Mathematical Methods in Reliability* Bordeaux, France. 509-512 (2000).
9. J. Huang and J. A. Wellner, Estimation of a monotone density or a monotone hazard under random censoring. *Scand. J. Statist.*, **22**, 3-33 (1995).
10. H. Ishwaran and L. F. James, Bayesian nonparametric methods for hazard mixture models using weighted gamma process approximations. Preprint. (2000).

11. L. F. James, Bayesian calculus for gamma processes with applications to semiparametric intensity models. Preprint. (2000).
12. L. Kuo, Computations of mixtures of Dirichlet processes. *SIAM J. Sci. Statist. Comput.*, **7**, 60-71 (1986).
13. P. W. Laud, A. F. M. Smith and P. Damien, Monte Carlo methods for approximating a posterior hazard rate process. *Stat. Comput.*, **6**, 77-83 (1996).
14. A. Y. Lo, Bayesian nonparametric statistical inference for Poisson point process, *Z. Wahrsch. Verw. Gebiete*, **59**, 55-66 (1982).
15. A. Y. Lo, L. J. Brunner and A. T. Chan, Weighted Chinese restaurant processes and Bayesian mixture models & Revision I. Research report. ISMT Department, HKUST, Hong Kong, (1996, 1998).
16. A. Y. Lo and C. S. Weng, On a class of Bayesian nonparametric estimates: II. Hazard rate estimates. *Ann. Instit. Statist. Math.*, **41**, 227-245 (1989).
17. S. N. MacEachern, Estimating normal means with a conjugate style Dirichlet process prior. *Communications in Statistics-Simulation*, **23**, 727-741 (1994).
18. J. Propp and D. Wilson, Exact sampling with coupled Markov chains and applications to statistical mechanics. *Random Structure and Algorithm*, **9**, 223-252 (1996).
19. R. L. Wolpert and K. Ickstadt, Poisson/gamma random field models for spatial statistics. *Biometrika*, **85**, 251-267 (1998).

CHAPTER 19

BAYESIAN SENSITIVITY ANALYSIS

Roger Cooke* and Daniel Lewandowski†
Department of Mathematics, Delft University of Technology
PO Box 5031, 2600 GA, Delft, The Netherlands
E-mail: *r.m.cooke@its.tudelft.nl,*
E-mail: †*D.Lewandowski@its.tudelft.nl*

We discuss Bayesian sensitivity analysis, how to measure sensitivity, and how to compute these measures efficiently. An illustration is taken from the Swedish Nuclear Inspectorate's two stage Bayesian model for processing field data from a population of similar but not identical plants[6].

1. Introduction

As Bayesian models become more popular and more complex, it issues of appraising model performance, and identifying important parameters receive more attention.

Classically, a parameter is identifiable if there is a function of the data which converges almost surely to the value of this parameter. A distinctive feature of Bayesian methods is that we can easily learn from data about parameters which in a classical sense are *not* identifiable. In particular, the posterior distribution can differ from the prior, without having a Dirac distribution as its limit. This simple remark means that sensitivity, sensitivity to data, and the consequences of sensitivity must be rethought from a Bayesian perspective.

In this paper we discuss a number of issues that arise in Bayesian model criticism and sensitivity analysis. Methods are suggested for identifying important parameters and for analysing the impact of data versus prior information.

These ideas are illustrated with an analysis of the two stage hierarchical model used the Swedish Nuclear Inspectorate's data processing[6] [a]. This model is designed to utilize failure data from similar plants to update failure rates at a given plant. Relevant questions concern the sensitivity to hyperparameters, the sensitivity to data, and the impact of information from other plants.

2. Sensitivity in Hierarchical Models

We are interested in characterizing the sensitivity of aspects of a model to the data. Loosely, if a parameter "does not listen to the data" then it is weakly identifiable in a Bayesian sense. This is relevant for model criticism for two reasons. First, if other quantities of interest depend on weakly identifiable parameters, then this might be a cause of concern as these quantities would remain strongly dependent on the prior distribution. Second, weakly identifiable parameters might be eliminated in a more parsimonious model.

We consider a generic two stage model with parameter vector $\lambda = \lambda_1, \ldots \lambda_n$, hyperparameter vector $\theta = (\alpha, \beta, \gamma, \ldots)$, data matrix $X = X_{ij}$ with row i associated with λ_i, and some distribution g over θ containing constants K. In general it is assumed that the components of λ are independent given θ, and that given λ_i the data $X_{i,\cdot}$ is independent of $(\theta, X_{j,\cdot}), j \neq i$.

Among the questions of interest are:

- Which parameters are sensitive to the data?
- Which parameters are sensitive for a given parameter?
- Can the computation of sensitivities be done efficiently?

Which parameters are sensitive to the data? This question can be answered by computing the relative information of a parameter's posterior with respect to its prior. Letting $(\lambda_i|X)$ denote a random variable with distribution equal to the conditional distribution of λ_i given X, and assuming densities exist, we could compute

$$I((\lambda_i|X); \lambda_i) = \int f(\lambda_i|X) \ln \left(\frac{f(\lambda_i|X)}{f(\lambda_i)} \right) d\lambda_i. \tag{1}$$

[a]Prof. R.E. Barlow was instrumental in the development of this model and served on the PhD thesis committee

Note that if λ_i follows the improper prior $d\lambda_i$, then

$$-I((\lambda_i|X);\lambda_i) = H(\lambda_i|X) = -\int f(\lambda_i|X)\,\ln f(\lambda_i|X)\,d\lambda_i \tag{2}$$

where H denotes the entropy.

Which parameters are sensitive for a given parameter? Having determined the sensitivity of parameters to data, we now examine the sensitivity of parameters to each other. There are many ways of doing this and the best method will depend on the specific case at hand. The simplest idea is to regress the parameter of interest, say λ_i onto other parameters, indexed as ψ:

$$\lambda_i = \sum_{\psi} \frac{\rho(\lambda_i,\psi)\sigma_{\lambda_i}}{\sigma_\psi}\psi + \text{Error} \tag{3}$$

In many cases the relationship between λ_i and ψ is exponential rather than linear and (3) might be replaced by log linear regression:

$$\ln(\lambda_i) = \sum_{\psi} \frac{\rho(\ln(\lambda_i),\psi)\sigma_{\ln(\lambda_i)}}{\sigma_\psi}\psi + \text{Error}; \tag{4}$$

or by rank regression:

$$F_i(\lambda_i) = \sum_{\psi} \rho(F_i(\lambda_i), F_\psi(\psi))F_\psi(\psi) + \text{Error}; \tag{5}$$

where F_* denotes the cumulative distribution function of its argument. Equation (5) regresses the quantile function of λ_i on the quantile functions of the ψ's. The quantile function of a continuous random variable is of course uniform on $[0,1]$ with variance $1/12$. The *rank correlation* $\rho(F_i(\lambda_i), F_\psi(\psi))$ is also denoted $\rho_r(\lambda_i,\psi)$.

The expressions (3) (4) and (5) can be evaluated either in the prior or in the posterior joint distributions. The former two can be evaluated on-the-fly, as their computation involves only moments. The rank regression coefficients cannot be evaluated on-the-fly as we must first determine the cumulative distribution function, and this is typically known only at the end of a simulation.

A more general notion of sensitivity. We consider a function $G = G(X,Y)$ of random vectors X and Y with $\sigma^2_G < \infty$. In analogy with the above, we may ask for which function $f(X)$ with $\sigma^2_{f(X)} < \infty$ is $\rho^2(G, f(X))$ maximal? The answer is given in the following

Proposition 2.1: *Let $G - G(X,Y)$ with $\sigma^2_G < \infty$; then*

(i) $Cov(G, E(G|X)) = \sigma^2_{E(G|X)}$,

(ii) $max_{f;\, \sigma^2_{f(X)} < \infty}\ \rho^2(G, f(X)) = \rho^2(G, E(G|X)) = \frac{\sigma^2_{E(G|X)}}{\sigma^2_G}$

Proof:

(i) $Cov(G, E(G|X)) = E(E(GE(G|X)|X)) - EGE(E(G|X)) = E(E^2(G|X)) - E^2(E(G|X))$.

(ii) Let $\delta(X)$ be any function with finite variance. Put $A = \sigma^2_{E(G|X)}$, $B = \mathrm{Cov}(E(G|X), \delta(X))$, $C = \sigma^2_G$, and $D = \sigma^2_\delta$. Then

$$\rho^2(G, E(G|X) + \delta(X)) = \frac{(A+B)^2}{C(A+D+2B)};$$

$$\frac{\sigma^2_{E(G|X)}}{\sigma^2_G} = \frac{A}{C}.$$

$$\frac{(A+B)^2}{C(A+D+2B)} \leq \frac{A}{C} \iff B^2 \leq AD.$$

The latter inequality follows from the Cauchy-Schwarz inequality. This is similar to a result by Whittle[7].

□

The quantity $\frac{\sigma^2_{E(G|X)}}{\sigma^2_G}$ is called the *correlation ratio*, and may be taken as the general sensitivity of G to X. Note that the correlation ratio is always positive, and hence gives no information regarding the direction of the influence. The following propositions explore some properties of the correlation ratio. The first is straightforward, the second uses Proposition 2.1.

Proposition 2.2: *Let $G(X,Y) = f(X) + h(Y)$ with $\sigma^2_f < \infty$, $\sigma^2_h < \infty$ and X, Y are not both simultaneously constant ($\sigma^2_G > 0$). If X and Y are independent then :*

$$\rho^2(G, E(G|X)) + \rho^2(G, E(G|Y)) = 1$$

Proposition 2.3: *Let $G = G(X,Y)$ with $Cov(E(G|X), E(G|Y)) = 0$; then*

$$\rho^2(G, E(G|X)) + \rho^2(G, E(G|Y)) \leq 1$$

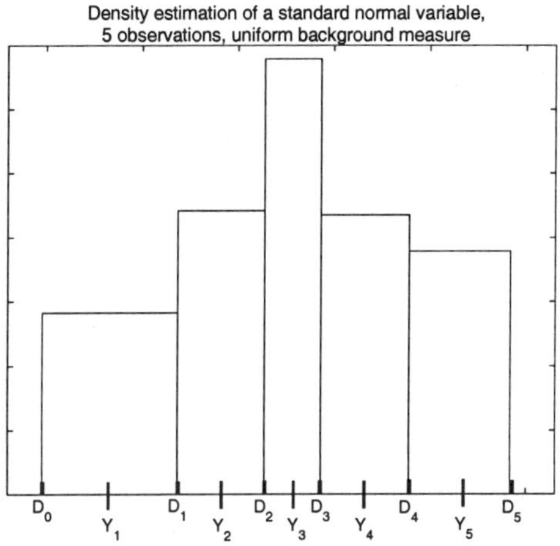

Fig. 1. Density approximation with 5 observations

Proof:

$$\rho(E(G|X), G - E(G|Y)) = \frac{Cov(E(G|X), G - E(G|Y))}{\sigma_{E(G|X)}\sqrt{\sigma_G^2 - \sigma_{E(G|Y)}^2}}$$

$$= \frac{\sigma_{E(G|X)}}{\sqrt{\sigma_G^2 - \sigma_{E(G|Y)}^2}} \leq 1$$

$$\sigma_{E(G|X)}^2 + \sigma_{E(G|Y)}^2 \leq \sigma_G^2$$

□

Can the computations be done efficiently? The computations frequently use Monte Carlo methods. Efficiency in this context usually means on-the-fly. That is, we would like to perform all necessary calculations on a sample, then discard the sample and proceed to the next sample. A computation which involves operations on the entire sample is not efficient.

For reasons of simplicity we first discuss an on-the-fly method of computing the entropy (2). Values for λ_i are generated by simulation and are plotted on the real line as $Y_1 \ldots Y_N$ (see Figure 1)[b]. We approximate the

[b]Although we must retain the ordered sample in memory, we do not perform operations

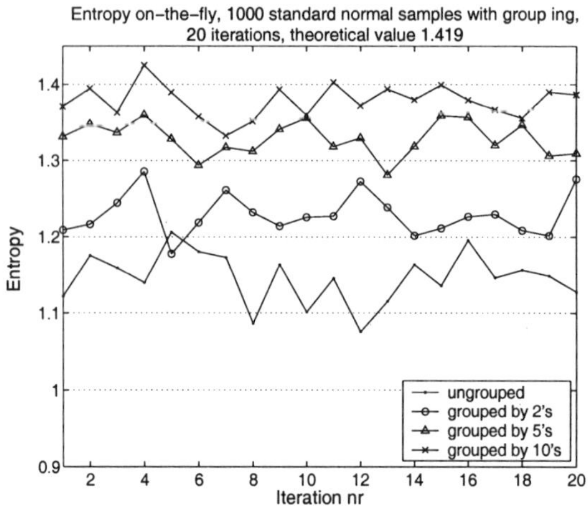

Fig. 2. Entropy, 1000 standard normals, 20 iterations

density as a step function whose steps occur at the midpoints D_i between Y_i, and Y_{i+1}. We set $D_0 = Y_1 - (Y_2 - Y_1)/2$, and $D_N = Y_N + (Y_N - Y_{N-1})/2$.

Each point has equal probability, namely $1/N$. The density above the point Y_i is therefore estimated as:

$$P_i = \frac{1}{N(D_i - D_{i-1})}. \tag{6}$$

The entropy is then computed as

$$H \approx (-1/N) \sum_{i=1}^{N} \ln \left(\frac{1}{N(D_i - D_{i-1})} \right). \tag{7}$$

The relative information (1) is computed in a similar fashion. The prior density of λ_i is approximated as a step function with value $f(Y_i)$ between points D_{i-1}, D_i, where f is the prior density.

$$I((\lambda_i|X); \lambda_i) \approx (1/N) \sum_{i=1}^{N} \ln \left(\frac{1}{N(D_i - D_{i-1}) f(Y_i)} \right). \tag{8}$$

These quantities can be computed on-the-fly. If the computation has been done for N samples, adding one additional sample requires only a

---

on the whole sample; in this sense we are still on-the-fly.

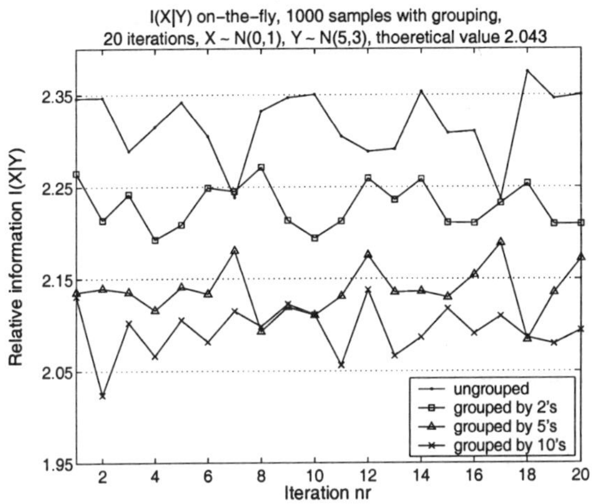

Fig. 3. $I(X|Y)$ 1000 samples, 20 iterations, X=N(0,1), Y = N(5,3)

local adjustment. We can also group the samples, such that we consider k adjacent samples and estimate the density over such a k-tuple starting with Y_i as $k/(N(D_{i+k} - D_{i-1}))$. Grouping the samples in this way a larger "local adjustment" in (8), but gives better results.

To illustrate, the entropy of 1000 samples from the standard normal distribution has been computed on 20 iterations. The results are shown as computed above, and also after grouping the data points by 2's, by 5's and by 10's (see Figure 2). The Entropy of the standard normal is 1.419. If we truncate the standard normal to $[-3.3]$ corresponding roughly the above density approximation with 1000 samples, the entropy integral is 1.4056.

We see that the results are stable, though the ungrouped data tend to be a bit jittery, thus producing a lower entropy value than the theoretical value.

The same pattern emerges in computing the relative information on the fly. Figure 3 shows 20 iterations of 1000 samples for calculating $I(X|Y)$, where X is standard normal, and Y is normal with mean 5 and standard deviation 3. The distribution of X is treated as the posterior in (8). The theoretical value is 2.043. Again, the estimator with ungrouped data is stable but overestimates the relative information due to sample jitter.

A. O'Hagan (personal communication) has recently proved that the bias in computing entropy with ungrouped data is asymptotically equal to $\gamma - 1 + ln(2)$, where γ is Euler's constant. Thus, for a sufficiently large sample, we can remove this bias and obtain better results.

Computing correlation ratio may be difficult in some cases. However, if we can sample Y' from the conditional distribution $(Y|X)$ independently of Y, and if the evaluation of G is not too expensive, then the following simple algorithm may be applied[4]:

(1) Sample (x, y) from (X, Y),
(2) Compute $G(x, y)$,
(3) Sample y' from $(Y|X = x)$ independent of $Y = y$,
(4) Compute $G' = G(x, y')$,
(5) Store $Z = G * G'$,
(6) Repeat

The average value of Z will approximate $E(E^2(G|X))$, from which the correlation ratio may be computed as

$$\frac{E(E^2(G|X)) - E^2(G)}{\sigma_G^2}$$

Of course, if Y and X are independent, then this algorithm poses no problems. If Y and X are not independent, then it may be difficult to sample from $(Y|X)$. In this case there is no alternative to the "pedestrian" method: save a large sample, compute $E(G|X = x_i \pm \varepsilon)$ for suitable $x_1, ...x_n$, and compute the variance of these conditional expectations. To do this for a large number of variables can be slow.

3. Example, the SKI model

A Bayesian model for dealing with plant-to-plant variability has been adopted by the Swedish Nuclear Inspectorate SKI[6]. This model has been reviewed by Cooke[1] and discussed by Hofer and Peschke[2] and Meyer and Hennings[5]. Hora and Iman[3] describe a similar model. We consider a collection of classes of components (plants). Each class consists of components which are considered identical for the purposes of lifetime estimation, and which are used in a specific plant under plant-specific conditions. Different plant specific conditions lead to different Rate of Occurrence of Failure (ROCOF). Since we would like to use data from given plants to make inference about the

ROCOF in another plant, we have to assume something about the underlying relationship between the ROCOF's of the various plants. In Pörn's model, these ROCOF's are treated as independent realisations of random quantities with the same distribution.

Specifically:

(1) The pattern of failures at each plant is supposed to follow a Poisson process. At plant i, we have x_i failures in an operating time of T_i. The plant specific ROCOF is λ_i, which is a realisation of the random variable Λ.

(2) Λ follows a mixed gamma distribution

$$P(\lambda_i|\theta) = G(\lambda_i|\alpha, \beta)(1 - c) + G(\lambda_i|1/2, 1)c. \tag{9}$$

α and β are unknown shape and scale parameters, and c is a random variable taking values in $[0, 1]$. According to [6], c is a contamination parameter mixing $G(\lambda_i|\alpha, \beta)$ with a relatively vague gamma to "add a pinch of uncertainty". The uncertainty over values of θ is modelled by assuming that θ is a random vector.

(3) $\lambda_1, \ldots, \lambda_n$ are independent realisations of Λ.

(4) Given $(\lambda_1, \ldots, \lambda_n)$, $(x_1, \ldots, x_n)$ are independent.

(5) Given λ_i, x_i and $(\theta, \lambda_1, \ldots \lambda_{i-1}, \lambda_{i+1}, \ldots, \lambda_n)$ are independent.

A consequence of the dependence structure here is that data (x_i, T_i) from plant i can only influence our beliefs on the value of λ_j $(j \neq i)$ through its influence on our beliefs on the value of θ.

Inference under Pörn's model Suppose that we have data (x_i, T_i) from plants $i = 1, \ldots, n + 1$. Suppose that we have chosen a prior distribution for the hyperparameters $P(\alpha, \beta, c)$, and that we wish to have an updated distribution of λ_{n+1}. Note that we have information on λ_{n+1} from two directions:

(1) The influence of all of the data on our beliefs of the value of θ, and

(2) The influence of the plant-specific data (x_{n+1}, T_{n+1}) on our beliefs over the value of λ_{n+1}.

It will be noted that the data (x_{n+1}, T_{n+1}) is used twice in this procedure. According to Pörn[6] the effect of this double counting is small.

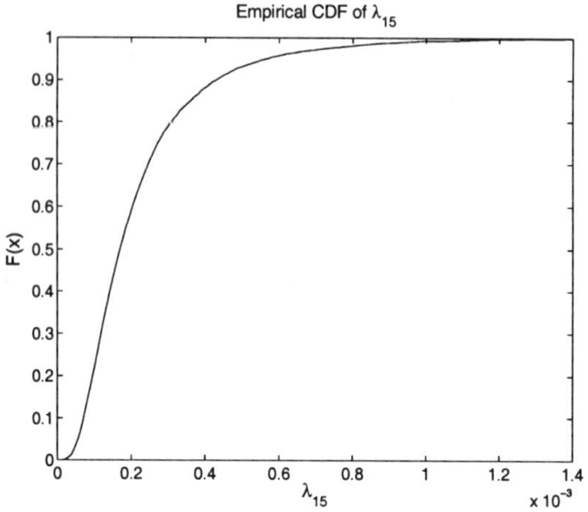

Fig. 4. Cumulative distribution function of λ_{15}

Under the assumptions of Pörn's model, it can be shown that the likelihood function of $\theta = (\alpha, \beta, c)$ given the data (x_i, T_i) $(i = 1, \ldots, n)$ is proportional to

$$\prod_{i=1}^{n} [A\,(1-c) + B\,c\,] \tag{10}$$

$$A = \frac{\Gamma(x_i+\alpha)}{\Gamma(x_i+1)\Gamma(\alpha)} \left(\frac{\beta}{\beta+T_i}\right)^{\alpha} \left(\frac{T_i}{\beta+T_i}\right)^{x_i}$$

$$B = \frac{\Gamma(x_i+1/2)}{\Gamma(x_i+1)\Gamma(1/2)} \left(\frac{1}{1+T_i}\right)^{1/2} \left(\frac{T_i}{1+T_i}\right)^{x_i}$$

The terms under the product operator in formula (10) will be denoted p_i. Each p_i represents an update of the hyperpriors based only on the data from plant i. The likelihood of λ_{n+1} given x_{n+1} is easily seen to be

$$P(x_{n+1}|\theta) = e^{-\lambda_{n+1}T_{n+1}} \frac{(\lambda_{n+1}T_{n+1})^{x_{n+1}}}{\Gamma(x_{n+1}+1)}$$

Thus, the posterior distribution of λ_{n+1} given data (x_i, T_i) $(i = 1, ..., n+1)$ is

$$G(\alpha + x_{n+1}, \beta + T_{n+1})(1-c) + G(1/2 + x_{n+1}, 1 + T_{n+1})c \tag{11}$$

Table 1. Swedish Nuclear Plant Centrifugal Pump Data

Plant	Nr Failures	Operating Hours
1	2	17600
2	1	17600
3	3	10700
4	1	10700
5	0	29500
6	0	29500
7	4	15000
8	1	15000
9	1	22000
10	1	22000
11	0	4600
12	0	4600
13	1	5600
14	0	5600
15	3	5000

where $\theta = (\alpha, \beta, c)$ follows a distribution proportional to the product of the prior $P(\alpha, \beta, c)$ and (10).

Table 2. Results for λ_{15}

	λ_{15} [failures/hour]
5%	8.0×10^{-5}
50%	2.3×10^{-4}
95%	6.3×10^{-4}
mean	2.8×10^{-4}

This model may be computed by analytical methods, or with Monte Carlo integration. In the later case we sample the hyperpriors and apply acceptance-rejection to produce samples of (10).

Figure 4 shows posterior cumulative distribution function of λ_{15} which was obtain using data from all plants.

Data in Table 1 is analysed in [6] and in a mathematical review [1], to which we refer for the prior distributions and details on the inference model. Table 2 shows the results in [6] for the updated distribution of the failure rate for plant 15 using the data of Table 1.

Table 3. Correlation ratios for λ_{15}

X	$\rho(\lambda_{15}, X)$	$\rho_r(\lambda_{15}, X)$	$\rho^2(\lambda_{15}, X)$	$\rho_r^2(\lambda_{15}, X)$	$\rho^2(\lambda_{15}, E(\lambda_{15}\|X))$
p_1	-0.19	-0.14	0.0361	0.0196	0.0558
p_2	-0.47	-0.55	0.2209	0.3025	0.234
p_3	0.28	0.44	0.0784	0.1936	0.113
p_4	-0.33	-0.31	0.1089	0.0961	0.119
p_5	0.02	-0.04	0.0004	0.0016	0.0185
p_6	0.02	-0.04	0.0004	0.0016	0.0185
p_7	0.29	0.46	0.0841	0.2116	0.123
p_8	-0.44	-0.50	0.01936	0.25	0.208
p_9	-0.48	-0.59	0.2304	0.3481	0.247
p_{10}	-0.48	-0.59	0.2304	0.3481	0.247
p_{11}	-0.29	-0.36	0.0841	0.1296	0.102
p_{12}	-0.29	-0.36	0.0841	0.1296	0.102
p_{13}	-0.04	0.05	0.0016	0.0025	0.0162
p_{14}	-0.27	-0.35	0.0729	0.1225	0.1415
α	-0.33	-0.45	0.1089	0.2025	0.153
β	-0.39	-0.59	0.1521	0.3481	0.251
c	0.09	0.03	0.0081	0.0009	0.0207

4. Sensitivity Results

We are interested in λ_{15} after updating on the data from all plants. In particular, we are interested in how sensitive λ_{15} is to the data from plant 15 (i.e. p_{15}), how sensitive it is to the data from other plants, and how sensitive it is to the hyperparameters. In these calculations, the posterior distribution was obtained with acceptance-rejection sampling. All sensitivity results concern the posterior distribution. Based on 9521 posterior samples the posterior mean and variance of λ_{15} are (compare Table 2):

i) $E(\lambda_{15}) =$ 2.22E-04
ii) $Var(\lambda_{15}) =$ 3.2629E-08

The acceptance-rejection method rendered the on-the-fly algorithm impractical, and correlation ratios were computed with the pedestrian method.

Table 3 presents sample based correlations, rank correlations, and correlation ratio's. All these refer to the posterior distribution. Evidently, the information from some plants are more important than from others. Plants $2, 8, 9, 10$ have the biggest influence on the function λ_{15}. The posterior parameter β is associated with the largest correlation ratio. Note that the correlation ratio is always greater than ρ^2; the size of this difference is an index of the non-linearity of the regression function $E(\lambda_{15}|X)$.

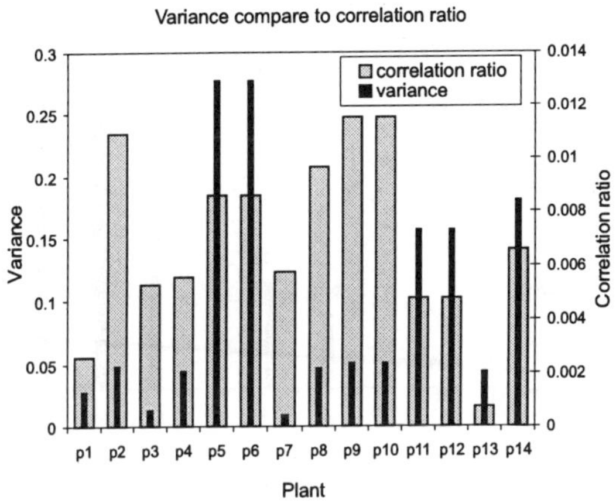

Fig. 5. Variance of p_i compared to correlation ratio between λ_{15} and p_i

We ask, which features of the data at plant i are driving these results. The variance of p_i is directly related to the amount of data available at plant i. Figure 5 shows no obvious relation between the correlation ratio with λ_{15} and variance of p_i $(i = 1, \ldots, n)$. Note that the posterior variance will be inversely proportional to the square of T_i, so that p_5, p_6 have the highest posterior variances.

The key to understanding the sensitivities of Table 3 is given in Figure 6. The 'mean time to failure' MTTF of a plant i is the operating time T_i divided by the number of failures x_i. If this time is bigger than the inverse expectation of λ_{15} (4444) then the function p_i is strongly negatively correlated with λ_{15}. The greater this difference the greater is the absolute value of the correlation (both product moment correlation and rank correlation). Notice, that if plant i has a MTTF smaller than MTTF of λ_{15} then the correlation is positive. Although the mean lifetime at plant 3 is only a little smaller than 4444, p_3 and λ_{15} are quite strongly correlated. We know that $\rho^2(\lambda_{15}, p_i) \leq \rho^2(\lambda_{15}, E(\lambda_{15}|p_i))$ (Proposition 2.1). Thus if a MTTF of a plant i is much different from the inverse of the expectation of λ_{15} then p_i has greater influence on λ_{15}. Note the effect of the sign of the correlation.

α, β and c are hyperparameters. We use data from the plants to update our belief in these variables. In the posterior distribution, we see that β

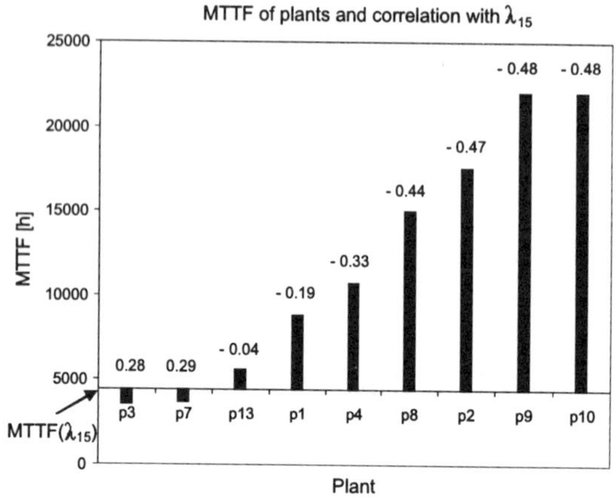

Fig. 6. The MTTF of the plant i compared to the inverse of expectation of λ_{15} (MTTF of λ_{15}) and the correlation between λ_{15} and p_i

is the most important parameter, and α is more important than 10 of the 14 plants. This indicates that the final result is still strongly driven by the hyperparameters. It must be noted that the distributions of the p_i also depend on the hyperparameters; however we should hope that their influence dies off rapidly as the data from all plants is gathered. After all, the data represents a total of 188,600 operating hours. The parameter c, on the other hand, does not play a significant role. The persistence of hyperparameters α and β observed here confirms the conclusion of Cooke[1], obtained with a much more laborious analysis.

The rank correlation $\rho_r(\lambda_{15}, \beta)$ is negative. Thus, small values of β tend to arise in combination with large values of λ_{15}. This is illustrated in Figure 8 which shows that the regression $E(\lambda_{15}|\beta)$ is decreasing.

The variable c is applied to add "a pinch of uncertainty" and it is a pinch indeed. c does't play a significant role in this model.

Consider Figure 7. β is a function inter alia of α. Thus, α influences λ_{15} through itself and β, and we might expect that α has greater influence on λ_{15} than β. However, $Var(E(\lambda_{15}|\alpha))$ is less than $Var(E(\lambda_{15}|\beta))$.

Figures 7 and 8 show that the regression of λ_{15} on α and β is not linear. This recommends the use of the correlation ratio as a measure of dependence. The relative unimportance of c is illustrated in Figure 9, where

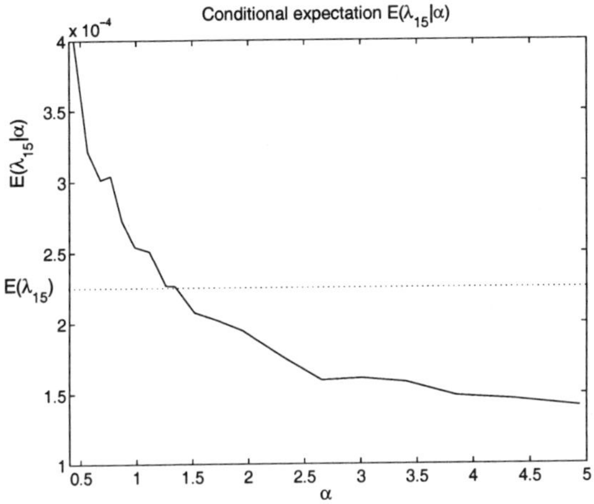

Fig. 7. Dependence between $E(\lambda_{15}|\alpha)$ and α compared to $E(\lambda_{15})$

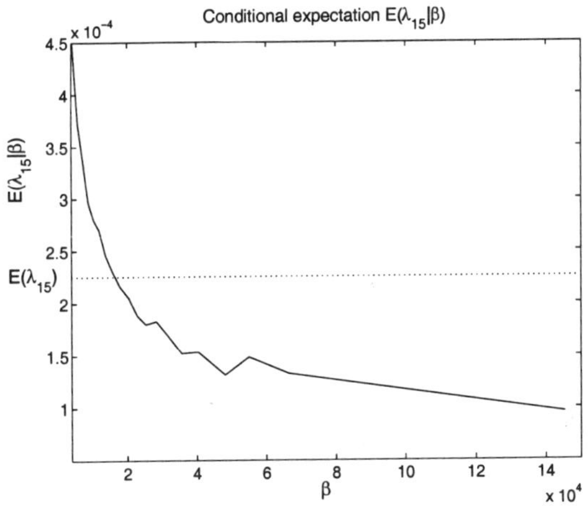

Fig. 8. Dependence between $E(\lambda_{15}|\beta)$ and β compared to $E(\lambda_{15})$

we see that the conditional expectation of λ_{15} given c does not differ greatly from the unconditional expectation of λ_{15}.

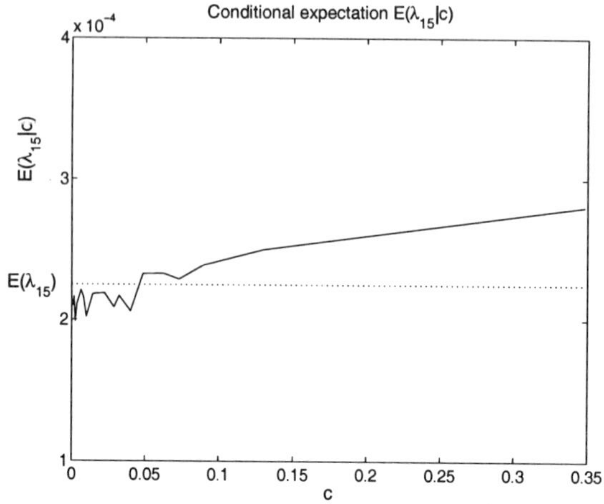

Fig. 9. Dependence between $E(\lambda_{15}|c)$ and c compared to $E(\lambda_{15})$

5. Conclusions

Sensitivity analysis and model criticism are active topics at the moment. The Bayesian approach allows these issues to be ràised and analysed in a natural way. By analysing the sensitivity of a parameter of interest to data and to prior parameters, we can judge the relative importance of prior assumptions.

There are many ways to quantify sensitivity using entropy based concepts or regression based concepts. We have argued that the correlation ratio is particularly attractive in this regard, although it cannot always be computed on-the-fly, and may be difficult to compute analytically.

The SKI two stage Hierarchical Bayes model is a very interesting case because, (i) it is an important application (ii) it has been studied and reviewed extensively, and (iii) it is complicated enough that quantitative measures of sensitivity greatly contribute to understanding the model. The main conclusions regarding the persistence of hyperparameters α and β reached by Cooke[1] by laborious re-computations, are obtained quite simply in Table 3. Moreover, we also gain insight into the features of the data which drive the parameter of interest (Figure 5).

Acknowledgments

We gratefully acknowledge support from the European Union and helpful discussions with Tony O'Hagan.

References

1. R.M. Cooke, J. Dorrepaal and T.J. Bedford, *Review of SKI Data Processing Methodology, SKi Report 95:2*, (1995), also published in abridged form as *Mathematical review of Swedish Bayesian methodology for nuclear plant reliability data bases* in *The Practice of Bayesian Analysis* (French, S. and Smith J.Q. eds) Arnold London, p. 25-55, (1997).
2. E. Hofer and J. Peschke, *Bayesian modeling of failure rates and initiating event frequencies* in *Safety and Reliability* (Schuëller and Kafka eds) Balkema, Rotterdam, p. 887-887, (1999).
3. S.C. Hora and R.L. Iman, *Bayesian modelling of initiating event frequencies at nuclear power plants* in *Risk Analysis* Vol 10 Nr 1 p. 102-109, (1990).
4. T. Ishigami T. Homma, *An importance quantification technique in uncertainty analysis for computer models, Proceedings of the ISUMA '90 First International Symposium on Uncertianty Modelling and Analysis*, University of Maryland, USA, p. 398-403, (1990).
5. W. Meyer and W. Hennings, *Prior distributions in two-stage Bayesian estimation of failure rates* in *Safety and Reliability* (Schuëller and Kafka eds) Balkema, Rotterdam, p. 893-899, (1999).
6. K. Pörn, *On Empirical Bayesian Inference Applied to Poisson Probability Models*, Linkoping Studies in Science and Technology Dissertation No.234, Linkoping, (1990).
7. P. Whittle, *Probability via Expectation*, Springer Verlag, New York, (1992).

CHAPTER 20

BAYESIAN SAMPLING ALLOCATIONS TO SELECT THE BEST NORMAL POPULATION WITH DIFFERENT SAMPLING COSTS AND KNOWN VARIANCES

Stephen E. Chick
Department of Industrial and Operations Engineering
University of Michigan, Ann Arbor, MI 48109-2117, U.S.A.
E-mail: sechick@engin.umich.edu

Masaru Hashimoto
School of Business Administration
University of Michigan, Ann Arbor, MI 48109-1234, U.S.A.
E-mail: mhashimo@umich.edu

Koichiro Inoue
Department of Industrial and Operations Engineering
University of Michigan, Ann Arbor, MI 48109-2117, U.S.A.
E-mail: Koichiro_Inoue@i2.com

Selection procedures are proposed for finding the population with the largest mean, from among k independent normal populations with known but potentially different variances. Suppose that a first stage of sampling is completed, and the cost of samples from each population may differ. The problem is how to allocate further observations to each population for second stage sampling. Both budget constrained and unconstrained allocations are considered. Allocations that minimize the expected linear loss or maximize the probability of correct selection are unknown except for special cases. However, we identify suboptimal allocations that have attractive asymptotic properties, are easy to compute, can be used in multistage procedures, and perform well in a numerical experiment.

1. Introduction

Reliability, engineering system design, medicine, agriculture, and business face the problem of selecting the best of several populations, when the precise value of each alternative is unknown[1,7,9,10]. Often, a few samples from each population are observed in a preliminary stage, and the decision-maker measures the evidence that a given population is best. If the evidence is insufficient, an additional stage of sampling is performed to obtain more information about the identity of the best population.

Suppose there are k different populations, and that the population with the largest mean is to be identified. The means $\mathbf{w} = (w_1, \ldots, w_k)$ are presumed unknown, and their values are to be inferred from statistical sampling $x_{i,j}$ (for $i = 1, \ldots, k$; $j = 1, 2, \ldots$). Given $\mathbf{w}$, the samples are presumed to have a jointly independent normal distribution

$$X_{i,j} \sim \mathcal{N}(w_i, \lambda_i), \text{ jointly independent, for } i = 1, \ldots, k;\ j = 1, 2, \ldots$$

for some known precision λ_i (precision = one over the variance) that may vary from population to population. Denote the vector of precisions by $\boldsymbol{\lambda} = (\lambda_1, \ldots, \lambda_k)$. The sampling costs $\mathbf{c} = (c_1, \ldots, c_k)$ for each population may be different, so that an additional $\mathbf{r} = (r_1, \ldots, r_k)$ samples incurs a cost $\mathbf{cr}^T$. The goal is to determine a sampling allocation $\mathbf{r}$ that will be effective in some sense for identifying the population with the largest mean. Let $\mathbf{X}_{r_i} = (X_{i,1}, \ldots, X_{i,r_i})$ be the random samples that are to be observed during additional sampling from population i, let $\mathbf{x}_{r_i}$ denote their realization, and let $\mathbf{x}_\mathbf{r} = (\mathbf{x}_{r_1}, \ldots, \mathbf{x}_{r_k})$ denote all observations.

This paper uses a Bayesian formulation to derive sampling allocations that reduce the risk of incorrect selections. To that end, let the joint distribution ζ for the unknown means $\mathbf{W}$ be independent and normal, with

$$W_i \sim \mathcal{N}(\mu_i, \tau_i), \text{ for } i = 1, \ldots, k.$$

This distribution is intended to reflect both prior information and any previously observed samples, such as from a first stage of observations. Multistage sampling allocations can be obtained by using the posterior distribution from a given stage of sampling as the prior distribution for the next stage of sampling.

Sampling allocations for both the linear loss and 0-1 loss functions are presented. When population i is selected and the mean vector is $\mathbf{w}$, these losses are defined as

$$\mathcal{L}_{o.c.}(i, \mathbf{w}) = \max_j w_j - w_i \tag{1}$$

$$\mathcal{L}_{0-1}(i, \mathbf{w}) = \begin{cases} 0 \text{ when } w_i = \max_j w_j \\ 1 \text{ otherwise} \end{cases} \tag{2}$$

Each allocation presumes that the population selected as best is the population with the highest posterior mean given $\mathbf{x_r}$, the so-called natural decision rule, $\delta^N(\mathbf{x_r})$. With these assumptions, the cost of sampling plus the expected loss of selecting a population is

$$\rho^*(\mathbf{r}) \stackrel{\text{def}}{=} \mathbf{cr}^T + E_{\mathbf{X_r}}\left[E_{\mathbf{W}|\mathbf{X_r}}[\mathcal{L}\left(\delta^N(\mathbf{X_r}), \mathbf{W}\right) \mid \mathbf{X_r}]\right]. \tag{3}$$

when $\mathcal{L}$ is the appropriate loss function. The natural decision-rule is optimal for the linear loss function, and though it is suboptimal for the 0-1 loss function, it has natural appeal and leads to a tractable analysis.

Both budget constrained and unconstrained allocations are presented. The unconstrained budget allocations attempt to balance the expected value of information to be gained with the cost of sampling. The optimal *unconstrained* allocation solves

$$\begin{aligned} &\min_{\mathbf{r}} \ \rho^*(\mathbf{r}) \\ &\text{s.t. } \ r_i \geq 0 \ \text{ for } i = 1, \ldots, k, \end{aligned} \tag{4}$$

The optimal budget *constrained* allocation solves Eq. 4 subject to $\mathbf{cr}^T = B$, where B is a sampling budget.

An optimal allocation in the general setting is computationally challenging to resolve, as a closed-form solution has proven difficult to obtain for $k \geq 2$. But structural results are available for special cases. Gupta and Miescke[8] show that when $k = 2$; $c_1 = c_2$; $r_1 + r_2 = B$, the optimal policy is to minimize the absolute difference in the posterior precision for the mean of each population. When $k \geq 2$ and $c_1 = \ldots = c_k$, $\mathbf{cr}^T = B$, Gupta and Miescke[9] give an optimal allocation for a single observation ($B = 1$).

On the other hand, there is interest in easily computed allocations that have appealing asymptotic properties. In computer simulation experiments, for example, the cost of samples might be measured in CPU time. A good but easily computed allocation would leave more time for observing samples than an optimal allocation that requires significant computation.

This paper therefore derives suboptimal but easily computed allocations that asymptotically minimize a bound on the expected loss. Chick and

Inoue[5] is a companion article that derives sampling allocations in a similar manner when the variance is unknown. They also present preliminary empirical evidence that suggests that the approach compares favorably with two indifference-zone procedures (both Rinott[15] and the combined screening and selection procedure of Nelson *et al.*[14]), at least for the unknown variance case. Inoue *et al.*[11] provide further empirical evidence (for the unknown variance case) that the method of deriving allocations presented here may lead to significant sampling efficiencies when compared to some indifference-zone results.

2. Linear Loss

This section derives sampling allocations $\mathbf{r}$ when the linear loss function is used. Consider first the case of an unconstrained sampling budget before including a budget constraint. In this case, the goal is to balance the cost of sampling with the expected information gain.

Using the notation introduced above, well-known results[6] imply that the posterior distribution of W_i given $\mathbf{x}_{r_i}$ is $\mathcal{N}\left(\mu_i(\mathbf{x}_{r_i}), \tau_i + r_i\lambda_i\right)$, where

$$\mu_i(\mathbf{x}_{r_i}) = (\tau_i\mu_i + r_i\lambda_i\bar{x}_i)/(\tau_i + r_i\lambda_i),$$

and $\bar{x}_i = \sum_{j=1}^{r_i} x_{i,j}/r_i$. Further, the prior predictive distribution of the posterior mean $Z_i \stackrel{\text{def}}{=} \mu_i(\mathbf{X}_{r_i})$ is normally distributed with mean μ_i and precision $\tau_i(\tau_i + r_i\lambda_i)/(r_i\lambda_i)$.

Define the permutation (i) so that $\mu_{(1)} > \mu_{(2)} \geq \cdots \geq \mu_{(k)}$ is nonincreasing. When the μ_i contain prior information and data from a first stage of sampling, ties occur with probability 0 in nondegenerate cases. Then the prior distribution of $W_{(i)} - W_{(j)}$ is $\mathcal{N}\left(\mu_{(i)} - \mu_{(j)}, \tau_{i,j}\right)$, where

$$\tau_{i,j} = \left(\frac{\tau_{(i)}\tau_{(j)}}{\tau_{(i)} + \tau_{(j)}}\right).$$

Further, the prior predictive distribution of $Z_{(i)} - Z_{(j)}$ is

$$Z_{(i)} - Z_{(j)} = \mu_{(i)}(\mathbf{X_r}) - \mu_{(j)}(\mathbf{X_r}) \sim \mathcal{N}\left(\mu_{(i)} - \mu_{(j)}, \tau_{\{i,j\}}\right),$$

where

$$\tau_{\{i,j\}} = \left(\frac{r_{(i)}\lambda_{(i)}}{\tau_{(i)}[\tau_{(i)} + r_{(i)}\lambda_{(i)}]} + \frac{r_{(j)}\lambda_{(j)}}{\tau_{(j)}[\tau_{(j)} + r_{(j)}\lambda_{(j)}]}\right)^{-1}.$$

Finally, define $\Psi(s) = \int_s^\infty (x-s)\phi(x)dx = \phi(s) - s(1-\Phi(s))$, where $\phi(s)$ and $\Phi(s)$ are the density function and the cumulative distribution function, respectively, of the standard normal random variable. Intuitively, $\Psi(s)$ is the expected linear loss when the value of a standard normal random variable is asserted to be less than s.

Theorem 1: Assume that the $X_{i,j}$ are conditionally independent with $\mathcal{N}(w_i, \lambda_i)$ distribution, given the mean w_i and known precision λ_i, and denote by ζ the joint prior distribution of $\mathbf{W}$, with independent $W_i \sim \mathcal{N}(\mu_i, \tau_i)$, for $i = 1, \ldots, k$. Let the loss function be the linear loss of Eq. 1, and let $\mathbf{c}$, (i), $\tau_{i,j}$, and $\tau_{\{i,j\}}$ be as above. Let $\mathcal{A}_i = \{\mathbf{z} \mid z_i = \max_j z_j\}$ be the event that the posterior mean of population i is maximal. Then:

- The natural decision rule, $\delta^N(\mathbf{x_r}) = i$, when $z_i \geq \max_j z_j$, is optimal.
- The expected total risk is:

$$\rho^*_{o.c.}(\mathbf{r}) = \mathbf{c}\mathbf{r}^T + E_\zeta[max_j w_j - w_{(1)}] - \sum_{i=2}^{k} p_{G_\mathbf{r}}\left(\mathcal{A}_{(i)}\right) E_{G_\mathbf{r}|\mathcal{A}_{(i)}}[z_{(i)} - z_{(1)}], \tag{5}$$

  and $\rho^*_{o.c.}(\mathbf{r})$ is bounded above by:

$$\mathbf{c}\mathbf{r}^T + E_\zeta[max_j w_j - w_{(1)}] - \max_{i\neq 1}\left\{ {\tau_{\{1,i\}}}^{-1/2}\Psi\left[{\tau_{\{1,i\}}}^{1/2}(\mu_{(i)} - \mu_{(1)})\right]\right\} \tag{6}$$

  and below by:

$$\hat{\rho}^*_{o.c.}(\mathbf{r}) \stackrel{\text{def}}{=} \mathbf{c}\mathbf{r}^T + E_\zeta[max_j w_j - w_{(1)}] - \sum_{i=2}^{k} {\tau_{\{1,i\}}}^{-1/2}\Psi\left[{\tau_{\{1,i\}}}^{1/2}(\mu_{(i)} - \mu_{(1)})\right] \tag{7}$$

- In the limit $c_i \to 0$ (small cost of sampling), the optimal population sizes r_i^* to minimize $\hat{\rho}^*_{o.c.}(\mathbf{r})$ are asymptotically:

$$\tilde{r}^*_{(1)} = \left(\sum_{j=2}^{k} \frac{(\tau_{1,j})^{1/2}\phi\left[(\tau_{1,j})^{1/2}(\mu_{(j)} - \mu_{(1)})\right]}{2c_{(1)}\lambda_{(1)}}\right)^{1/2} - \frac{\tau_{(1)}}{\lambda_{(1)}} \tag{8}$$

$$\tilde{r}^*_{(i)} = \left(\frac{(\tau_{1,i})^{1/2}\phi\left[(\tau_{1,i})^{1/2}(\mu_{(i)} - \mu_{(1)})\right]}{2c_{(i)}\lambda_{(i)}}\right)^{1/2} - \frac{\tau_{(i)}}{\lambda_{(i)}}, \text{ for } i \neq 1$$

Proof: That the natural decision rule is optimal for linear loss has been shown elsewhere[8]. To show that the expected total loss $\rho^*_{o.c.}(\mathbf{r})$ is as in Eq. 5, it is useful to consider the modified loss function:

$$\mathcal{L}^*_{o.c.}(i, \mathbf{w}) = \mathcal{L}_{o.c.}(i, \mathbf{w}) - \mathcal{L}_{o.c.}((1), \mathbf{w}) = w_{(1)} - w_i \tag{9}$$

Subtracting $\mathcal{L}_{o.c.}((1), \mathbf{w})$ from $\mathcal{L}_{o.c.}(i, \mathbf{w})$ does not change the optimal decision[6]. Thus $E[\mathcal{L}^*_{o.c.}(i, \mathbf{w}) \mid \mathbf{x_r}] = z_{(1)} - z_i$, the difference of the posterior means. Further, i is selected if and only if the event $\mathcal{A}_i$ obtains. Take the expectation over all experiments, add $E_\zeta[\mathcal{L}_{o.c.}((1), \mathbf{w})]$ to compensate for subtracting $\mathcal{L}_{o.c.}((1), \mathbf{w})$ earlier, and add the cost $\mathbf{cr}^T$ of the experiment to obtain Eq. 5.

The upper bound of Eq. 6 is determined by considering only losses incurred by each pairwise comparison. In a comparison of two distinct populations (1) and (i) alone,

$$E[\mathcal{L}^*_{o.c.}((i), \mathbf{w}) \mid \mathbf{x_r}] \leq \begin{cases} z_{(1)} - z_{(i)} & \text{when } z_{(1)} < z_{(i)} \\ 0 & \text{otherwise} \end{cases} \tag{10}$$

Bracken and Schleifer[4] indicate that the expected loss for this pairwise comparison is $E[E[\mathcal{L}^*_{o.c.}(i, \mathbf{w}) \mid \mathbf{X_r}]] = \tau_{\{1,i\}}{}^{-1/2}\Psi\left[\tau_{\{1,i\}}{}^{1/2}(\mu_{(i)} - \mu_{(1)})\right]$. Since the loss when comparing k populations is at least as great as for 2 populations, the claimed upper bound holds.

The lower bound of Eq. 7 is obtained by summing the losses from each of $k-1$ pairwise comparisons. Define $\mathcal{B}_i = \{\mathbf{z} \mid z_i \geq z_{(1)}\}$ so that $\mathcal{A}_i \subseteq \mathcal{B}_i$, and let $\chi_{\mathcal{A}_i}(\mathbf{z})$ be the indicator function. The lower bound follows from:

$$\begin{aligned}
\sum_{i=2}^{k} p_{G_\mathbf{r}}(\mathcal{A}_{(i)}) E_{G_\mathbf{r}|\mathcal{A}_{(i)}}[z_{(i)} - z_{(1)}] &= \sum_{i=2}^{k} E_{G_\mathbf{r}}[\chi_{\mathcal{A}_{(i)}}(\mathbf{z})(z_{(i)} - z_{(1)})] \\
&\leq \sum_{i=2}^{k} E_{G_\mathbf{r}}[\chi_{\mathcal{B}_{(i)}}(\mathbf{z})(z_{(i)} - z_{(1)})] \\
&= \sum_{i=2}^{k} \tau_{\{1,i\}}{}^{-1/2}\Psi\left[\tau_{\{1,i\}}{}^{1/2}(\mu_{(i)} - \mu_{(1)})\right].
\end{aligned}$$

To minimize $\hat{\rho}^*_{o.c.}(\mathbf{r})$, assume that $\mathbf{r}$ is continuous solve $\partial\hat{\rho}^*_{o.c.}(\mathbf{r})/\partial r_i = 0$, to obtain optimality conditions:

$$c_{(i)} = \frac{(\tau_{\{1,i\}})^{1/2}\phi[(\tau_{\{1,i\}})^{1/2}(\mu_{(i)} - \mu_{(1)})]\lambda_{(i)}}{2(\tau_{(i)} + r_{(i)}\lambda_{(i)})^2} \quad \text{for } i \neq 1 \tag{11}$$

$$c_{(1)} = \sum_{i=2}^{k} \frac{(\tau_{\{1,i\}})^{1/2}\phi[(\tau_{\{1,i\}})^{1/2}(\mu_{(i)} - \mu_{(1)})]\lambda_{(1)}}{2(\tau_{(1)} + r_{(1)}\lambda_{(1)})^2} \tag{12}$$

By the monotonicity of the value of information, $r_{(i)} \to \infty$ as all $c_{(i)} \to 0$, as more samples can be observed for the same price. But $\tau_{\{1,i\}} \to \tau_{1,i}$ as $r_{(1)}, r_{(i)} \to \infty$. Solving for $r_{(i)}$ from Eq. 11 and Eq. 12, substituting the limiting value $\tau_{1,i}$ for $\tau_{\{1,i\}}$, gives the desired $\tilde{r}^*_{(i)}$. □

Miescke[12] provides the bounds in Eq. 6 and Eq. 7 for the special case of a common known precision. The derivation of $\tilde{r}^*_{(i)}$ does not require that $E_\zeta[max_j w_j - w_{(1)}]$ be calculated, as it does not depend on $\tilde{r}^*_{(i)}$. Further, the same allocation is obtained by minimizing the surrogate objective function:

$$\mathbf{cr}^T + \sum_{i=2}^{k} \left\{ \tau_{1,i}{}^{-1/2}\Psi\left[\tau_{1,i}{}^{1/2}(\mu_{(i)} - \mu_{(1)})\right] - \tau_{\{1,i\}}{}^{-1/2}\Psi\left[\tau_{\{1,i\}}{}^{1/2}(\mu_{(i)} - \mu_{(1)})\right]\right\}, \tag{13}$$

which has the sum of expected pairwise losses between the 'current' best and the others. Since the Bonferroni inequality is also a sum of pairwise losses (albeit for the 0-1 loss), the minimization of the lower bound for Bayes risk that leads to Eq. 8 corresponds in some sense to optimizing a Bonferroni-type bound.

Turn now to a sampling allocation that is subject to a budget constraint.

Corollary 2: *The solution to* $\min \hat{\rho}^*_{o.c.}(\mathbf{r})$ *subject to* $\mathbf{cr}^T = B$ *(but unconstrained* r_i*) for asymptotically large* B *is:*

$$\tilde{r}^*_{i,B} = \frac{B + \sum_{j=1}^{k} \frac{c_j \tau_j}{\lambda_j}}{\sum_{j=1}^{k} \left(\frac{c_j c_i \lambda_i \eta_j}{\lambda_j \eta_i}\right)^{1/2}} - \frac{\tau_i}{\lambda_i} \tag{14}$$

where the factor $\eta_{(i)}$ *is given by*

$$\eta_{(i)} = \begin{cases} (\tau_{1,i})^{1/2}\phi\left[(\tau_{1,i})^{1/2}(\mu_{(1)} - \mu_{(i)})\right] & \textit{for } (i) \neq (1) \\ \sum_{j|j\neq(1)} \eta_j & \textit{for } (i) = (1) \end{cases} \tag{15}$$

Proof: Use of Lagrange multipliers to minimize $\hat{\rho}^*_{o.c.}(\mathbf{r})$ subject to the budget constraint $\mathbf{cr}^T = B$, together with the observation that $\tau_{\{1,i\}} \to \tau_{1,i}$

when B is large, leads to

$$\frac{(\tau_{(i)} + \tilde{r}^*_{(i)}\lambda_{(i)})^2}{(\tau_{(j)} + \tilde{r}^*_{(j)}\lambda_{(j)})^2} = \frac{\lambda_{(i)}\eta_{(i)}/c_{(i)}}{\lambda_{(j)}\eta_{(j)}/c_{(j)}}$$

Recall that $\mathbf{cr}^T = B$ and solve for $\tilde{r}^*_{(i)}$ to obtain the desired result. □

What happens when the constraint $r_i \geq 0$ is added to the optimization problem of Corollary 2? Suppose that one or more $\tilde{r}^*_{i,B} < 0$, as determined by Eq. 14, and let $\mathcal{S} = \{i \mid \tilde{r}^*_{i,B} \geq 0\}$. One cannot make a negative number of samples, so $\tilde{r}^*_{i,B}$ must be 0 for those $i \notin \mathcal{S}$. Resetting $\tilde{r}^*_{i,B} = 0$, however, causes the total sampling budget to exceed B, which indicates the number of samples for other populations should decrease. Further, the precision of $Z_{(i)} - Z_{(j)}$ is better approximated by $\tau_{(j)}$ than by $\tau_{i,j}$ when $\tilde{r}^*_{(j),B}$ is large and $\tilde{r}^*_{(i),B} = 0$. Eq. 14 should therefore be applied again to recalculate each $\tilde{r}^*_{(i),B}$, while summing only over the $i \in S$ and making appropriate adjustments for the asymptotic approximation for the precision. Since recalculating may further violate the nonnegativity constraint, this process may need repeating, as in the following algorithm.

Allocation $\mathcal{LL}(\mathcal{B})$.

(1) Initialize the set of populations considered for second-stage sampling, $\mathcal{S} = \{1, \ldots, k\}$.

(2) Compute a tentative number of additional samples for each population $(i) \in \mathcal{S}$,

$$r_{(i)} = \frac{B + \sum_{j\in\mathcal{S}} \frac{c_j\tau_j}{\lambda_j}}{\sum_{j\in\mathcal{S}} \left(\frac{c_j c_{(i)}\lambda_{(i)}\eta_j}{\lambda_j\eta_{(i)}}\right)^{1/2}} - \frac{\tau_{(i)}}{\lambda_{(i)}}$$

where

$$\eta_{(i)} = \begin{cases} (\tau_{1,i})^{1/2}\phi\left[(\tau_{1,i})^{1/2}(\mu_{(1)} - \mu_{(i)})\right] & \text{for } (i) \neq (1) \\ \sum_{j|j\neq(1)} \eta_j & \text{for } (i) = (1) \end{cases}$$

(3) If all allocations $r_{(i)}$ are nonnegative, then continue to Step 4. Otherwise, remedy the nonnegativity constraint violation: (a) For each $(i) \in \mathcal{S}$ such that $r_{(i)} \leq 0$, remove (i) from $\mathcal{S}$ and set $r_{(i)} = 0$, (b)

For each (i) still in $\mathcal{S}\backslash\{(1)\}$, reassign $\tau_{1,i}$ as

$$\tau_{1,i} = \begin{cases} \left(\frac{\tau_{(1)}\tau_{(i)}}{\tau_{(1)}+\tau_{(i)}}\right) & \text{if } (1),(i) \in \mathcal{S} \\ \tau_{(1)} & \text{if } (i) \notin \mathcal{S}; (1) \in \mathcal{S}, \\ \tau_{(i)} & \text{if } (1) \notin \mathcal{S}; (i) \in \mathcal{S}, \end{cases}$$

and (c) return to Step 2.

(4) Round the r_i to integers.

When the budget B is large, rounding is likely to have no significant effect. Since the posterior distribution of the unknown means is normal, as is the prior, the allocation can be repeated to provide a multistage procedure, **Allocation $\mathcal{LL}(\mathcal{S})$**, that sequentially allocates some number of observations per stage.

3. Zero-One Loss

The reasoning for the linear loss in Section 2 also applies to the 0-1 loss function. The 0-1 loss is related to probability of correct selection P(CS) by $p\left(\mathbf{w} \mid w_i = \max_{j=1,\dots,k} w_j\right) = 1 - E[\mathcal{L}_{0-1}\left(i, \mathbf{w}\right)]$. Although the optimal decision to maximize P(CS) is *not* $\delta^N(\mathbf{x_r})$, the decision rule $\delta^N(\mathbf{x_r})$ has intuitive appeal, is tractable, and we conjecture that the sampling allocation derived below tends to avoid situations when the decision rule is not optimal (e.g., when the mean of the 'best' population is known with a much lower precision than the mean of the 'second best'[2]).

Denote the posterior precision of $W_{(j)} - W_{(i)}$, given $\mathbf{x_r}$ by

$$\tilde{\tau}_{i,j} = \left(\frac{1}{\tau_{(i)} + r_{(i)}\lambda_{(i)}} + \frac{1}{\tau_{(j)} + r_{(j)}\lambda_{(j)}}\right)^{-1}.$$

First consider the case of $k = 2$ populations.

Lemma 3: *Represent by ζ the prior distribution of $\mathbf{W} = (W_1, W_2)$, with $W_i \sim \mathcal{N}\left(\mu_i, \tau_i\right)$ jointly independent; and assume that the $x_{i,j}$ are jointly independent with $\mathcal{N}\left(w_i, \lambda_i\right)$ distribution, given w_i and a known precision λ_i. Let $G_\mathbf{r}$ be the prior predictive distribution for the unknown posterior means $\mathbf{Z}$. WLOG, assume $\mu_1 > \mu_2$. Then the expected total risk is*

$$\rho_{0-1}(\delta^N(\mathbf{X_r})) = \mathbf{c}\mathbf{r}^T + \Phi\left[{\tau_{1,2}}^{1/2}(\mu_2 - \mu_1)\right] + \Phi\left[{\tau_{\{1,2\}}}^{1/2}(\mu_2 - \mu_1)\right] \cdot \left(1 - 2E\left[\Phi\left[\tilde{\tau}_{1,2}^{1/2}(Z_2 - Z_1)\right] \mid Z_2 \geq Z_1\right]\right) \quad (16)$$

Proof: Add $-\mathcal{L}_{0-1}(1,\mathbf{w})$ to the loss function to obtain:

$$\mathcal{L}^*_{0-1}(i,\mathbf{w}) = \mathcal{L}_{0-1}(i,\mathbf{w}) - \mathcal{L}_{0-1}(1,\mathbf{w}) = \begin{cases} 0 \text{ for } i=1 \\ -1 \text{ for } i=2, w_1 < w_2 \\ 1 \text{ for } i=2, w_2 < w_1 \end{cases} \quad (17)$$

$$E[\mathcal{L}^*_{0-1}(1,\mathbf{w}) \mid \mathbf{x}] = 0$$

$$E[\mathcal{L}^*_{0-1}(2,\mathbf{w}) \mid \mathbf{x}] = p(w_1 \text{ best} \mid \mathbf{x}) - p(w_2 \text{ best} \mid \mathbf{x}) = 1 - 2\Phi\left[(\tilde{\tau}_{1,2})^{1/2}(\mu_2(\mathbf{x}) - \mu_1(\mathbf{x}))\right], \quad (18)$$

Population 2 is selected when its posterior mean is greater than the posterior mean for population 1. So population 2 is selected with probability $\Phi\left[\tau_{\{1,2\}}{}^{1/2}(\mu_2 - \mu_1)\right]$. The expected total risk is obtained by conditioning on which population has the highest posterior mean, then adding the cost of replications and $E_\zeta[\mathcal{L}_{0-1}(1,\mathbf{w})] = \Phi\left[\tau_{1,2}{}^{1/2}(\mu_2 - \mu_1)\right]$. □

Consider the behavior of Eq. 16 at extremes. First, $\tilde{\tau}_{1,2} \to \infty$ as r_1, r_2 grow without bound, so $E\left[\Phi\left[\tilde{\tau}_{1,2}^{1/2}(Z_2 - Z_1)\right] \mid Z_2 \geq Z_1\right] \to 1$. Further, $\tau_{\{1,2\}} \to \tau_{1,2}$, $G_\mathbf{r} \to \zeta$, and

$$\rho_{0-1}(\delta^N(\mathbf{X_r})) - \mathbf{cr}^T \to \Phi\left[\tau_{1,2}{}^{1/2}(\mu_2 - \mu_1)\right] + \Phi\left[\tau_{1,2}{}^{1/2}(\mu_2 - \mu_1)\right](1 - 2\cdot 1) = 0$$

as expected, given the value of perfect information. On the other hand, when there are no additional observations ($\mathbf{r} \to (0,0)$), then $\tilde{\tau}_{1,2} \to \tau_{1,2}$, $\tau_{\{1,2\}} \to \infty$, and

$$\rho_{0-1}(\delta^N(\mathbf{X_r})) - \mathbf{cr}^T = \Phi\left[\tau_{1,2}{}^{1/2}(\mu_2 - \mu_1)\right] + 0(1-2a) = \text{prior probability 2 is best} = 1 - \text{prior } P(CS),$$

for some $a \in (1/2, 1]$, as desired.

Because $E\left[\Phi\left[\tilde{\tau}_{1,2}^{1/2}(Z_2 - Z_1)\right] \mid Z_2 \geq Z_1\right] \to 1$ when the c_i are small (large r_i), and because easily calculable allocations are desired, the following asymptotically tight bound for $\rho_{0-1}(\delta^N(\mathbf{X_r}))$ is used for deriving an allocation.

$$\rho_{0-1}(\delta^N(\mathbf{X_r})) \geq \hat{\rho}^*_{0-1}(\mathbf{r}) \stackrel{\text{def}}{=} \mathbf{cr}^T + \Phi\left[\tau_{1,2}{}^{1/2}(\mu_2 - \mu_1)\right] - \Phi\left[\tau_{\{1,2\}}{}^{1/2}(\mu_2 - \mu_1)\right].$$

Theorem 4 argues that an asymptotically good approximation to the $\mathbf{r}$ that minimizes this bound is

$$\tilde{r}_i^* = \left(\frac{(\tau_{1,2})^{3/2}(\mu_1 - \mu_2)\phi\left[(\tau_{1,2})^{1/2}(\mu_2 - \mu_1)\right]}{2c_i\lambda_i} \right)^{1/2} - \frac{\tau_i}{\lambda_i}. \tag{19}$$

Adding the constraint $\mathbf{c}\mathbf{r}^T = B$, assuming that B is sufficiently large, gives the same allocation as for linear loss for this special case (because $\eta_1 = \eta_2$ when $k = 2$).

$$\tilde{r}_{i,B}^* = \frac{B + \sum_{j=1}^{2} \frac{c_j\tau_j}{\lambda_j}}{\sum_{j=1}^{2} \left(\frac{c_j c_i \lambda_i}{\lambda_j}\right)^{1/2}} - \frac{\tau_i}{\lambda_i}.$$

These results are special cases of the following general theorem, which gives different allocations than the linear case when $k > 2$.

Theorem 4: Presume the setup of Theorem 1, but use the 0-1 loss of Eq. 2 rather than the linear loss function. Then:

- The total risk of the natural decision rule is:

$$\rho_{0-1}(\delta^N(\mathbf{X_r})) = \mathbf{c}\mathbf{r}^T + \left(1 - p_\zeta\left(\mathcal{A}_{(1)}\right)\right) + \sum_{i=2}^{k} p_{G_\mathbf{r}}\left(\mathcal{A}_{(i)}\right) \cdot \left(1 - 2E\left[\Phi\left[\tilde{\tau}_{1,i}^{1/2}(Z_{(i)} - Z_{(1)})\right] \mid \mathcal{A}_{(i)}\right]\right)$$

  is bounded below by:

$$\hat{\rho}_{0-1}^*(\mathbf{r}) \stackrel{\text{def}}{=} \mathbf{c}\mathbf{r}^T + \left(1 - p_\zeta\left(\mathcal{A}_{(1)}\right)\right) - \sum_{i=2}^{k} \Phi\left[\tau_{\{1,k\}}{}^{1/2}(\mu_{(i)} - \mu_{(1)})\right] \tag{20}$$

- Let $\Delta_i = \mu_{(1)} - \mu_{(i)}$. As costs $c_i \to 0$, the r_i^* that minimize $\hat{\rho}_{0-1}^*(\mathbf{r})$ asymptotically approach:

$$\tilde{r}_{(1)}^* = \left(\sum_{i=2}^{k} \frac{(\tau_{1,i})^{3/2}\Delta_i\phi\left[(\tau_{1,i})^{1/2}\Delta_i\right]}{2c_{(1)}\lambda_{(1)}} \right)^{1/2} - \frac{\tau_{(1)}}{\lambda_{(1)}} \tag{21}$$

$$\tilde{r}_{(i)}^* = \left(\frac{(\tau_{1,i})^{3/2}\Delta_i\phi\left[(\tau_{1,i})^{1/2}\Delta_i\right]}{2c_i\lambda_i} \right)^{1/2} - \frac{\tau_{(i)}}{\lambda_{(i)}}, \text{ for } i \neq 1$$

Proof: The proof that the expected risk is as given repeats the arguments of the proof of Theorem 1 (except that here, $\delta^N(\mathbf{X})$ is not necessarily optimal). Similarly, the proof that Eq. 20 is a lower bound for the expected risk is obtained by adding the pairwise losses quantified by Lemma 3, then noting that $1 - 2\Phi\left[(\tilde{\tau}_{1,i})^{1/2}(\mu_1(\mathbf{x}) - \mu_i(\mathbf{x}))\right] \geq -1$.

Treat $\mathbf{r}$ as continuous and solve for $\partial \hat{\rho}^*_{0-1}(\mathbf{r})/\partial r_i = 0$, to get optimality conditions:

$$c_{(1)} = \sum_{j=2}^{k} \frac{(\tau_{\{1,j\}})^{3/2}(\mu_{(1)} - \mu_{(j)})\phi[(\tau_{\{1,j\}})^{1/2}(\mu_{(j)} - \mu_{(1)})]\lambda_{(1)}}{2(\tau_{(1)} + r_{(1)}\lambda_{(1)})^2} \tag{22}$$

$$c_{(i)} = \frac{(\tau_{\{1,i\}})^{3/2}(\mu_{(1)} - \mu_{(i)})\phi[(\tau_{\{1,i\}})^{1/2}(\mu_{(i)} - \mu_{(1)})]\lambda_{(i)}}{2(\tau_{(i)} + r_{(i)}\lambda_{(i)})^2}, \text{ for } i \neq 1$$

Solving for $r_{(i)}$ in Eq. 22 and substituting the limiting value $\tau_{1,i}$ for $\tau_{\{1,i\}}$ as each $c_i \to 0$, gives the claimed number of replications $\tilde{r}^*_{(i)}$. □

Minimizing the following surrogate objective function

$$\mathbf{c}\mathbf{r}^T + \sum_{i=2}^{k} \Phi\left[{\tau_{1,i}}^{1/2}(\mu_{(i)} - \mu_{(1)})\right] - \Phi\left[{\tau_{\{1,i\}}}^{1/2}(\mu_{(i)} - \mu_{(1)})\right] \tag{23}$$

gives the same solution $\tilde{r}^*_{(i)}$ of Eq. 21 for minimizing Eq. 20. Neglecting the cost of replications, this surrogate function represents sum of approximations of pairwise probabilities incorrect selection. For the special case of no replications, the sum in Eq. 23 becomes $\sum_{i=2}^{k} \Phi\left[{\tau_{1,i}}^{1/2}(\mu_{(i)} - \mu_{(1)})\right]$, an upper bound on the probability of incorrect selection when the population (1) is selected without additional samples, and corresponds exactly the Bonferroni inequality applied to the prior P(CS).

The sampling budget constraint is handled like Corollary 2 above.

Corollary 5: *The solution to* $\min \hat{\rho}^*_{0-1}(\mathbf{r})$ *subject to* $\mathbf{c}\mathbf{r}^T = B$ *(unconstrained* r_i*) for asymptotically large* B *is:*

$$\tilde{r}^*_{i,B} = \frac{B + \sum_{j=1}^{k} \frac{c_j \tau_j}{\lambda_j}}{\sum_{j=1}^{k}\left(\frac{c_j c_i \lambda_i \gamma_j}{\lambda_j \gamma_i}\right)^{1/2}} - \frac{\tau_i}{\lambda_i}$$

where

$$\gamma_{(i)} = \begin{cases} (\tau_{1,i})^{3/2}(\mu_{(1)} - \mu_{(i)})\phi\left[(\tau_{1,i})^{1/2}(\mu_{(1)} - \mu_{(i)})\right], & \textit{for } i \neq 1 \\ \sum_{j=2}^{k} \gamma_j, & \textit{for } i = 1 \end{cases}$$

Proof: The same as for Corollary 2, except that $\gamma_{(i)}$ replaces $\eta_{(i)}$. □

Allocation 0-1($\mathcal{B}$), the 0-1 loss analog to Allocation $\mathcal{LL}(\mathcal{B})$, can be used to constrain the number of observations to be nonnegative, assuming that $\gamma_{(i)}$ replaces $\eta_{(i)}$ from Eq. 15. **Allocation 0-1**($\mathcal{S}$) is the analogous multistage procedure that sequentially allocates a given number of observations per stage.

4. Numerical Example

Bickel and Doksum[3] consider the identification of the socio-economic subgroup with the highest cholesterol level from among $k = 3$ choices: $\mathcal{P}_1$ (high end), $\mathcal{P}_2$ (middle), $\mathcal{P}_3$ (low end). They presume samples to be normally distributed, that the means are unknown, and that there is a common precision of samples $\lambda_i = 1/\sigma^2 = 1/3386$. Gupta and Miescke[9] consider this problem, and suppose that $\mu_i = 266, \tau_i = 1/1508$, and $c_i = 1$. Data from a first stage of observations are available with 5 samples from $\mathcal{P}_1$ with $\bar{x}_1 = 315.6$, 10 samples from $\mathcal{P}_2$ with $\bar{x}_2 = 320.1$, and 6 samples from $\mathcal{P}_3$ with $\bar{x}_3 = 302.3$. The posterior given this data is used as the prior for the second stage of sampling.

Gupta and Miescke[9] indicate that it is optimal to observe the next sample from $\mathcal{P}_1$ when the linear loss is used, and that determining the optimal allocation of more than one sample at a time is computationally more intensive. Allocation $\mathcal{LL}(\mathcal{B})$, which also assumes the linear loss function, agrees with the allocation for the first observation, and readily gives allocations for larger budgets as shown in Table 1. Also presented are the analogous allocations for the 0-1 loss function.

Miescke[13] describes several Bayesian sampling allocations, including some easily computable suboptimal allocations, and evaluates their performance numerically. One of those procedures, called $\mathcal{LAH}$, repeatedly allocates one sample at a time as if it were the last sample to be allocated before a selection, doing so until a total number of samples has been observed. Although the $\mathcal{LAH}$ procedure is suboptimal, it performs reasonably well and is easy to compute.

The sequential procedure $\mathcal{LAH}$ is compared here with the two new two-stage budget-constrained allocations ($\mathcal{LL}(\mathcal{B})$ and 0-1($\mathcal{B}$)), as well as two analogous sequential procedures ($\mathcal{LL}(\mathcal{S})$ and 0-1($\mathcal{S}$)) in an empirical test for the cholesterol level problem.

Table 1. Sampling Allocations for a Cholesterol Study.

		$\mathcal{P}_1$	$\mathcal{P}_2$	$\mathcal{P}_3$
Stage One Sample Mean		315.6	320.1	302.3
Optimal (linear loss)	$B = 1$	1	0	0
Allocation $\mathcal{LL}(\mathcal{B})$	$B = 1$	1	0	0
(linear loss)	$B = 10$	4	3	3
	$B = 100$	31	41	28
	$B = 1000$	303	414	283
Allocation 0-1($\mathcal{B}$)	$B = 1$	0	0	1
(0-1 loss)	$B = 10$	2	4	4
	$B = 100$	25	41	34
	$B = 1000$	253	416	331

Each allocation is compared with respect to three performance metrics: the empirical fraction of correct selections (EFCS), the expected Bonferroni bound for the posterior probability of correct selection (EBPCS), and the expected Bonferroni bound for the posterior expectation of the linear loss, or opportunity cost (EBOC). The EBPCS is computed after observing all samples by (a) using the posterior distribution for each unknown mean, given all data, (b) reordering the sample statistics (i) to account for the potential changes in ordering for the posterior means $\mu_i(\mathbf{x_r})$, and (c) computing EBPCS as 1 minus the Bonferroni bound for incorrect selection obtained by setting $\mathbf{r} = 0$ in Eq. 23. EBOC is computed similarly by summing the expected linear loss for each pairwise comparison in Eq. 13.

Table 2 presents a summary of the effectiveness of $\mathcal{LAH}$ and the four new allocations (for the linear and 0-1 loss functions; two-stage and sequential). As with $\mathcal{LAH}$, the new sequential procedures allocation one observation at a time until B are observed. Each performance metric is measured as a function of the total number of samples ($B = 1, 10, 100, 1000$) and is estimated with simulation using 3000 macroreplications (repeated applications of the selection procedure). In each macroreplication, the unknown means $\mathbf{w}$ are sampled independently from the distribution for the unknown means described in the first paragraph of this section, and then samples are obtained conditional on that mean.

When $B = 1$ sample is allocated, $\mathcal{LAH}$, $\mathcal{LL}(\mathcal{B})$ and $\mathcal{LL}(\mathcal{S})$ perform within error bounds for all three metrics, as expected, since they each allocate one sample to $\mathcal{P}_1$. (When $B = 1$, the standard error for EFCS is 0.009; for EBPCS is 0.0016; and for EBOC is 0.040).

The asymptotic approximations used to derive allocations $\mathcal{LL}(\mathcal{B})$ and

0-1($\mathcal{B}$) seem to deleteriously affect their performance when $B = 10$ for this problem. This is particularly true for 0-1($\mathcal{B}$), which makes one more approximation than $\mathcal{LL}(\mathcal{B})$. However, the performance of all four new allocations, relative to $\mathcal{LAH}$, improves as the number of samples increases. Each clearly outperforms $\mathcal{LAH}$ when $B = 1000$. This seems to indicate that studies with large sampling budgets, as is common in computer simulation experiments, would particularly benefit from using $\mathcal{LL}(\mathcal{B})$, 0-1($\mathcal{B}$), $\mathcal{LL}(\mathcal{B})$ or 0-1($\mathcal{B}$) rather than $\mathcal{LAH}$.

Allocation $\mathcal{LL}(\mathcal{S})$ is either within error bounds or better than $\mathcal{LAH}$ for each figure of merit and for all sampling budgets. It appears to be a particularly good choice for an easy to compute allocation.

Table 2. Performance of sampling allocations $\mathcal{LAH}$ (Miescke[13]), new two-stage procedures $\mathcal{LL}(\mathcal{B})$ (linear loss) and 0-1($\mathcal{B}$) (0-1 loss), and new sequential procedures $\mathcal{LL}(\mathcal{S})$ and 0-1($\mathcal{B}$) for the empirical fraction of correct selections (EFCS), the expected Bonferroni bound for the posterior probability of correct selection (EBPCS), and the expected Bonferroni bound for the posterior expectation of the linear loss, or opportunity cost (EBOC).

Setup		Allocation				
Metric	B	$\mathcal{LAH}$	$\mathcal{LL}(\mathcal{B})$	0-1($\mathcal{B}$)	$\mathcal{LL}(\mathcal{S})$	0-1($\mathcal{S}$)
EFCS	1	0.5413	0.5463	0.5190	0.5397	0.5261
	10	0.6303	0.5983	0.5900	0.6218	0.5753
	100	0.8083	0.8183	0.7950	0.8176	0.7959
	1000	0.8886	0.9333	0.9316	0.9371	0.9458
EBPCS	1	0.4055	0.4052	0.3940	0.4041	0.3968
	10	0.5268	0.5018	0.4801	0.5231	0.4786
	100	0.7722	0.7865	0.7618	0.7901	0.7704
	1000	0.8851	0.9321	0.9229	0.9352	0.9306
EBOC	1	10.16	10.18	10.49	10.20	10.45
	10	6.81	7.48	8.06	6.94	8.15
	100	1.96	1.59	1.91	1.54	1.81
	1000	0.598	0.185	0.216	0.157	0.177

5. Comments

The problem of multiple selection has a rich history and a number of important structural results, such as the work of Gupta and Miescke[9]. Still, the optimal allocation for an additional stage of sampling has resisted closed-form solution in a more general setting, such as when there are $k > 2$ populations with potentially different variances and sampling costs. For this reason, reasonable allocations that can be efficiently computed are of great interest. This may be particularly true in computer simulation experiments, where a good, easy-to-compute allocation is better than an optimal but difficult to compute allocation, because CPU time can be used to obtain additional samples, rather than computing an optimal allocation.

The approach taken here is to minimize a lower bound on the expected risk (which gives the same allocation suggested by the cost of samples plus the sum of expected pairwise losses), then look at allocations with attractive asymptotic properties. Companion articles[5,11] indicate that similar allocations that assume an unknown variance perform well when compared to the indifference-zone formulation, even when the number of alternatives is much larger ($k = 100$) than in the numerical experiment above.

The new two-stage allocations ($\mathcal{LL}(\mathcal{B})$ and 0-1($\mathcal{B}$)) outperformed another easy-to-compute allocation, $\mathcal{LAH}$, in a numerical experiment when the total number of samples is large, even though $\mathcal{LAH}$ allocates sequentially. Sequential analogs of the new allocations perform even better, particularly $\mathcal{LL}(\mathcal{S})$.

The new two-stage allocations are somewhat inferior when the total number of samples is small. This is explained by the asymptotic approximations used to derive $\mathcal{LL}(\mathcal{B})$ and 0-1($\mathcal{B}$). On the other hand, the new allocations also have the flexibility to account for differing sampling costs. However, the merits of that cost accounting is not explored here.

A potential area of further research is the exploration of multiplicative bounds (e.g., Slepian's inequality) rather than the loose Bonferroni-type additive bound for the loss function. This may lead to improvements in sampling efficiency.

References

1. Bechhofer, R. E., T. J. Santner, and D. M. Goldsman (1995). *Design and Analysis for Statistical Selection, Screening, and Multiple Comparisons.* New York: John Wiley & Sons, Inc.
2. Berger, J. O. (1988). A Bayesian approach to ranking and selection of related means with alternatives to analysis-of-variance methodology. *Journal of the American Statistical Association* **83**(402), 364–373.
3. Bickel, P. J. and K. A. Doksum (1977). *Mathematical Statistics: Basic Ideas and Selected Topics.* San Francisco: Holden-Day.
4. Bracken, J. and A. Schleifer (1964). *Tables for Normal Sampling with Unknown Variance.* Boston: Graduate School of Business Administration,

Harvard University.

5. Chick, S. E. and K. Inoue (2001). New two-stage and sequential procedures for selecting the best simulated system. *Operations Research* **49**(5), to appear.
6. de Groot, M. H. (1970). *Optimal Statistical Decisions.* New York: McGraw-Hill, Inc.
7. Goldsman, D. and B. L. Nelson (1998). Statistical screening, selection, and multiple comparisons in computer simulation. In D. J. Madeiros, E. F. Watson, J. S. Carson, and M. S. Manivannan (Eds.), *Proceedings of the Winter Simulation Conference*, Piscataway, New Jersey, pp. 159–166. Institute of Electrical and Electronics Engineers, Inc.
8. Gupta, S. S. and K. J. Miescke (1994). Bayesian look ahead one stage sampling allocations for selecting the largest normal mean. *Statistical Papers* **35**, 169–177.
9. Gupta, S. S. and K. J. Miescke (1996). Bayesian look ahead one-stage sampling allocations for selecting the best population. *Journal of Statistical Planning and Inference* **54**, 229–244.
10. Hsu, J. C. (1996). *Multiple Comparisons: Theory and Methods.* New York: Chapman & Hall.
11. Inoue, K., S. E. Chick, and C.-H. Chen (1999). An empirical evaluation of several methods to select the best system. *ACM Transactions on Modeling and Computer Simulation* **9**(4), 381–407.
12. Miescke, K. J. (1979). Bayesian subset selection for additive and linear loss functions. *Commun. Statist.-Theor. Meth.* **A8**(12), 1205–1226.
13. Miescke, K. J. (1999). Bayes sampling designs for selection procedures. In S. Ghosh (Ed.), *Multivariate, Design, and Sampling.* New York: M. Dekker.
14. Nelson, B. L., J. Swann, D. Goldsman, and W. Song (2001). Simple procedures for selecting the best simulated system when the number of alternatives is large. *Operations Research*, to appear.
15. Rinott, Y. (1978). On two-stage selection procedures and related probability-inequalities. *Communications in Statistics* **A7**, 799–811.

CHAPTER 21

BAYES ESTIMATES OF FLOOD QUANTILES USING THE GENERALISED GAMMA DISTRIBUTION

Jan M. van Noortwijk

HKV Consultants
P.O. Box 2120, NL-8203 AC Lelystad, The Netherlands;
Department of Mathematics, Delft University of Technology
P.O. Box 5031, NL-2600 GA Delft, The Netherlands
E-mail: j.m.van.noortwijk@hkv.nl

In this paper, a Bayesian approach is proposed to estimate flood quantiles while taking statistical uncertainties into account. Predictive exceedance probabilities of annual maximum discharges are obtained using the three- and four-parameter generalised gamma distribution (without and with location parameter respectively). The parameters of this distribution are assumed to be random quantities rather than deterministic quantities and to have a prior joint probability distribution. On the basis of observations, this prior joint distribution is then updated to the posterior joint distribution by using Bayes' theorem.

An advantage is that the generalised gamma distribution fits well with the stage-discharge rating curve being an approximate power law between water level and discharge. Furthermore, since the generalised gamma distribution has three or four parameters, it is flexible in fitting data. Many well-known probability distributions which are commonly used to estimate quantiles of hydrological random quantities are special cases of the generalised gamma distribution. As an example, a Bayesian analysis of annual maximum discharges of the Rhine River at Lobith is performed to determine flood quantiles including their uncertainty intervals. The generalised gamma distribution can also be applied in lifetime and reliability analysis.

1. Introduction

The Dutch river dikes have to withstand water levels and discharges with an average return period up to 1250 years, where a downstream water level can be determined on the basis of the upstream discharge by using a river flow simulation model[33,34]. Therefore, a design discharge is defined as the annual maximum river discharge for which the probability of exceedance is 1/1250 per year. A common problem in obtaining design discharges is that there is a limited amount of observations available (for example, with respect to the Dutch Rhine River there are only 98 annual maximum discharges available).

The design discharge is usually estimated by extrapolating a probability distribution which is fitted to the observed discharges. The two-parameter gamma distribution (Pearson type III distribution) is among the distributions commonly used. When more shape flexibility is needed to fit the data, Ashkar & Ouarda[2] suggest the three-parameter generalised gamma distribution. In order to take the uncertainties into account when there is a small amount of data, they developed approximate confidence intervals for quantiles of the generalised gamma distribution. These confidence intervals have been obtained by applying techniques from classical statistics.

In this paper, an alternative way to take the statistical uncertainties into account is proposed: regarding the statistical parameters as being random quantities rather than deterministic quantities. On the basis of the observed annual maximum discharges, the prior density of these random quantities can be updated to the posterior density by using Bayes' theorem. In order to describe the a priori 'lack of knowledge', we use the non-informative Jeffreys prior for the scale and shape parameters of the three-parameter generalised gamma distribution. This Bayesian analysis is extended to estimating the location parameter of the four-parameter generalised gamma distribution as well.

To fit a generalised gamma distribution to discharge data, we refer to the relation between this distribution and the stage-discharge rating curve (an approximate power law in which the discharge is expressed in terms of the water level). Although it was not possible to derive the generalised gamma distribution solely on the basis of this physical law, the reason for using this distribution is in the spirit of the so-called 'engineering probability' (see Barlow[3] and Mendel[24]). According to Mendel[24] "engineering probability concerns the derivation of probabilistic models from the physical laws, geometric constraints, and other engineering knowledge concerning an engineering system". The observations that have been analysed are the annual maximum discharges of the Dutch Rhine River at Lobith (near the

Dutch-German border).

This paper is set out as follows. In Section 2, an approximate rating curve is derived for the Rhine River at Lobith. The mathematical procedure to apply a Bayesian analysis to the observed annual maximum discharges can be found in Section 3. The Jeffreys prior of the three-parameter generalised gamma distribution is derived in Section 4. The posterior density of the scale and shape parameters of the three-parameter generalised gamma distribution is obtained in Section 5. Section 6 is devoted to taking account of the statistical uncertainty in the location parameter of the four-parameter generalised gamma distribution. Results for the Rhine River are presented in Section 7 and conclusions are in Section 8.

2. Stage-discharge Rating Curve

Although a flood wave is a gradually varied unsteady non-uniform flow, the uniform-flow condition is frequently assumed in the computation of flow in rivers. The results obtained from this assumption are recognised to be approximate and general, but they offer a relatively simple and satisfactory solution to many practical problems.

Under the condition of uniform open-channel flow, the average discharge can be determined using Manning's equation:

$$q = v\tilde{a} = \frac{\tilde{r}^{2/3}\tilde{a}\sqrt{s}}{n_m} = \frac{\tilde{a}^{5/3}\sqrt{s}}{\tilde{p}^{2/3}\,n_m},$$

where q = discharge [m^3/s], v = mean velocity [m/s], $\tilde{a}$ = cross-sectional flow area [m^2], $\tilde{p}$ = wetted perimeter [m], $\tilde{r} = \tilde{a}/\tilde{p}$ = hydraulic radius or hydraulic mean depth [m], s = channel slope [-], and n_m = Manning roughness coefficient [s/m$^{1/3}$] (see e.g. Chow[9] or Chow, Maidment & Mays[10]).

In the event of extreme discharges, a river breaks its banks and inundates the flood plain. Since the total width of the flood plain of the Rhine at Lobith is about 1150 m (including the main channel), the inundated flood plain can approximately be regarded as a wide rectangular cross-section with a width w [m] much larger than its water depth $\tilde{d}$ [m]. In this situation, the hydraulic radius can be approximated by

$$\tilde{r} = \frac{\tilde{a}}{\tilde{p}} = \frac{w\tilde{d}}{w + 2\tilde{d}} \approx \tilde{d}$$

and, accordingly,

$$q \approx \frac{w\tilde{d}^{5/3}\sqrt{s}}{n_m}. \tag{1}$$

The stage-discharge rating curve (1) suggests that the rating curve can be approximated by the following power law between water level and discharge (see Shaw[29] and Chow, Maidment & Mays[10]):

$$q - q_0 \approx \alpha \left[h - h_0\right]^{\beta}, \qquad q \geq q_0,\ h \geq h_0,\ \alpha, \beta > 0, \tag{2}$$

where q = discharge [m^3/s], q_0 = threshold value discharge [m^3/s], h = water level [m +NAP], h_0 = threshold value water level [m +NAP], $\tilde{d} = h - h_0$ [m], and 1 m +NAP means 1 m above 'normal Amsterdam level'. By applying rating curve (2), regularly observed or continuously recorded water levels can be converted to corresponding discharge estimates. Hydrological statistical analyses are mostly performed on the basis of discharge data. As a result of Eq. (2), and by assuming the location parameter q_0 to be known, we fit a probability distribution to those discharges which are higher than the threshold $q_0 = 1750$ m^3/s (or, in other words, water levels which are higher than $h_0 = 9.0$ m +NAP). The threshold q_0 has been determined by maximising the marginal density of the observations in Eq. (5).

With the aid of a least-squares method, α and β in Eq. (2) have been fitted to measurements and extrapolations of the Dutch Ministry of Transport, Public Works, and Water Management[26]. As it can be seen in Figure 1, the estimated values $\alpha = 207.9$ and $\beta = 1.917$ result in a good approximation of the actual rating curve of the Rhine at Lobith, especially for extreme discharges.

3. Bayesian Analysis of Discharges

In this section, it is shown that the generalised gamma distribution fits well with the stage-discharge rating curve in terms of the power law between water level and discharge. As a matter of fact, if the discharge is assumed to have a generalised gamma distribution and if the power law holds, then the water level has a generalised gamma distribution as well (and this also applies in reverse). Furthermore, we regard the scale and shape parameters of the three-parameter generalised gamma distribution to be unknown, having a joint probability distribution.

Because the rating curve (2) approximately holds for the thresholds q_0 and h_0, the following question must be answered. What is the probability

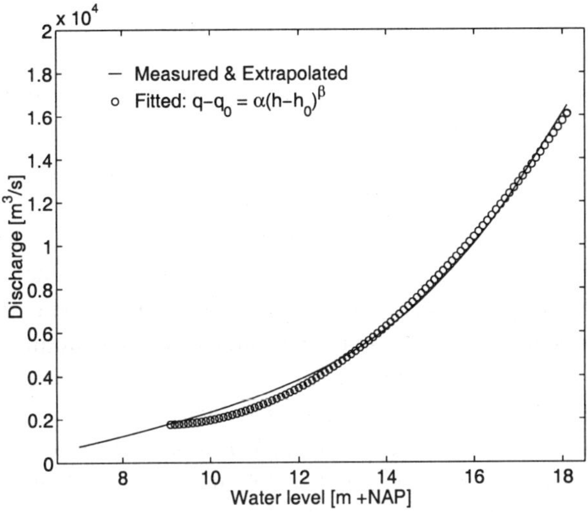

Fig. 1. Actual (measured and extrapolated) and fitted rating curve of the Rhine River at Lobith.

distribution of the annual maximum discharge Q for $Q \geq q_0$ while taking the statistical uncertainties into account? (or, in other words, what is the probability distribution of the annual maximum water level H for $H \geq h_0$?) This question can be answered using Bayesian statistics[6,8].

The probability distribution of the annual maximum discharge at Lobith is chosen so that the type of distribution remains the same when the transformation (2) is applied from discharge to water level. If the probability density functions of the discharge Q and the water level H are given by $p(q - q_0)$ and $\tilde{p}(h - h_0)$, respectively, then

$$\tilde{p}\,(h - h_0) = \alpha\beta\,[h - h_0]^{\beta-1}\, p\left(\alpha\,[h - h_0]^{\beta}\right)$$

for $h \geq h_0$ or, in terms of the discharge,

$$p\,(q - q_0) = \tfrac{1}{\beta}\alpha^{-\frac{1}{\beta}}\,[q - q_0]^{\frac{1}{\beta}-1}\,\tilde{p}\left(\alpha^{-\frac{1}{\beta}}\,[q - q_0]^{\frac{1}{\beta}}\right)$$

for $q \geq q_0$. A probability density function of Q for which $p(\cdot)$ and $\tilde{p}(\cdot)$ belong to the same type of distribution is the three-parameter generalised gamma distribution (see Stacy[31], Parr & Webster[27], Hager & Bain[13] and Johnson, Kotz & Balakrishnan[17]):

$$\ell\,(q - q_0|\, a, b, c) = \mathrm{Gga}\,(q - q_0|\, a, b, c) =$$

$$= \frac{c}{b\Gamma(a)} \left[\frac{q-q_0}{b}\right]^{ca-1} \exp\left\{-\left[\frac{q-q_0}{b}\right]^c\right\} I_{(0,\infty)}(q-q_0) \qquad (3)$$

with parameters $a, b, c > 0$. Indeed, the probability density function of h is also a generalised gamma distribution:

$$\ell\left(h-h_0 \,\middle|\, a, [b/\alpha]^{1/\beta}, \beta c\right) = \mathrm{Gga}\left(h-h_0 \,\middle|\, a, [b/\alpha]^{1/\beta}, \beta c\right).$$

The three-parameter generalised gamma distribution has been successfully fitted to annual maximum discharges by Ashkar & Ouarda[2]. Since it has three parameters, it includes many well-known probability distributions like the exponential distribution ($a = 1$ and $c = 1$), the gamma distribution ($c = 1$), the chi-square distribution ($a = t/2$, $b = 2$, and $c = 1$), the Weibull distribution ($a = 1$), the l_c-isotropic distribution ($a = 1/c$), the Rayleigh distribution ($a = 1$ and $c = 2$), and the Maxwell distribution ($a = 3/2$ and $c = 2$). In addition, the lognormal distribution is a limiting special case when a tends to infinity. The proof follows by noting that the logarithm of a random quantity with a generalised gamma distribution has a generalised Gompertz distribution. Ahuja & Nash[1] proved that the so-obtained generalised Gompertz distribution is asymptotically normal as $a \to \infty$. An interesting physical characterisation of generalised gamma distributions on the basis of statistical mechanics can be found in Lienhard & Meyer[21]. As a special case, Lienhard[20] derived a generalised gamma distribution (with $c = 2$) to describe rainfall run-off from a watershed.

Subsequently, we determine the moments and the exceedance probability of the generalised gamma distribution. For this purpose, the incomplete gamma function is defined as

$$\Gamma(x,y) = \int_{t=y}^{\infty} t^{x-1}e^{-t}\,dt, \qquad x > 0,\ y \geq 0.$$

By means of the gamma function, $\Gamma(x) = \Gamma(x,0)$, the nth moment of $Q - q_0$ can be written as

$$E\left([Q-q_0]^n \mid a,b,c\right) = b^n\, \Gamma\left(\tfrac{n}{c}+a\right)/\,\Gamma(a)$$

for $q \geq q_0$ and $n \geq 0$. Similarly, the conditional exceedance probability follows:

$$\Pr\{Q > q \mid a,b,c\} = 1 - \Gamma\left(a, \left[\frac{q-q_0}{b}\right]^c\right)\Big/\,\Gamma(a) \qquad (4)$$

for $q \geq q_0$. Note that the generalised gamma distribution takes its name from the fact that $Y = [(Q-q_0)/b]^c$ has a gamma distribution with shape parameter a and scale parameter 1. A random quantity X has a gamma distribution with shape parameter $\nu > 0$ and scale parameter $\mu > 0$ if its probability density function is given by:

$$\mathrm{Ga}(x|\,\nu,\mu) = [\mu^{\nu}/\Gamma(\nu)]\, x^{\nu-1} \exp\{-\mu x\}\, I_{(0,\infty)}(x).$$

Recall that n annual maximum discharges have been observed. Since these discharges are conditionally independent when the values of the parameters a, b and c are given, the likelihood function of the observations q_j, $j = 1,\ldots,n$, can be written as

$$\ell\,(\mathbf{q} - q_0|\, a,b,c) = \textstyle\prod_{j=1}^{n} \ell\,(q_j - q_0|\, a,b,c)\,,$$

where $\mathbf{q} = (q_1,\ldots,q_n)'$. In order to quantify the uncertainty in the parameters a, b, and c, we assume that they have a prior probability distribution. The marginal density $\pi(\mathbf{q} - q_0)$ of the observations $\mathbf{q}$ is obtained by integrating the likelihood over (a,b,c):

$$\pi(\mathbf{q} - q_0) = \textstyle\int_{a=0}^{\infty}\int_{b=0}^{\infty}\int_{c=0}^{\infty} \ell(\mathbf{q} - q_0|\, a,b,c)\pi(a,b,c)\, da db dc. \qquad (5)$$

4. Non-informative Jeffreys Prior

For the purpose of flood prevention, we would like the observations to 'speak for themselves', especially in comparison to the prior information. This means that the prior distribution should describe a certain 'lack of knowledge' or in other words, should be as 'vague' as possible. For this purpose, so-called non-informative priors have been developed. A disadvantage of most non-informative priors is that these priors can be improper; that is, they often do not integrate to unity. This disadvantage can be resolved by focussing on the posterior distributions rather than the prior distributions. As a matter of fact, formally carrying out the calculations of Bayes' theorem by combining an improper prior with observations often results in a proper posterior.

The pioneer in using non-informative priors was Bayes[4] who considered a uniform prior. However, the use of uniform priors is criticised because of a lack of invariance under one-to-one transformations. For example, let us consider an unknown parameter θ and suppose the problem has been parameterised in terms of $\phi = \exp\{\theta\}$. This is a one-to-one transformation, which should have no bearing on the ultimate result. The Jacobian of this transformation is given by $d\theta/d\phi = d\log\phi/d\phi = 1/\phi$. Hence, if the non-informative prior for θ is chosen to be uniform (constant), then the non-informative prior for ϕ should be proportional to $1/\phi$ to maintain consistency. Unfortunately, we cannot maintain consistency and choose both the non-informative priors for θ and ϕ to be constant.

An illustration of the danger of using uniform distributions as non-informative priors for the parameters (a, b, c) of the generalised gamma distribution is given below. Let us make a change of variables from (a, b, c) to $(\theta, \lambda, \phi) = (a^{-1}, ac, b^{-c})$ and assume the non-informative prior of (θ, λ, ϕ) to be $\pi(\theta, \lambda, \phi) \propto \phi^{-1}$. Hence, the marginal priors of θ, λ, and $\log \phi$ are uniform. Accounting for the Jacobian of this transformation, the corresponding non-informative prior of (a, b, c) has the form $\pi(a, b, c) \propto (c/a) \cdot b^{-1}$. The marginal prior of c is all but non-informative!

The physicist Sir Jeffreys[16] was the first to produce an alternative to solely using uniform non-informative priors. His main reason for deriving non-informative priors (now known as Jeffreys priors) were invariance requirements for one-to-one transformations. In a multi-parameter setting, the Jeffreys prior takes account of dependence between the parameters. For decades, there has been discussion going on as to whether the multivariate Jeffreys rule is appropriate[19]. We concur with the following statement made by Dawid[1]: "we do not consider it as generally appropriate to use other improper priors than the Jeffreys measure for purposes of 'fully objective' formal model comparison". When a location parameter, which is bounded from below or above, is involved, it may be recommended to modify the Jeffreys prior (see Section 6). The main advantage of the Jeffreys prior is that it is always both dimensionless and invariant under transformations.

As an example, the multivariate Jeffreys prior for the normal model with unknown mean μ and unknown standard deviation σ is

$$J(\mu, \sigma)\, d\mu d\sigma = \frac{\sqrt{2}}{\sigma^2}\, d\mu d\sigma.$$

It can be easily seen that the above prior is dimensionless: i.e., $d\mu$, $d\sigma$, and σ have the same units. For another example, see the Jeffreys prior of the three-parameter generalised gamma distribution in Eq. (7). Because the non-dimensionality argument is sound (from a physics point of view), the multivariate Jeffreys prior is used as a non-informative prior for the three-parameter generalised gamma distribution.

To explain the derivation of non-informative Jeffreys priors, see Box & Tiao[7]. Let $\mathbf{x} = (x_1, \ldots, x_n)'$ be a random sample from a multi-parameter probability distribution with likelihood function $\ell(x|\boldsymbol{\theta})$ with parametric vector $\boldsymbol{\theta}$. When the probability distribution obeys certain regularity conditions, then for sufficiently large n, the posterior density function of $\boldsymbol{\theta}$ is approximately normal and remains approximately normal under mild one-to-one transformations of $\boldsymbol{\theta}$. As a consequence, the prior distribution for $\boldsymbol{\theta}$ is approximately non-informative if it is taken proportional to the square root of Fisher's information measure for a single observation. In mathematical terms, the elements of this matrix are

$$I_{ij}(\boldsymbol{\theta}) = E\left(-\frac{\partial^2 \log \ell(X|\boldsymbol{\theta})}{\partial\theta_i \partial\theta_j}\right), \qquad i,j = 1,\ldots,r,$$

where r is the number of parameters. The corresponding non-informative Jeffreys prior is defined by the square root of the determinant of Fisher's information matrix for a single observation:

$$J(\boldsymbol{\theta}) = \sqrt{|I(\boldsymbol{\theta})|} = \sqrt{\det I_{ij}(\boldsymbol{\theta})}, \qquad i,j = 1,\ldots,r.$$

According to Hager & Bain[13], the Fisher information matrix of the three-parameter generalised gamma distribution is found to be

$$I(a,b,c) = \begin{pmatrix} \psi'(a) & \dfrac{c}{b} & -\dfrac{\psi(a)}{c} \\ \dfrac{c}{b} & \dfrac{c^2 a}{b^2} & -\dfrac{1+a\psi(a)}{b} \\ -\dfrac{\psi(a)}{c} & -\dfrac{1+a\psi(a)}{b} & \dfrac{1+2\psi(a)+a\psi'(a)+a[\psi(a)]^2}{c^2} \end{pmatrix}. \qquad (6)$$

After straightforward algebra, the determinant of Fisher's information matrix can be written as

$$|I(a,b,c)| = \frac{[a\psi'(a)]^2 - \psi'(a) - 1}{b^2}.$$

Hence, the Jeffreys prior of the three-parameter generalised gamma distribution is

$$J(a,b,c) = \sqrt{|I(a,b,c)|} = \frac{\sqrt{[a\psi'(a)]^2 - \psi'(a) - 1}}{b} = \frac{J(a)}{b}. \qquad (7)$$

The function $\psi'(a)$ is the first derivative of the digamma function:

$$\psi'(a) = \frac{\partial\psi(a)}{\partial a} = \frac{\partial^2 \log\Gamma(a)}{\partial a^2}$$

for $a > 0$. It is called the trigamma function. The digamma function and the trigamma function can be accurately computed using algorithms developed by Bernardo[5] and Schneider[28], respectively.

5. Posterior Density

On the basis of the observations, we obtain the posterior conditional probability density function of b when the values of a and c are given, as well as the posterior joint probability density function of a and c. On the basis of the Jeffreys prior for the generalised gamma distribution, the posterior distribution of b, given $(a, c, \mathbf{q} - q_0)$, can be expressed in explicit form:

$$\pi(b|\, a, c, \mathbf{q} - q_0) = \tag{8}$$
$$= \frac{\left(\sum_{j=1}^{n} [q_j - q_0]^c\right)^{na}}{\Gamma(na)} c b^{-nca-1} \exp\left\{-\sum_{j=1}^{n}\left[\frac{q_j - q_0}{b}\right]^c\right\}.$$

In a similar manner, the marginal density of the observations, given a and c, can be expressed in explicit form

$$\pi(\mathbf{q} - q_0|\, a, c) = \tag{9}$$
$$= \int_{b=0}^{\infty} \ell(\mathbf{q} - q_0|\, a, b, c)\, b^{-1}\, db = c^{n-1} \cdot \frac{\Gamma(na)}{[\Gamma(a)]^n} \cdot \frac{\prod_{j=1}^{n} [q_j - q_0]^{ca-1}}{\left(\sum_{j=1}^{n} [q_j - q_0]^c\right)^{na}}.$$

Since the posterior joint probability density function of a and c cannot be expressed in explicit form, we have to resort to approximations. For this, it is convenient to define discrete distributions in the same way as they were applied to quantify the uncertainty in the shape parameter of a Weibull distribution in Soland[30] and Mazzuchi & Soyer[22,23]. Using Eq. (9) and Bayes' theorem, it follows that

$$p(a_h, c_i|\, \mathbf{q} - q_0) = \frac{\pi(\mathbf{q} - q_0|\, a_h, c_i)\, p(a_h)\, p(c_i)}{\sum_{h=1}^{k}\sum_{i=1}^{m} \pi(\mathbf{q} - q_0|\, a_h, c_i)\, p(a_h)\, p(c_i)}, \tag{10}$$

where

$$a_h = a_L + [(2h-1)/2] \cdot [(a_U - a_L)/k], \qquad h = 1, \ldots, k,$$
$$c_i = c_L + [(2i-1)/2] \cdot [(c_U - c_L)/m], \qquad i = 1, \ldots, m,$$

with a_L and a_U being the lower and upper bounds for a, and c_L and c_U being the lower and upper bounds for c. Suitably wide integration bounds may be determined on the basis of an approximate posterior density such as a (transformed) normal density with a mean equal to the maximum-likelihood estimator $\hat{\boldsymbol{\theta}}$ and a covariance matrix equal to $[nI(\hat{\boldsymbol{\theta}})]^{-1}$, being the inverse Fisher information matrix for a sample of n observations evaluated at $\hat{\boldsymbol{\theta}}$. Convergence to normality of the approximate posterior density can often be improved by transformation (e.g., by taking the logarithm of a non-negative parameter).

The non-informative prior probability functions of a and c are based on the Jeffreys prior (7); that is,

$$p(a_h) = \Pr\{A = a_h\} = J(a_h)/[\textstyle\sum_{h=1}^{k} J(a_h)], \qquad h = 1, \ldots, k,$$
$$p(c_i) = \Pr\{C = c_i\} = 1/m, \qquad i = 1, \ldots, m.$$

Alternative forms of the prior distributions of a, b, and c are also possible; they can be informative rather than non-informative as well.

Finally, the predictive expected exceedance probability of the annual maximum discharge can be determined by integrating Eq. (4) over the random quantities a, b, and c:

$$\Pr\{Q > q\} = \sum_{h=1}^{k}\sum_{i=1}^{m}\int_{b=0}^{\infty} \Pr\{Q > q|\, a_h, c_i, b\}\, p\,(a_h, c_i, b|\, \mathbf{q} - q_0)\; db, \quad (11)$$

where

$$p\,(a_h, c_i, b|\, \mathbf{q} - q_0) = \pi\,(b|\, a_h, c_i, \mathbf{q} - q_0)\, p\,(a_h, c_i|\, \mathbf{q} - q_0) \qquad (12)$$

(the product of Eqs. (8) and (10), respectively) for $h = 1, \ldots, k$ and $i = 1, \ldots, m$. Note that the prior independence between the random quantities a, b, and c converts to posterior dependence given the observations.

6. Location Parameter

This section discusses how to extend the Bayesian analysis from the three-parameter generalised gamma distribution (without location parameter) to the four-parameter generalised gamma distribution (with location parameter). The likelihood function of the four-parameter generalised gamma distribution with scale parameter b, shape parameters a and c, and location parameter d is defined as[14]

$$\ell\,(q|\, a, b, c, d) = \mathrm{Gga}\,(q - d|\, a, b, c) =$$
$$= \frac{c}{b\Gamma(a)}\left[\frac{q-d}{b}\right]^{ca-1} \exp\left\{-\left[\frac{q-d}{b}\right]^{c}\right\} I_{(d,\infty)}(q). \qquad (13)$$

For determining the Jeffreys prior in the presence of location parameter d, we have to extend Fisher's information matrix with seven elements (see the Appendix) of which three are equal to each other (due to the symmetry of Fisher's information matrix). Theoretically speaking, the Jeffreys prior of the four-parameter generalised gamma distribution with location parameter d may be obtained by taking the square root of the determinant of the four-dimensional Fisher information matrix (whose elements are given in Eq. (6) and in the Appendix). Unfortunately, a serious disadvantage of the so-obtained Jeffreys prior is that it is only defined for $ac > 2$ (see the Appendix) thus possibly excluding valuable distributions such as the exponential and Rayleigh distribution. For this reason, maximum-likelihood estimation for the four-parameter generalised gamma distribution is called *regular* if and only if the product of the two shape parameters is greater than two. For the definition of regular estimation problems, see e.g. Cox & Hinkley[11].

The problem of a non-existing Jeffreys prior notably arises when parameters of different kinds are considered simultaneously. Jeffreys[16] pointed out that his multi-parameter rule must be applied with caution, especially where scale and location parameters co-exist. To counter this problem Jeffreys suggested: "We can then deal with location parameters, on the hypothesis that the scale and numerical parameters are irrelevant to them, by simply taking their prior probability uniform". In deriving Jeffreys priors, serious problems may occur in situations where a location parameter is bounded from below or above (e.g., is non-negative).

For the four-parameter generalised gamma distribution with a location parameter being bounded from below, we follow Jeffreys' recommendation and assume the location parameter d to be a priori independent of the scale and shape parameters (a, b, c) and take the uniform prior as a non-informative prior for d and the Jeffreys prior (7) as a non-informative prior for (a, b, c); that is,

$$J(a, b, c, d) = J(a, b, c) = \frac{J(a)}{b}. \tag{14}$$

On the basis of the Jeffreys prior for the four-parameter generalised gamma distribution, the posterior distribution of (a, b, c, d) given the observations $\mathbf{q}$ remains to be calculated. This posterior can be determined by extending the formulas in Section 5 with a summation over the discretised

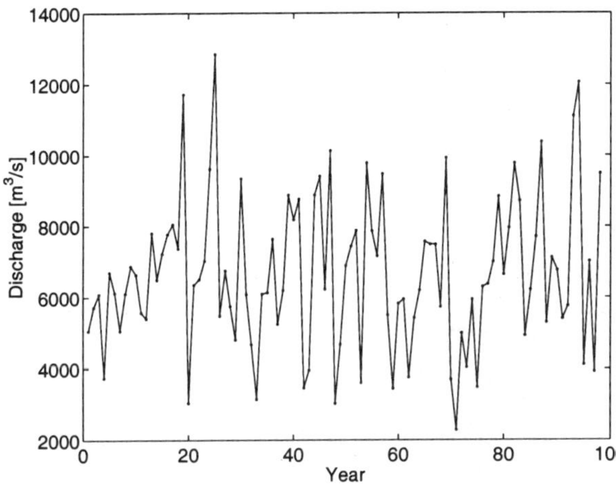

Fig. 2. Observed annual maximum discharges of the Rhine River at Lobith during 1901-1998.

prior distribution of d, which is defined by

$$p(d_j) = \Pr\{D = d_j\} = 1/l, \quad j = 1, \ldots, l,$$

where

$$d_j = d_L + [(2j-1)/2] \cdot [(d_U - d_L)/l], \quad j = 1, \ldots, l,$$

with d_L and d_U being the lower and upper bounds for d. Obviously, river flow physics suggests the lower bound of d to be zero.

7. Results: Design Discharge of the Rhine River

The Bayesian updating approach has been applied to the annual maximum discharges of the Rhine River at Lobith during the period 1901-1998 (see Figure 2). The four largest discharges in descending order are 12,849 m^3/s (1926), 12,060 m^3/s (1995), 11,712 m^3/s (1920), and 11,100 m^3/s (1993). The Bayesian analysis of the annual maxima has been performed with respect to both the three-parameter generalised gamma distribution (without location parameter) and the four-parameter generalised gamma distribution (with location parameter).

The prior and posterior probability functions of the shape parameters a and c for the three-parameter gamma distribution are shown in Figures 3

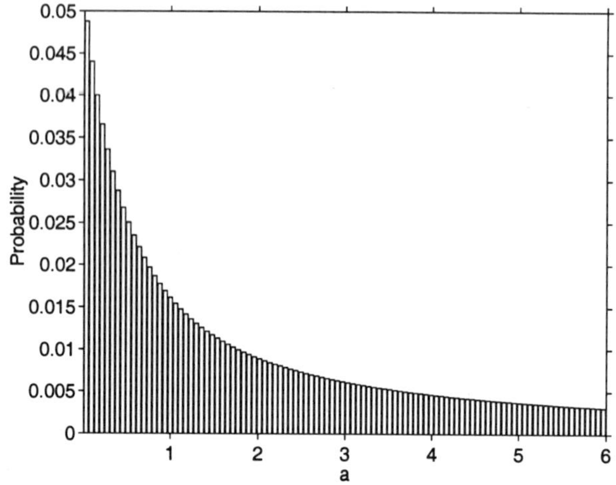

Fig. 3. Prior probability function of a: $p(a_h)$, $h = 1, \ldots, k$.

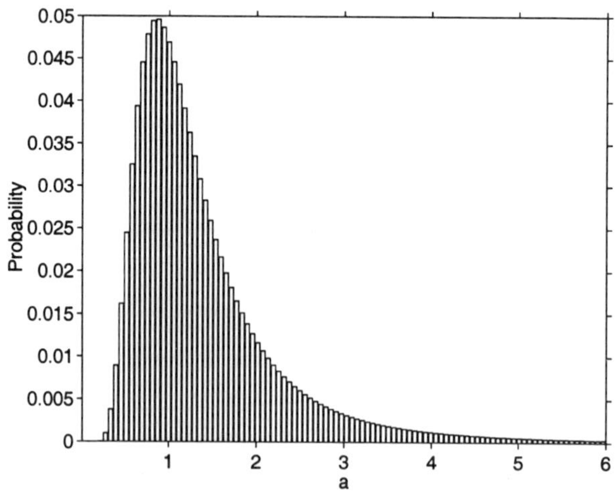

Fig. 4. Posterior probability function of a: $p(a_h \,|\, \mathbf{q} - q_0)$, $h = 1, \ldots, k$.

to 6. The parameters of the prior probability distributions of a and c, as well as the posterior expectations of a, b, and c can be found in Table 1. Figure 7 presents the empirical cumulative probability distribution, the pre-

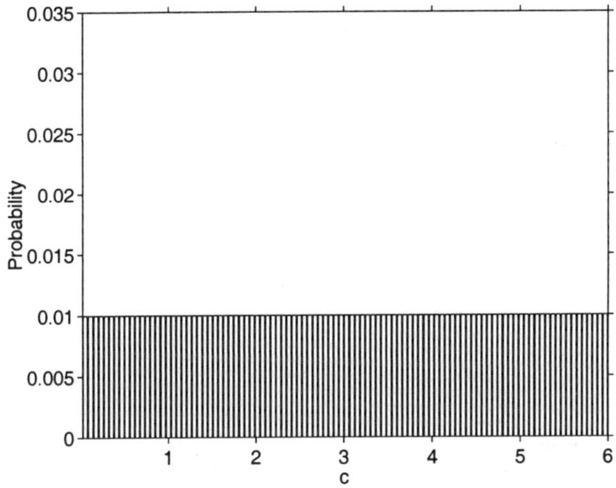

Fig. 5. Prior probability function of c: $p(c_i)$, $i = 1, \ldots, m$.

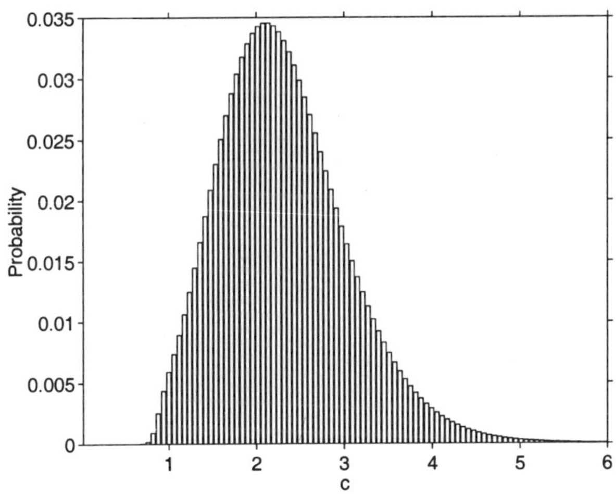

Fig. 6. Posterior probability function of c: $p(c_i | \mathbf{q} - q_0)$, $i = 1, \ldots, m$.

dictive cumulative probability distribution, and the 90 per cent uncertainty intervals (Bayesian credible sets or regions). These 90 per cent uncertainty intervals (the 5th and 95th percentile) are determined using Monte Carlo

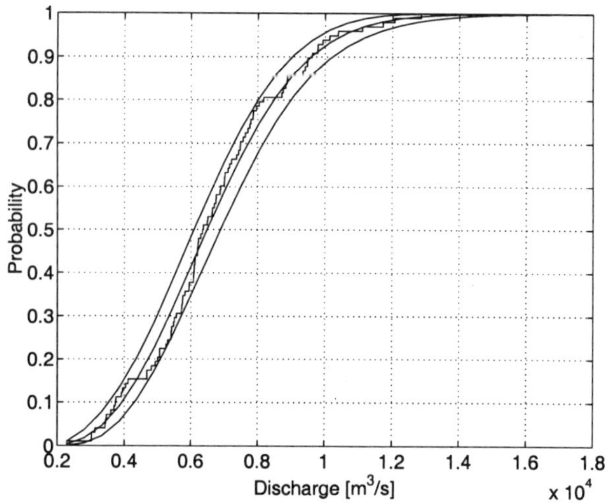

Fig. 7. Empirical and predictive cumulative probability distribution of the annual maximum discharge of the Rhine River at Lobith, including their 90 per cent uncertainty interval, for the three-parameter generalised gamma distribution.

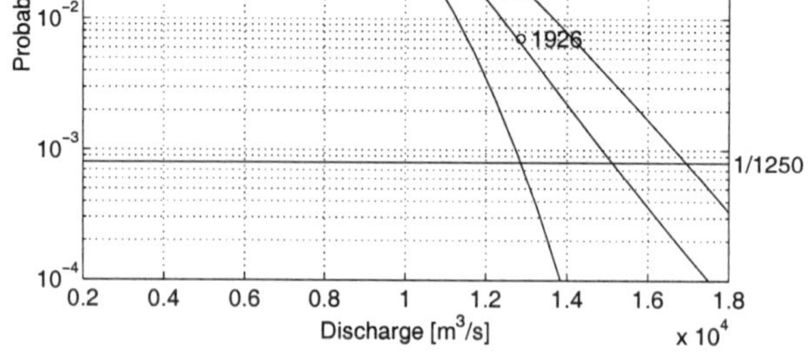

Fig. 8. Empirical and predictive probability of exceedance of the annual maximum discharge of the Rhine River at Lobith, including their 90 per cent uncertainty interval, for the three-parameter generalised gamma distribution.

Table 1. Cross-sectional parameters, as well as Bayes estimates of a, b, and c for the three-parameter generalised gamma distribution.

parameter	description	value	dimension
w	width main channel and flood plain	1147	m
α	parameter rating curve	207.9	-
β	parameter rating curve	1.917	-
q_0	threshold value discharge	1750	m^3/s
h_0	threshold value water level	9.015	m +NAP
n	number of observations	98	-
a_L	lower bound for a	0.01	-
a_U	upper bound for a	6	-
k	number of subdivisions for a	100	-
c_L	lower bound for c	0.01	-
c_U	upper bound for c	6	-
m	number of subdivisions for c	100	-
$E(A\vert \mathbf{q}-q_0)$	posterior expectation of a	1.380	-
$E(B\vert \mathbf{q}-q_0)$	posterior expectation of b	4936	m^3/s
$E(C\vert \mathbf{q}-q_0)$	posterior expectation of c	2.310	-

simulation on the basis of 10,000 samples. The empirical and predictive probabilities of exceedance, according to Eq. (11), are displayed in Figure 8 including their 90 per cent uncertainty intervals. Note that the empirical probabilities of exceedance are calculated with the aid of Chegodayev's formula[10]: $\Pr\{Q > q_{z-u+1:z}\} = [u - 0.3]/[z + 0.4]$, where $q_{z-u+1:z}$ is the uth largest observation among the sample of z observations ordered by descending magnitude. For the three-parameter generalised gamma distribution, the Bayes estimate of the design discharge (i.e. the discharge with an exceedance probability of 1/1250) is 15,150 m^3/s with a 90 per cent uncertainty interval of (12,950; 16,950).

In order to take account of the uncertainty in the location parameter, the four-parameter generalised gamma distribution has been studied as well. The posterior probability functions of the shape parameters a and c, and the location parameter d are displayed in Figures 9 to 11, as well as the empirical and predictive exceedance probability and its 5th and 95th percentiles in Figure 12. The Bayes estimates of the four parameters are summarised in Table 2. For the four-parameter generalised gamma distribution, the Bayes estimate of the design discharge is 15,200 m^3/s with a 90 per cent uncertainty interval of (13,000; 16,950). Surprisingly, the uncertainty in the location parameter does not seem to have a big influence. It should be noted however, that it is not yet known whether the generalised gamma

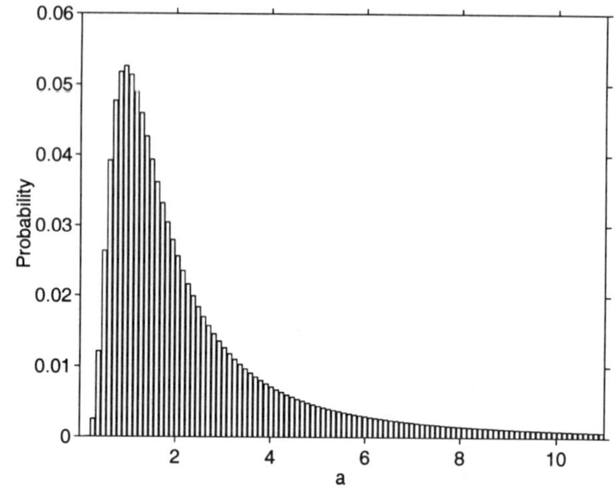

Fig. 9. Posterior probability function of a: $p(a_h \mid \mathbf{q})$, $h = 1, \ldots, k$.

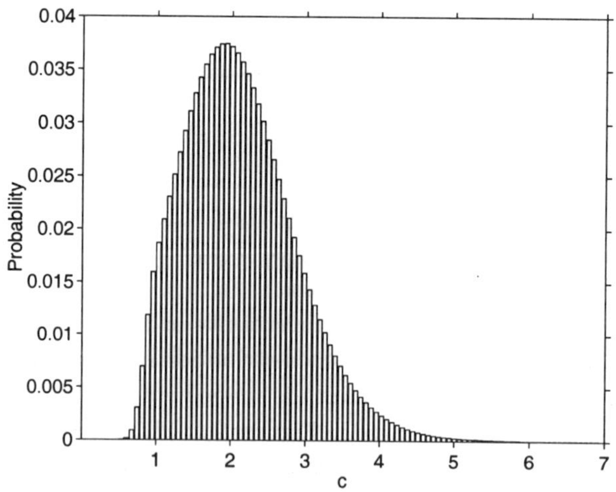

Fig. 10. Posterior probability function of c: $p(c_i \mid \mathbf{q})$, $i = 1, \ldots, m$.

distribution performs the best in fitting the observed annual maximum discharges. This is currently being investigated using Bayes factors[18].

Using the maximum-likelihood method, the estimate of the design dis-

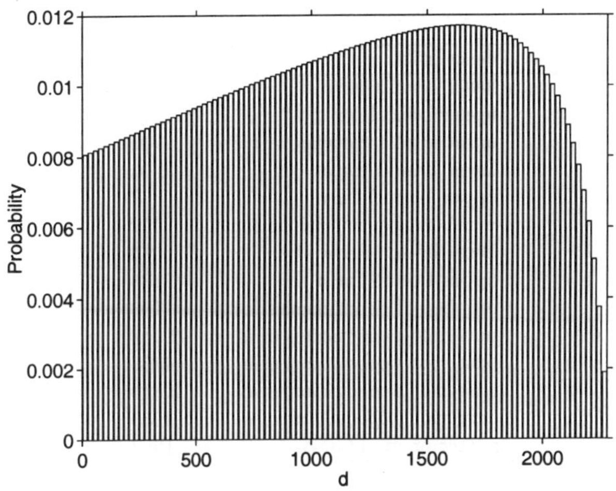

Fig. 11. Posterior probability function of d: $p(d_j \,|\, \mathbf{q})$, $j = 1, \ldots, l$.

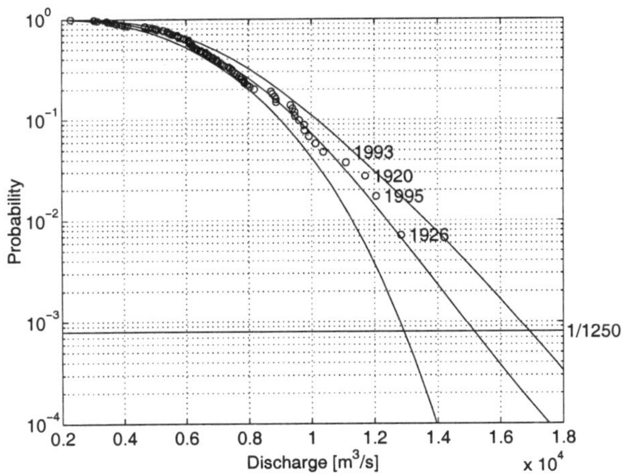

Fig. 12. Empirical and predictive probability of exceedance of the annual maximum discharge of the Rhine River at Lobith, including their 90 per cent uncertainty interval, for the four-parameter generalised gamma distribution.

charge decreases to 14,285 m^3/s for the three-parameter generalised gamma distribution and 14,289 m^3/s for the four-parameter generalised gamma dis-

Table 2. Bayes estimates of a, b, c, and d for the four-parameter generalised gamma distribution.

parameter	description	value	dimension
a_L	lower bound for a	0.01	-
a_U	upper bound for a	11	-
k	number of subdivisions for a	100	-
c_L	lower bound for c	0.01	-
c_U	upper bound for c	7	-
m	number of subdivisions for c	100	-
d_L	lower bound for d	0	m^3/s
d_U	upper bound for d	2280	m^3/s
l	number of subdivisions for d	100	-
$E(A\mid \mathbf{q})$	posterior expectation of a	2.348	-
$E(B\mid \mathbf{q})$	posterior expectation of b	4363	m^3/s
$E(C\mid \mathbf{q})$	posterior expectation of c	2.113	-
$E(D\mid \mathbf{q})$	posterior expectation of d	1155	m^3/s
d^*	posterior mode of d	1653	m^3/s

tribution. As expected, taking account of parameter uncertainty results in larger design discharges.

Finally, I would like to mention that more or less the same results have been obtained with the Markov chain Monte Carlo method[8,32]. The Metropolis algorithm—developed by Metropolis et al.[25] and generalised by Hastings[15]—was used to determine the predictive exceedance probabilities and their 5th and 95th percentiles. As a proposal or jumping density, the symmetric three- or four-dimensional normal density was chosen with mean equal to the parameter values at the current state and covariance matrix equal to the inverse Fisher information matrix for a sample of n observations evaluated at the maximum-likelihood estimator. Probably due to the possible non-regularity of the four-parameter generalised gamma distribution, numerical integration in combination with Monte Carlo performed better than Markov chain Monte Carlo.

8. Conclusions

In this paper, the discharge of the Rhine at Lobith with an average return period of 1250 years has been determined on the basis of a Bayesian analysis. Both the three-parameter generalised gamma distribution (without location parameter) and the four-parameter generalised gamma distribution (with location parameter) were used to obtain predictive exceedance probabilities of annual maximum discharges. In order to take account of the statistical uncertainties, the parameters of the generalised gamma distribution are assumed to be unknown and to have a prior joint probability distribution.

As prior density, the non-informative Jeffreys prior has been used. On the basis of the observed annual maximum discharges at Lobith, this prior density has been updated to the posterior density by using Bayes' theorem. An advantage is that the generalised gamma distribution fits well with the stage-discharge rating curve being an approximate power law between water level and discharge. Furthermore, since the generalised gamma distribution has three or four parameters, it is flexible in fitting data. Many well-known probability distributions which are commonly used to estimate quantiles of hydrological random quantities, are special cases of the generalised gamma distribution.

Appendix

In this Appendix, the four extra elements of the symmetric Fisher information matrix are derived when the three-parameter generalised gamma distribution (3) is extended to the four-parameter generalised gamma distribution (13) with scale parameter b, shape parameters a and c, and location parameter d; that is, when $\boldsymbol{\theta} = (a, b, c, d)'$. By applying the transformation $T = [(Q - d)/b]^c$ and using

$$\int_{t=0}^{\infty} t^a e^{-t} \log t \, dt = \Gamma'(a+1) = \Gamma(a) + a\Gamma'(a), \qquad a > 0,$$

the four elements of Fisher's information matrix for location parameter d are

$$E\left(-\frac{\partial^2 \log \ell(Q|\boldsymbol{\theta})}{\partial a \partial d}\right) = \frac{c}{b} E\left(T^{-\frac{1}{c}}\right) = \frac{c\Gamma(a - c^{-1})}{b\Gamma(a)},$$

$$E\left(-\frac{\partial^2 \log \ell(Q|\boldsymbol{\theta})}{\partial b \partial d}\right) = \frac{c^2}{b^2} E\left(T^{1-\frac{1}{c}}\right) = \frac{c(ca-1)\Gamma(a - c^{-1})}{b^2\Gamma(a)},$$

$$E\left(-\frac{\partial^2 \log \ell(Q|\boldsymbol{\theta})}{\partial c \partial d}\right) = \frac{1}{b} E\left(aT^{-\frac{1}{c}} - T^{1-\frac{1}{c}} - T^{1-\frac{1}{c}} \log T\right) =$$

$$= \frac{(1-c)\Gamma(a - c^{-1}) - (ca-1)\Gamma'(a - c^{-1})}{bc\Gamma(a)},$$

$$E\left(-\frac{\partial^2 \log \ell(Q|\boldsymbol{\theta})}{\partial d^2}\right) = E\left(\frac{ca-1}{b^2} T^{-\frac{2}{c}} + \frac{c(c-1)}{b^2} T^{1-\frac{2}{c}}\right) =$$

$$= \frac{(c^2 a - 2c + 1)\Gamma(a - 2c^{-1})}{b^2 \Gamma(a)}. \qquad \text{(A.1)}$$

The other elements of Fisher's information matrix can be found in Eq. (6). Because $\Gamma(a - 2c^{-1})$ in Eq. (A.1) only exists for $ac > 2$, the determinant of Fisher's information matrix for a single observation is only finite for $ac > 2$.

References

1. J.C. Ahuja and S.W. Nash. The generalized Gompertz-Verhulst family of distributions. *Sankya: The Indian Journal of Statistics, Series A*, **29**:141–156 (1967).
2. F. Ashkar and T.B.M.J. Ouarda. Approximate confidence intervals for quantiles of gamma and generalized gamma distributions. *Journal of Hydrologic Engineering*, **3**(1):43–51 (1998).
3. R.E. Barlow. *Engineering Reliability.* (American Statistical Association (ASA) & Society for Industrial and Applied Mathematics (SIAM), Philadelphia, 1998).
4. T. Bayes. An essay towards solving a problem in the doctrine of chances. *Philosophical Transactions of the Royal Society of London*, **53**:370–418 (1763).
5. J.M. Bernardo. Algorithm AS 103: Psi (digamma) function. *Applied Statistics*, **25**:315–317 (1976).
6. J.M. Bernardo and A.F.M. Smith. *Bayesian Theory.* (John Wiley & Sons, Chichester, 1994).
7. G.E.P. Box and G.C. Tiao. *Bayesian Inference in Statistical Analysis.* (John Wiley & Sons, New York, 1973), Section 1.3.
8. B.P. Carlin and T.A. Louis. *Bayes and Empirical Bayes Methods for Data Analysis.* (Chapman & Hall, London, 2000).
9. V.T. Chow. *Open-Channel Hydraulics.* (McGraw-Hill, Singapore, 1959), Chapters 5–6.
10. V.T. Chow, D.R. Maidment, and L.W. Mays. *Applied Hydrology.* (McGraw-Hill, Singapore, 1988), Chapters 2,9,12.
11. D.R. Cox and D.V. Hinkley. *Theoretical Statistics.* (Chapman & Hall, London, 1974), pages 106–113.
12. A.P. Dawid. The trouble with Bayes factors. Technical Report No. 202, Department of Statistical Science, University College London (1999).
13. H.W. Hager and L.J. Bain. Inferential procedures for the generalized gamma distribution. *Journal of the American Statistical Association*, **65**(332):1601–1609 (1970).
14. H.L. Harter. Maximum-likelihood estimation of the parameters of a four-parameter generalized gamma population from complete and censored samples. *Technometrics*, **9**(1):159–165 (1967).

15. W.K. Hastings. Monte Carlo sampling methods using Markov chains and their applications. *Biometrika*, **57**(1):97–109 (1970).
16. H.J. Jeffreys. *Theory of Probability; Third Edition.* (Clarendon Press, Oxford, 1961), Chapters 3–4.
17. N.L. Johnson, S. Kotz, and N. Balakrishnan. *Continuous Univariate Distributions, Volume 1; Second Edition.* (John Wiley & Sons, New York, 1994), Chapter 17.
18. R.E. Kass and A.E. Raftery. Bayes factors. *Journal of the American Statistical Association*, **90**(430):773–795 (1995).
19. R.E. Kass and L. Wasserman. The selection of prior distributions by formal rules. *Journal of the American Statistical Association*, **91**(435):1343–1370 (1996).
20. J.H. Lienhard. A statistical mechanical prediction of the dimensionless unit hydrograph. *Journal of Geophysical Research*, **69**(24):5231 (1964).
21. J.H. Lienhard and P.L. Meyer. A physical basis for the generalized gamma distribution. *Quarterly of Applied Mathematics*, **25**(3):330–334 (1967).
22. T.A. Mazzuchi and R. Soyer. Adaptive Bayesian replacement strategies. In J.M. Bernardo, J.O. Berger, A.P. Dawid, and A.F.M. Smith, editors, *Bayesian Statistics 5*, pages 667–674. (Oxford University Press, Oxford, 1996).
23. T.A. Mazzuchi and R. Soyer. A Bayesian perspective on some replacement strategies. *Reliability Engineering and System Safety*, **51**:295–303 (1996).
24. M. Mendel. The case for engineering probability. In R. Cooke, M. Mendel, and H. Vrijling, editors, *Engineering Probabilistic Design and Maintenance for Flood Protection*, pages 1–22. (Kluwer Academic Publishers, Dordrecht, 1997).
25. N. Metropolis, A.W. Rosenbluth, M.N. Rosenbluth, A.H. Teller, and E. Teller. Equation of state calculations by fast computing machines. *Journal of Chemical Physics*, **21**(6):1087–1092 (1953).
26. Ministry of Transport, Public Works, and Water Management. *Year-book Monitoring Rivers and Canals 1993 [Jaarboek Monitoring Rijkswateren 1993 (in Dutch)]*. The Hague, The Netherlands (1994).
27. V.B. Parr and J.T. Webster. A method for discriminating between failure density functions used in reliability predictions. *Technometrics*, **7**(1):1–10 (1965).
28. B.E. Schneider. Algorithm AS 121: Trigamma function. *Applied Statistics*, **27**:97–99 (1978).
29. E.M. Shaw. *Hydrology in Practice; Second Edition.* (Chapman & Hall, London, 1988), Chapter 6.
30. R.M. Soland. Bayesian analysis of the Weibull process with unknown scale and shape parameters. *IEEE Transactions on Reliability*, **18**(4):181–184 (1969).
31. E.W. Stacy. A generalization of the gamma distribution. *Annals of Mathematical Statistics*, **33**:1187–1192 (1962).

32. L. Tierney. Markov chains for exploring posterior distributions (with Discussion). *Annals of Statistics*, **22**(4):1701–1762 (1994).
33. W.E. Walker, A. Abrahamse, J. Bolten, J.P. Kahan, O. van de Riet, M. Kok, and M. den Braber. A policy analysis of Dutch river dike improvements: trading off safety, cost, and environmental impacts. *Operations Research*, **42**(5):823–836 (1994).
34. Waterloopkundig Laboratorium (WL) and European American Center (EAC) for Policy Analysis/RAND. Investigating basic principles of river dike improvement; Supporting Volume 2: Design loads [Toetsing uitgangspunten rivierdijkversterkingen; Deelrapport 2: Maatgevende belastingen (in Dutch)]. Delft, The Netherlands (1993).

CHAPTER 22

BAYESIAN OPERATIONAL APPROACH FOR REGRESSION MODELS IN FINITE POPULATIONS

Heleno Bolfarine

Departamento de Estatistica, Universidade de São Paulo
Caixa Postal 66281, CEP 05315-970, São Paulo, Brasil
E-mail: hbolfar@ime.usp.br

Pilar Iglesias[a]

Departamento de Estadistica, Pontificia Universidad Católica de Chile
Casilla 306, Santiago 22, Chile
E-mail: pliz@mat.puc.cl

Loretta Gasco

Seccion de Matematicas, Universidad Católica de Perú
Avenida Universitaria, Cuadra 18, Lima 32, Peru
E-mail: lgasco@pucp.edu.pe

In this paper we discuss the operational Bayesian approach for inference in finite populations. It is assumed that the distribution of the observable quantities is invariant under an orthogonal group of transformations. The quantities of interest are introduced as operational parameters, which depended only on observable quantities. Interest centers on the population total and on the finite population regression coefficient although predictors for the finite population variance are also considered. An operational likelihood function is defined which is a function of the operational parameters. Bayes estimators for the operational parameters are obtained by using the operational likelihood functions under representable prior distributions yielding conjugate priors. As shown, the Pearson type II distribution plays an important role on deriving the main results.

[a]This research was partally supported by FONDECYT grant 8000004-Chile and grants from FAPESP and CNPq-Brasil.

1. Introduction

Let $\mathcal{P} = \{1, \ldots, N\}$ denote a finite population where the number N of units is known. We consider that associated with unit k of $\mathcal{P}$ there are $p+1$ quantities, which we denote by $Y_k, X_{k1}, \ldots, X_{kp}$, $k = 1, \ldots, N$. Further, let $\mathbf{Y} = (Y_1, \ldots, Y_N)'$ and $\mathbf{X}_k = (X_{k1}, \ldots, X_{kp})'$, so that in matrix notation, we write $\mathbf{X} = [\mathbf{1}_N \; \mathbf{X}_1 \ldots \mathbf{X}_N]'$, where $\mathbf{1}_N$ is a N-dimensional vector of ones. Prediction of quantities $\theta(\mathbf{Y})$ like the population total $T = \sum_{i=1}^{N} Y_i$ or the finite population regression coefficient, $\mathbf{B}_N = (\mathbf{X}'\mathbf{X})^{-1}\mathbf{X}'\mathbf{Y}$ has been the subject of a great attention in the recent statistical literature. Comprehensive reviews can be found in Bolfarine and Zacks (1992). Most of the literature associated with the subject consider that the quantities Y_k and $\mathbf{X}_k$ are related through the linear relation

$$Y_k = \beta_0 + \mathbf{X}_k^t \boldsymbol{\beta} + \epsilon_k, \tag{1.1}$$

$k = 1, \ldots, N$, where $\boldsymbol{\beta} = (\beta_1, \ldots, \beta_p)'$, a p-dimensional vector of fixed and unknown parameters and $\boldsymbol{\epsilon} = (\epsilon_1, \ldots, \epsilon_N)'$ is a vector of random errors, typically satisfying $E[\boldsymbol{\epsilon}] = \mathbf{0}$ and $Var[\boldsymbol{\epsilon}] = \sigma^2 \mathbf{I}_N$, with σ^2 unknown and $\boldsymbol{I}_N$ the identity matrix of dimension N. To obtain information on quantities $\theta(\mathbf{Y})$ like T or $\mathbf{B}_N$, a sample $\mathbf{s}$ is selected from $\mathcal{P}$ according to some specified sampling plan. Notice that $\boldsymbol{\beta}$ denotes the superpopulation model regression coefficient (non operational) and $\mathbf{B}$ denotes the finite population regression coefficient.

The unobserved part of $\mathcal{P}$ is denoted by $\mathbf{r} = \mathcal{P} - \mathbf{s}$. Given $\mathbf{s}$, we denote by $\mathbf{Y}_s = (Y_1, \ldots, Y_n)'$ and $\mathbf{Y}_r = (Y_{n+1}, \ldots, Y_N)'$, the observed and unobserved parts of $\mathbf{Y}$, respectively, with the corresponding partition $\mathbf{X}_s$ and $\mathbf{X}_r$ of the matrix $\mathbf{X}$.

The superpopulation (or model based) approach to the prediction problem consider the model (or superpopulation) parameters $(\beta_0, \boldsymbol{\beta}, \sigma^2)$ as the main connection between $\mathbf{s}$ and $\mathbf{r}$, no matter which sampling plan is used and inference should only be based on the superpopulation model (1.1). Thus, under the perspective of the superpopulation approach, according to the conditionality principle (Basu, 1971), the sampling plan is not relevant for inference. The design-model based approach utilizes model (1.1) only in the definition of estimators and to propose convenient sampling plans. However, the merits of an estimator is totally judged by its performance with respect to the sampling plan.

In this paper, we focus on the prediction of the population total from a pure predictivistic approach. The main assumption is that the distribution of $\mathbf{Y}$ is $O_N(M)$-invariant which, as a consequence, implies the Pearson type II representation of the distribution of projections of $\mathbf{Y}$ given in Proposition 2.1 and Example 2.2. The notation $O_N(M)$ is used to denote the space of all orthogonal transformations which leaves the elements of the subspace M invariant (see Section 2.1). One important consequence of those results when M is generated by columns of $\mathbf{X}$ is that $\mathbf{Y}_s$ and $\mathbf{X}_s$ are related through the linear model

$$\mathbf{Y}_s = \mathbf{X}_s\mathbf{B}_N + \mathbf{e}_s, \tag{1.2}$$

with $\mathbf{B}_N$ as above and the distribution of $\mathbf{e}_s = (e_1, \ldots, e_n)'$ given $\mathbf{B}_N$ and $\mathbf{S} = \mathbf{Y}'\mathbf{Q}_M\mathbf{Y}$ is the multivariate Pearson type II distribution which we denote by

$$MPII(\mathbf{0}, \mathbf{S}(\mathbf{I}_n - \mathbf{X}_s(\mathbf{X}_s'\mathbf{X}_s)^{-1}\mathbf{X}_s', \frac{N-m-n-2}{2}),$$

where $m = p+1$ is the dimension of M and $\mathbf{Q}_M$ is the projection matrix on $M^{\perp}$. We call attention to the fact that the invariance condition determines all the elements of the Bayesian model, the parameters, however, being functions only of observable quantities. The marginal distribution of $\mathbf{Y}$ then becomes completely specified as soon as prior distributions are specified for the (finite population) model parameters. Operational Bayesian approach is the nomenclature typically associated with such approach (Mendel, 1994) and the finite population parameters (quantities of interest) are termed operational parameters. Another important aspect of the development is that it provides justification for the fact that the normal distribution is typically associated with the distribution of the error vector $\boldsymbol{\epsilon}$.

The paper is organized as follows. Section 2 discusses the construction of $O_N(M)$-invariant distributions in finite populations and presents also some representation theorems for the distribution of the observed part of the population, represented as projected measures of $O_N(M)$-invariant distributions. Following Mendel (1994), Section 3 presents a systematic approach for proposing operational parameters in finite populations. The operational parameter and the likelihood function are presented in Section 4. Finally Bayesian inference solutions are presented for the operational parameters in Sections 5. A representable class of prior distributions is considered yielding posterior distributions with the same form as the prior distributions.

2. $O_N(M)$-invariant Distributions in Finite Populations

In this section we consider the distribution of the population vector $\mathbf{Y}$ to be $O_N(M)$-invariant. Consequences of this assumption including characterizations in terms of maximal invariants and representation of such distributions as mixtures of uniform distributions defined on the orbits induced by the maximal invariants under the groups considered are discussed. Moreover, following some results in Diaconis et al. (1992), stochastic representations of such distributions in terms of the Pearson type II distributions are considered. These results will be used in Section 3 to obtain the operational likelihood for the prediction problem in finite populations.

2.1. *Construction of $O_N(M)$-invariant Distributions*

In this section invariant conditions are established which allow considering an operational regression model for finite populations. Let M denote an m-dimensional subspace of $\Re^N$. Moreover, let's denote by O_N the compact subgroup of all real orthogonal $N \times N$ matrices and by $O_N(M) = \{\mathbf{\Gamma} \in O_N; \mathbf{\Gamma x} = \mathbf{x}, \mathbf{x} \in M\}$. Let $\mathcal{L}_N$ be the set of all N-dimensional real vectors, and $\mathbf{Y}$ a N-dimensional random vector taking values in $\mathcal{L}_N$. The random vector $\mathbf{Y}$ is said to be $O_N(M)$-invariant if $\mathbf{Y}$ and $\mathbf{\Gamma Y}$ are identically distributed for all $\mathbf{\Gamma} \in O_N(M)$.

Definition 2.1. *Let $\mathbf{U}$ be a random matrix with values in $O_N(M)$. Then $\mathbf{U}$ is uniformly distributed on $O_N(M)$ provided $\mathbf{U}$ is distributed according to the probability measure ν, which is the unique invariant probability measure on $O_N(M)$. Furthermore,*

$$E[\mathbf{U}] = \mathbf{P}_M \quad \text{and} \quad Var[\mathbf{U}] = \frac{\mathbf{Q}_M \otimes \mathbf{Q}_M}{N-m}, \tag{2.1}$$

where $\mathbf{P}_M$ is the orthogonal projection matrix on M, $\mathbf{Q}_M = \mathbf{I}_N - \mathbf{P}_M$ is the projection matrix on $M^\perp$ and $\otimes$ denotes the Kroenecker product.

The action of $O_N(M)$ on $\mathcal{L}_N$ yields a partition of $\mathcal{L}_N$ into orbits as follows. If $\mathbf{y} \in \mathcal{L}_N$ then the orbit of $\mathbf{y}$, O_y, is the set of $\mathbf{z} \in \mathcal{L}_N; \mathbf{z} = \mathbf{\Gamma y}$, for some $\mathbf{\Gamma} \in O_N(M)$, that is,

$$O_y = \{\mathbf{z} \in \mathcal{L}_N; t(\mathbf{z}) = t(\mathbf{y})\},$$

where $t(.)$ is the maximal invariant with respect to action of the group $O_N(M)$.

Since O_y is compact in $\mathcal{L}_N$ (a locally compact Hausdorff space with an enumerable basis for the Euclidean topology in $\Re^N$) and $O_N(M)$ acts transitively on O_y it follows that there exists a unique $O_N(M)$-invariant probability measure, ν, defined on the orbits of the group (Nachbin, 1965). However, if $\mathbf{U}$ is a random matrix uniformly distributed over $O_N(M)$ and $\mathbf{y} \in \mathcal{L}_N$ then $\mathbf{Uy}$ is a random matrix with values in O_y and with a $O_N(M)$-invariant distribution generated by ν, which we denote by ν_y. Thus, uniqueness of this measure implies that it is the corresponding invariant probability measure on the orbit O_y. In this sense, it is said that the random matrix $\mathbf{Uy}$ has uniform distribution on O_y, which is the surface of an $(N-m)$-dimensional sphere. Note from (2.1) that

$$E[\mathbf{Uy}] = \mathbf{P}_M\mathbf{y} \quad \text{and} \quad Var[\mathbf{Uy}] = \frac{\mathbf{Q}_M \otimes \mathbf{y}'\mathbf{Q}_M\mathbf{y}}{N-m}.$$

Consequently, if P denotes the probability measure of a N-dimensional random vector $\mathbf{Y}$ with

$$P = \int \nu_y P(dy), \tag{2.2}$$

then $\mathbf{Y}$ has a $O_N(M)$-invariant distribution. The converse also holds. A simple situation is illustrated in this context in the following examples.

Example 2.1. $O_N(M)$*-invariance.* As before, let M denotes a m-dimensional subspace of $\Re^N$ and $O_N(M)$, the compact set of $N \times N$ orthogonal matrices leaving invariant the elements of M. Moreover, let $\mathbf{Y}$ a N-dimensional random vector with a O_N-invariant distribution P. In this case, $t(\mathbf{Y}) = (\mathbf{P}_M\mathbf{Y}, ||\mathbf{Y} - \mathbf{P}_M\mathbf{Y}||^2)$ with $\mathbf{Y} = (Y_1, \ldots, Y_N)'$, is a maximal invariant under the action of $O_N(M)$. As a consequence, the conditional distribution of $\mathbf{Y}$ given $\mathbf{P}_M\mathbf{Y} = \mathbf{c}$ and $||\mathbf{Y} - \mathbf{P}_M\mathbf{Y}||^2 = r^2$ is uniform over the set

$$S_N(\mathbf{c}, r) = \{\mathbf{Y} \in \Re^N; \mathbf{P}_M\mathbf{Y} = \mathbf{c}, ||\mathbf{Y} - \mathbf{P}_M\mathbf{Y}||^2 = r^2\},$$

$\mathbf{c} \in M$ and $r > 0$. Thus, the distribution P can be represented as a center-radial mixture of these uniform distributions. The measure in the mixture is the P-law of $t(\mathbf{Y}) = (\mathbf{P}_M\mathbf{Y}, ||\mathbf{Y} - \mathbf{P}_M\mathbf{Y}||^2)$. For example, if $\mathbf{Y} \sim N_N(\mathbf{m}, \mathbf{I}_N)$, then $\mathbf{P}_M\mathbf{Y}$ and $||\mathbf{Y} - \mathbf{P}_M\mathbf{Y}||^2$ are independent with $\mathbf{P}_M\mathbf{Y} \sim N_N(\mathbf{P}_M\mathbf{m}, \mathbf{P}_M)$ and $||\mathbf{Y} - \mathbf{P}_M\mathbf{Y}||^2 \sim \chi^2_{N-m}$.

Remark. 2.1. It follows from the example above that if M is the column space generated by $\mathbf{1}_N$, then the measure P can be represented as a mixture of uniform distributions over the set $S_N(a,r) = \{\mathbf{Y} \in \Re^N; \bar{Y} = a, \sum_{i=1}^N (Y_i - \bar{Y})^2 = r^2\}$, $a, r \in \Re$, $r > 0$, where $Y = \sum_{i=1}^n Y_i/n$. Generally, if M is the column space generated by a $n \times p$ matrix $\mathbf{X}$ with rank p ($p < N$), then P can be represented as a mixture of uniform distributions over the set $S_N(\mathbf{b}, r) = \{\mathbf{Y} \in \Re^N; \mathbf{B}_N = \mathbf{b}, ||\mathbf{Y} - \mathbf{X}\mathbf{B}_N||^2 = r^2\}$, where $\mathbf{B}_N$ is the finite population regression coefficient and $\mathbf{b} \in \Re^p$.

Let $\mathbf{V} = \mathbf{A}\mathbf{U}\mathbf{B}'$, where $\mathbf{U}$ is a $N \times N$ random matrix uniformly distributed over $O_N(M)$, and $\mathbf{A}$ and $\mathbf{B}$ are arbitrary matrices of dimensions $k \times N$ and $1 \times N$, respectively, with $k < N - m$, where $m = dim(M)$. Diaconis et al. (1922) show that the density of $\mathbf{V}$ is given by

$$f(\mathbf{v}) = \frac{\Gamma(\frac{n-m}{2})|r^2\mathbf{C}|^{-1/2}}{\Gamma(\frac{N-m-k}{2})\pi^{k/2}}\{1 - \frac{Q(\mathbf{v},\mathbf{B})}{r^2}\}^{\frac{\nu}{2}}, \tag{2.3}$$

where $\nu = N - m - k - 2$ and $Q(\mathbf{v},\mathbf{B}) = (\mathbf{v} - \mathbf{A}\mathbf{E})'\mathbf{C}^{-1}(\mathbf{v} - \mathbf{A}\mathbf{E})$, $\mathbf{C} = \mathbf{A}\mathbf{Q}_M\mathbf{A}'$, $\mathbf{E} = \mathbf{P}_M\mathbf{B}'$ and $r^2 = \mathbf{B}\mathbf{Q}_M\mathbf{B}'$, provided $Q(\mathbf{v},\mathbf{B}) \leq r^2$. The distribution of $\mathbf{V}$ in this case is known in the literature as the k-variate Pearson type II distribution with location vector $\boldsymbol{\mu} = \mathbf{A}\mathbf{P}_M\mathbf{B}'$ and scale matrix $\boldsymbol{\Sigma} = r^2\mathbf{C}$. It follows from (2.1) that $E[\mathbf{V}] = \mathbf{A}\mathbf{E}$ and $Var[\mathbf{V}] = (N-m)^{-1}r^2\mathbf{C}$. A formal definition of the k-variate Pearson type II distribution is presented next.

Definition 2.2. *A k-dimensional random vector $\mathbf{Y}$ is said to have a symmetric k-variate Pearson type II distribution with parameters $\boldsymbol{\mu}$, which is N-dimensional, and $\boldsymbol{\Sigma}$, a positive definite $N \times N$ matrix, if its density is given by*

$$f(\mathbf{y}) = \frac{\Gamma(\frac{k}{2} + \nu + 1)}{\Gamma(\nu + 1)\pi^{k/2}}|\boldsymbol{\Sigma}|^{-1/2}(1 - (\mathbf{y} - \boldsymbol{\mu})'\boldsymbol{\Sigma}^{-1}(\mathbf{y} - \boldsymbol{\mu}))^{\nu},$$

which we denote by $\mathbf{Y} \sim MPII_k(\boldsymbol{\mu}, \boldsymbol{\Sigma}, \nu)$ provided $0 \leq (\mathbf{y} - \boldsymbol{\mu})'\boldsymbol{\Sigma}^{-1}(\mathbf{y} - \boldsymbol{\mu}) \leq 1$, $\nu > -1$. Moreover, $E[\mathbf{Y}] = \boldsymbol{\mu}$ and $Var[\mathbf{Y}] = \frac{1}{k+2\nu+2}\boldsymbol{\Sigma}$.

Remark 2.2. As a direct consequence we have that if the k-dimensional random vector $\mathbf{Y}$ is such that $\mathbf{Y} \sim MPII_k(\boldsymbol{\mu}, \boldsymbol{\Sigma}, \nu)$ and $\mathbf{C}$ is a $m \times k$ matrix of known constants, then $\mathbf{W} = \mathbf{C}\mathbf{Y} \sim MPII_k(\mathbf{C}\boldsymbol{\mu}, \mathbf{C}\boldsymbol{\Sigma}\mathbf{C}', \nu)$.

In the following we provide representations for projections of probability measures which are O_N-invariant on $\Re^N$. Let $\mathbf{Y}$ be a N-dimensional random vector which is $O_N(M)$-invariant and $\mathbf{\Pi}$ is a $n \times N$ projection matrix, that is, $\mathbf{\Pi\Pi}' = \mathbf{I}_n$, with $n < N$. As before, denote by P the law of $\mathbf{Y}$ and by $P^{(n)}$ the law of $\mathbf{Y}_s = \mathbf{\Pi Y}$. The following result is a consequence of related results in Diaconis et al. (1992).

Proposition 2.1. *Let P and $P^{(n)}$ be as defined above. Then, for each $n < N - m$,*

$$P^{(n)} = \int_{M \times [0,\infty)} MPII_k(\mathbf{\Pi c}, v\mathbf{Q}_1, \nu) d\mu_n(\mathbf{c}, v), \tag{2.4}$$

where μ_n is the P-law of $\mathbf{t}(\mathbf{Y}) = (\mathbf{P}_M\mathbf{Y}, \mathbf{Y}'\mathbf{Q}_M\mathbf{Y})$, $\nu = (N - m - n - 2)/2$ and $\mathbf{Q}_1 = \mathbf{\Pi Q}_M\mathbf{\Pi}'$.

We present next an example to illustrate the results derived previously.

Example 2.2. $O_N(M)-$*invariance.* We consider now the case where M is the column space generated by the columns of the $N \times p$-dimensional matrix $\mathbf{X}$, where p is the rank of $\mathbf{X}$. We assume that the N-dimensional random vector $\mathbf{Y}$ is $O_N(M)$ invariant and consider $\mathbf{Y}_s = \mathbf{\Pi Y}$ and $\mathbf{X}_s = \mathbf{\Pi X}$, with rank p. Thus, it follows that $\mathbf{\Pi P}_M\mathbf{Y} = \mathbf{X}_s\mathbf{B}_N$, $\mathbf{\Pi Q}_M\mathbf{\Pi}' = \mathbf{X}_s(\mathbf{X}'\mathbf{X})^{-1}\mathbf{X}_s'$, where $\mathbf{B}_N = (\mathbf{X}'\mathbf{X})^{-1}\mathbf{X}'\mathbf{Y}$, as defined in the introduction. Hence, using Corollary 2.1, it follows for $1 \leq n < N - p$, that

$$P^{(n)} = \int_{\Re^p \times [0,\infty)} MPII_n(\mathbf{X}_s\mathbf{b}, v\mathbf{H}_s, \nu) d\mu_N(\mathbf{b}, v), \tag{2.5}$$

where μ_N is the P-law of the maximal invariant $t(\mathbf{Y}) = (\mathbf{B}_N, (\mathbf{Y} - \mathbf{XB}_N)'(\mathbf{Y} - \mathbf{XB}_N))$, $\mathbf{H}_s = \mathbf{X}_s(\mathbf{X}'\mathbf{X})^{-1}\mathbf{X}_s'$ and $\nu = (N - n - p - 2)/2$. The conditional model for $\mathbf{Y}_s$ which is implied by the invariance condition is traditionally represented in the form

$$\mathbf{Y}_s = \mathbf{X}_s\mathbf{b} + \mathbf{e}_s,$$

where $\mathbf{e}_s$ is a n-dimensional random vector with distribution $MPII_n(\mathbf{0}, \mathbf{H}_s, \nu)$. Moreover, if $\mathbf{B}_s = (\mathbf{X}_s'\mathbf{X}_s)^{-1}\mathbf{X}_s'\mathbf{Y}_s$, then conditional on $\mathbf{B}_N = \mathbf{b}$ and $(\mathbf{Y} - \mathbf{XB}_N)'(\mathbf{Y} - \mathbf{XB}_N) = v$, the distribution of $\mathbf{B}_s$ is $MPII_p(\mathbf{b}, v(\mathbf{X}'\mathbf{X})^{-1}, \frac{n-2p-2}{2})$ and $(\mathbf{B}_s - \mathbf{b})'\mathbf{X}'\mathbf{X}(\mathbf{B}_s - \mathbf{b})$ has the same distribution as the random variable vW, where $W \sim Beta(p/1, (n-p)/2)$.

Proposition 2.1 provides an exact representation for marginal distributions of $O_N(M)$-invariant distributions. Proximity (in terms of the variation distance) of such representation to a mixture of $O_N(M)$ invariant normal distributions has been studied in Diaconis et al. (1992) in the case $M = \{\mathbf{0}\}$. Such results are known in the literature as finite forms of de Finetti type theorems and are extended within a general context in Diaconis et al. (1992). The results considered previously and the multivariate extension of Smith (1981) are actually special cases of those general results.

3. The Operational Structure for Finite Populations

In the following, we consider a systematic approach for proposing operational parameters as presented in Iglesias (1993), Mendel (1994) and Mendel and Kempthorne (1996). In accordance with such results, if the distribution P of the vector $\mathbf{Y}$ is $O_N(M)$-invariant, then the maximal invariant $\theta(\mathbf{Y})$ typically is an operational parameter. In particular, as seen in Section 2, $t(\mathbf{Y}) = (\mathbf{P}_M\mathbf{Y}, \mathbf{Y}'\mathbf{Q}_M\mathbf{Y})$ is a maximal invariant under the action of $O_N(M)$ and then an operational parameter for that model. Moreover, provided that $P = \mathcal{L}(\mathbf{Y})$ is $O_N(M)$-invariant, we can write

$$\mathcal{L}(\mathbf{Y}) = \int_{M\times(0,\infty)} \nu_{\mathbf{c},r} Q(d\mathbf{c}, dr), \tag{3.1}$$

where $\nu_{c,r}$ is the uniform distribution on the orbit O_y with $\mathbf{y}$ such that $\mathbf{P}_M\mathbf{y} = \mathbf{c}$, $\mathbf{y}^t\mathbf{Q}_M\mathbf{y} = r^2$ and Q the P-law of $t(\mathbf{Y}) = (\mathbf{P}_M\mathbf{Y}, \mathbf{Y}'\mathbf{Q}_M\mathbf{Y})$, which is the operational parameter. Hence, the operational parameter provides an indexation of the uniform distributions $\nu_{t(\mathbf{y})}$, which is involved in the representation of $P = \mathcal{L}(\mathbf{Y})$. Moreover, the uniform distribution is common for representing any $O_N(M)$-invariant distribution P. What makes the representation unique (makes models distinguishable) is the mixing measure Q in (3.1), the measure associated with the operational parameter $t(\mathbf{Y}) = (\mathbf{P}_M\mathbf{Y}, \mathbf{Y}'\mathbf{Q}_M\mathbf{Y})$ and induced by the particular distribution $P = \mathcal{L}(\mathbf{Y})$ under consideration. Hence, with respect to the operational parameter $t(\mathbf{Y}) = (\mathbf{P}_M\mathbf{Y}, \mathbf{Y}'\mathbf{Q}_M\mathbf{Y})$, the operational likelihood function for the class of the $O_N(M)$-invariant distributions is given by the uniform distribution $\nu_{t(\mathbf{y})}$. Now, corresponding to the observed part $\mathbf{Y}_s$ of $\mathbf{Y}$, it follows from Proposition 2.1 that the law $P^{(n)} = \mathcal{L}(\mathbf{Y}_s)$ can be represented as

$$\mathcal{L}(\mathbf{Y}_s) = \int_{M\times[0,\infty)} MPII_n\left(\mathbf{\Pi}\hat{\mathbf{y}}, sA, \nu\right) Q(d\hat{\mathbf{y}}, ds),$$

where $MPII_n(\mathbf{\Pi}\hat{\mathbf{y}}, sA, \nu)$ is the n-variate Pearson type II distribution, with $s = \mathbf{y}'\mathbf{Q}_M\mathbf{y}$, $\hat{\mathbf{y}} = \mathbf{P}_M\mathbf{y}$, $\mathbf{\Pi}$ as in Section 2 and $\nu = (N - m - n - 2)/2$. Thus, the likelihood for $\mathbf{Y}_s$, is defined by this model.

As pointed out above, difference among $O_N(M)$-invariant models is captured by the measure Q of the maximal invariant $t(\mathbf{Y})$, which corresponds to the prior distribution in the Bayesian operational structure.

4. The Operational Parameter and the Likelihood Function

In this section, we describe in detail the operational parameter and the likelihood function for the finite population model. We begin by calling attention to the fact that the subspace M which describes the group $O_N(M)$ is generated by the columns of the matrix $\mathbf{X}$ of rank m.

- *4.1 The operational parameter*

After obtaining the projection of $\mathbf{Y}$ on the space M which is generated by the columns of the matrix $\mathbf{X}$, it follows that (see also Example 2.2) $t(\mathbf{Y}) = (\mathbf{B}_N, \sigma_N^2)$, where

$$\mathbf{B}_N = (\mathbf{X}'\mathbf{X})^{-1}\mathbf{X}'\mathbf{Y} \text{ and } \sigma_N^2 = \frac{SCE_N}{N - m},$$

with $SCE_N = (\mathbf{Y} - \mathbf{X}\mathbf{B}_N)'(\mathbf{Y} - \mathbf{X}\mathbf{B}_N)$, is a maximal invariant which is also, according to Section 3, an operational parameter for the operational model.

- *4.2. The operational likelihood function*

As defined before, $\mathbf{Y}_s$ is the observed part of the vector $\mathbf{Y}$ and being $n < N - m$ and $\mathbf{V}_s = \mathbf{\Pi}\mathbf{Q}_M\mathbf{\Pi}' = \mathbf{I}_n - \mathbf{X}_s(\mathbf{X}'\mathbf{X})^{-1}\mathbf{X}_s'$, it follows from Proposition 2.1 that the marginal distribution of $\mathbf{Y}_s$ is given by

$$P^{(s)} = \int_{\Re^p \times \Re_+} MPII_n\left(\mathbf{X}_s\mathbf{B}_N, \sigma_N^2\mathbf{\Sigma}, \nu\right) Q(d\mathbf{B}_N, d\sigma_N^2),$$

where the support of the n-variate Pearson type II distribution is

$$F = \{\mathbf{v} \in \Re^n \mid (\mathbf{v} - \mathbf{X}_s\mathbf{B}_N)'[\mathbf{\Sigma}]^{-1}(\mathbf{v} - \mathbf{X}_s\mathbf{B}_N) \leq \sigma_N^2\},$$

with Q the P-law of $(\mathbf{B}_N, \sigma_N^2)$ and $\mathbf{\Sigma} = (N - m)\mathbf{V}_s$, $\nu = (N - m - n - 2)/2$. Thus, as seen in Section 3, the corresponding density of $\mathbf{Y}_s$, which is the operational likelihood function, is given by

$$f(\mathbf{y}_s|\mathbf{B}_N,\sigma_N^2)=\frac{\Gamma(\frac{n}{2}+\nu+1)}{\Gamma(\nu+1)\pi^{\frac{n}{2}}}[\sigma_N^2\mathbf{\Sigma}]^{-1/2} \tag{4.1}$$

$$\times[1-(\mathbf{y}_s-\mathbf{X}_s\mathbf{B}_N)'[\sigma_N^2\mathbf{\Sigma}]^{-1}(\mathbf{y}_s-\mathbf{X}_s\mathbf{B}_N)]^{\nu}I_F(\mathbf{y}_s).$$

Using the observed sample $\mathbf{Y}_s$, corresponding to a sample of size n, we define

$$\mathbf{B}_s=(\mathbf{X}_s\mathbf{V}_s^{-1}\mathbf{X}_s)^{-1}\mathbf{X}_s'\mathbf{V}_s^{-1}\mathbf{y}_s$$

and

$$s_n^2=\frac{(\mathbf{y}_s-\mathbf{X}_s\mathbf{B}_s)'\mathbf{V}_s^{-1}(\mathbf{y}_s-\mathbf{X}_s\mathbf{B}_s)}{n-m},$$

so that we can write the operational likelihood function (4.1) as

$$f(\mathbf{y}_s|\mathbf{B}_N,\sigma_N^2)\propto\left(\frac{1}{\sigma_N^2}\right)^{\frac{n}{2}}\left[1-\frac{h(\mathbf{B}_N)}{\sigma_N^2}\right]^{\nu}I_{[h([\mathbf{B}_N),\infty)}(\sigma_N^2), \tag{4.2}$$

where

$$h(\mathbf{B}_N)=\frac{1}{N-m}[(n-m)s_n^2+(\mathbf{B}_N-\mathbf{B}_s)'\mathbf{C}(\mathbf{B}_N-\mathbf{B}_s)],$$

and $\mathbf{C}=\mathbf{X}_s'\mathbf{V}_s^{-1}\mathbf{X}_s$.

5. Inference for the Operational Parameters under Representable Priors

In finite population sampling, the quantities of interest can be linear functions of $\mathbf{Y}$ such as $\theta_l=\mathbf{c}'\mathbf{Y}$, with $\mathbf{c}$ a vector of dimension N, or quadratic, such as $\theta_Q=\mathbf{Y}'\mathbf{A}\mathbf{Y}$, where $\mathbf{A}$ is a matrix of dimension $N\times N$. Important linear functions are the populational total, $T=\mathbf{1}_N'\mathbf{Y}$, or the finite population regression coefficient $\mathbf{B}_N=(\mathbf{X}'\mathbf{X})^{-1}\mathbf{X}'\mathbf{Y}$. Straightforward algebraic manipulations show that we can write $T=\mathbf{c}'\mathbf{B}_N$ with $\mathbf{c}=(N,N\bar{X}_1,\ldots,N\bar{X}_p)'$, where $\bar{X}_i=\sum_{j=1}^{n}X_{ij}/n$, is the i-th column mean of the matrix $\mathbf{X}$. By considering that $\mathbf{Y}$ is $O_N(M)$-invariant and a prior density π for $(\mathbf{B}_N,\sigma_N^2)$ we have from (4.2), that the posterior density is given by

$$\pi(\mathbf{B}_N,\sigma_N^2|\mathbf{y}_s)\propto\left(\frac{1}{\sigma_N^2}\right)^{\frac{n}{2}}\left[1-\frac{h(\mathbf{B}_N)}{\sigma_N^2}\right]^{\frac{N-m-n-2}{2}} \tag{5.1}$$

$$\times I_{[h(\mathbf{B}_N),\infty)}(\sigma_N^2)\pi(\mathbf{B}_N,\sigma_N^2),$$

where $h(\mathbf{B}_N)$ is as in (4.2).

In this section we consider prior distributions which can be represented as

$$\pi(\mathbf{B}_N,\sigma_N^2) = \int_{\Re^p\times(0,\infty)} g_N(\mathbf{B}_N,\sigma_N^2|\boldsymbol{\beta},\sigma^2)\mu(d\boldsymbol{\beta},d\sigma^2),$$

where $g_N(.|\boldsymbol{\beta},\sigma^2)$ denotes the density function of $(\mathbf{B}_N,\sigma_N^2)$ under the assumption that $\mathbf{Y}\sim N_N(\mathbf{X}\boldsymbol{\beta},\sigma^2\mathbf{I}_N)$ and $\mu(.)$ is a probability measure on the Borel sets of $\Re^p\times(0,\infty)$. In the following it is shown that the posterior distribution obtained by using the above prior presents the same form as the prior distribution. In fact, notice that

$$\pi(\mathbf{B}_N,\sigma_N^2|\mathbf{y}_s) \propto f(\mathbf{y}_s|\mathbf{B}_N,\sigma_N^2)\int_{\Re^p\times(0,\infty)} g_N(\mathbf{B}_N,\sigma_N^2|\boldsymbol{\beta},\sigma^2)\mu(d\boldsymbol{\beta},d\sigma^2),$$

where $f(\mathbf{y}_s|\mathbf{B}_N,\sigma_N^2)$ is given in (4.2). Now, given that $N_N(\mathbf{X}\boldsymbol{\beta},\sigma^2\mathbf{I}_N)$ is $O_N(M)$-invariant, it follows from Proposition 2.1 and Example 2.2 for $\boldsymbol{\beta}\in\Re^p$, $\sigma>0$, that

$$f(\mathbf{y}_s|\mathbf{B}_N,\sigma_N^2) = f_N(\mathbf{y}_s|\mathbf{B}_N,\sigma_N^2),$$

where as before, $f_N(\mathbf{y}_s|\mathbf{B}_N,\sigma_N^2)$ denotes the conditional density of $\mathbf{y}_s$ given $(\mathbf{B}_N,\sigma_N^2)$ when $\mathbf{Y}\sim N_N(\mathbf{X}\boldsymbol{\beta},\sigma^2\mathbf{I}_N)$. Thus,

$$\pi(\mathbf{B}_N,\sigma_N^2|\mathbf{y}_s) = [f(\mathbf{y}_s)]^{-1}$$

$$\times\int_{\Re^p\times(0,\infty)} f_N(\mathbf{y}_s|\mathbf{B}_N,\sigma_N^2)g_N(\mathbf{B}_N,\sigma_N^2|\boldsymbol{\beta},\sigma^2)\mu(d\boldsymbol{\beta},d\sigma^2).$$

Furthermore,

$$g_N(\mathbf{B}_N,\sigma_N^2|\mathbf{y}_s,\boldsymbol{\beta},\sigma^2) = \frac{f_N(\mathbf{y}_s|\mathbf{B}_N,\sigma_N^2)g_N(\mathbf{B}_N,\sigma_N^2|\boldsymbol{\beta},\sigma^2)}{f_N(\mathbf{y}_s|\boldsymbol{\beta},\sigma^2)},$$

with

$$f_N(\mathbf{y}_s|\boldsymbol{\beta},\sigma^2) = \int f_N(\mathbf{y}_s|\mathbf{B}_N,\sigma_N^2)g_N(\mathbf{B}_N,\sigma_N^2|\boldsymbol{\beta},\sigma^2)d\mathbf{B}_N d\sigma_N^2,$$

leading to,

$$\pi(\mathbf{B}_N,\sigma_N^2|\mathbf{y}_s) = [f(\mathbf{y}_s)]^{-1}$$

$$\times \int_{\Re^p \times (0,\infty)} g_N(\mathbf{B}_N, \sigma_N^2 | \mathbf{y}_s, \boldsymbol{\beta}, \sigma^2) f_N(\mathbf{y}_s|\beta, \sigma^2)\mu(d\boldsymbol{\beta}, d\sigma^2).$$

On other hand,

$$f(\mathbf{y}_s) = \int f(\mathbf{y}_s|\mathbf{B}_N, \sigma_N^2)\pi(\mathbf{B}_N, \sigma_N^2)d\mathbf{B}_N d\sigma_N^2$$

$$= \int\int f_N(\mathbf{y}_s|\mathbf{B}_N, \sigma_N^2) g(\mathbf{B}_N, \sigma_N^2|\boldsymbol{\beta}, \sigma^2)\mu(d\boldsymbol{\beta}, d\sigma^2)d\mathbf{B}_N d\sigma_N^2$$

$$= \int f_N(\mathbf{y}_s)\mu(d\boldsymbol{\beta}, d\sigma^2),$$

implying that

$$\pi(\mathbf{B}_N, \sigma_N^2|\mathbf{y}_s) = \int_{\Re^p \times (0,\infty)} g_N(\mathbf{B}_N, \sigma_N^2|\mathbf{y}_s, \beta, \sigma^2)\mu(d\boldsymbol{\beta}, d\sigma^2|\mathbf{y}_s, \boldsymbol{\beta}, \sigma^2), \tag{5.2}$$

where

$$\mu(d\boldsymbol{\beta}, d\sigma^2|\mathbf{y}_s) = \frac{f_N(\mathbf{y}_s|\boldsymbol{\beta}, \sigma^2)\mu(d\boldsymbol{\beta}, d\sigma^2)}{\int f_N(\mathbf{y}_s|\boldsymbol{\beta}, \sigma^2)\mu(d\boldsymbol{\beta}', d\sigma'^2)}, \tag{5.3}$$

is the probability measure associated with the posterior distribution of $(\boldsymbol{\beta}, \sigma^2)$ given $\mathbf{y}_s$ when $\mathbf{y}_s|\boldsymbol{\beta}, \sigma^2 \sim N_n(\mathbf{X}_s\boldsymbol{\beta}, \sigma^2\mathbf{I}_n)$, and the prior distribution for $(\boldsymbol{\beta}, \sigma^2)$ is defined by $\mu(.)$. Thus, we arrive at

Proposition 5.1. *If* $\mathbf{Y}$ *is* $O_N(M)$*-invariant and*

$$\pi(\mathbf{B}_N, \sigma_N^2) = \int_{\Re^p \times [0,\infty)} g_N(\mathbf{B}_N, \sigma_N^2|\boldsymbol{\beta}, \sigma^2)\mu(d\boldsymbol{\beta}, d\sigma^2) \tag{5.4}$$

then

$$\pi(\mathbf{B}_N, \sigma_N^2|\mathbf{y}_s) = \int_{\Re^p \times [0,\infty)} g_N(\mathbf{B}_N, \sigma_N^2|\mathbf{y}_s, \boldsymbol{\beta}, \sigma^2)\mu(d\boldsymbol{\beta}, d\sigma^2|\mathbf{y}_s), \tag{5.5}$$

where $\mu(.|\mathbf{y}_s)$ *is defined in (5.3).*

Corollary 5.1. *Under the assumptions of Proposition 5.1 and provided that appropriate posterior moments are defined, then*

$$\hat{\mathbf{B}}_N = (\mathbf{X}'\mathbf{X})^{-1}\mathbf{X}_s'\mathbf{y}_s + (\mathbf{X}'\mathbf{X})^{-1}\mathbf{X}_r'\hat{\mathbf{y}}_r, \tag{5.6}$$

where $\hat{\mathbf{y}}_r^B = \mathbf{X}_r E[\boldsymbol{\beta}|\mathbf{y}_s]$,

$$V(\mathbf{B}_N|\mathbf{y}_s) = (\mathbf{X}'\mathbf{X})^{-1}\mathbf{X}_r^t\{E(\sigma^2|\mathbf{y}_s)\mathbf{I}_{N-n} + V(\mathbf{X}_r\boldsymbol{\beta}|\mathbf{y}_s)\}\mathbf{X}_r(\mathbf{X}'\mathbf{X})^{-1} \tag{5.7}$$

and

$$\hat{\sigma}_N^2 = E[\sigma^2|\mathbf{y}_s] + \frac{1}{N-m}\{\mathbf{y}_s'\mathbf{y}_s + E[\boldsymbol{\beta}'|\mathbf{y}_s]\mathbf{X}_r'\mathbf{X}_r E[\boldsymbol{\beta}|\mathbf{y}_s] - \hat{\boldsymbol{\beta}}'\mathbf{X}'\mathbf{X}\hat{\boldsymbol{\beta}}_N$$

$$+tr[Var(\mathbf{X}_r\boldsymbol{\beta}|\mathbf{y}_s) - Var(\mathbf{X}\boldsymbol{\beta}|\mathbf{y}_s)\}, \tag{5.8}$$

where $tr[\mathbf{A}]$ *denotes the trace of the matrix* $\mathbf{A}$ *and for any function* $w(\boldsymbol{\beta}, \sigma^2)$ *of* $(\boldsymbol{\beta}, \sigma^2)$, $E[w(\boldsymbol{\beta}, \sigma^2)|\mathbf{y}_s]$ *is computed with respect to the conditional measure* μ *defined in* 5.3.

Proof: It follows by noticing that $\hat{\mathbf{B}}_N$ can be written as

$$\hat{\mathbf{B}}_N = E[\mathbf{B}_N|\mathbf{y}_s] = (\mathbf{X}'\mathbf{X})^{-1}\mathbf{X}_s'\mathbf{y}_s + (\mathbf{X}'\mathbf{X})^{-1}\mathbf{X}_r'\hat{\mathbf{y}}_r,$$

and using the properties of the expectation operator.

Remark 5.1: Note that

$$g_N(\mathbf{B}_N, \sigma_N^2|\boldsymbol{\beta}, \sigma^2) = g_1(\mathbf{B}_N|\boldsymbol{\beta}, \sigma^2) g_2(\sigma_N^2|\sigma^2),$$

where g_1 is the density function of the $N_p(\boldsymbol{\beta}, \sigma^2(\mathbf{X}'\mathbf{X})^{-1})$ and g_2 is the density function of the $\sigma^2\chi^2_{N-m}/(N-m)$. Thus,

$$\pi(\mathbf{B}_N, \sigma_N^2) \propto (\sigma_N^2)^{\frac{N-m-1}{2}} \tag{5.10}$$

$$\times \int (\sigma^2)^{-\frac{n}{2}+1} \exp\{-\frac{1}{2\sigma^2}\{Q(\mathbf{B}_N, \boldsymbol{\beta}) + (N-m)\sigma_N^2\}\}\mu(d\boldsymbol{\beta}, d\sigma^2),$$

where, $Q(\mathbf{B}_N, \boldsymbol{\beta}) = (\mathbf{B}_N - \boldsymbol{\beta})'\mathbf{X}'\mathbf{X}(\mathbf{B}_N - \boldsymbol{\beta})$.

Similarly, to determine $g_N(\mathbf{B}_N, \sigma_N^2|\mathbf{y}_s, \boldsymbol{\beta}, \sigma^2)$, we note from the invariance property of the normal model that

$$g_N(\mathbf{y}_s|\mathbf{B}_N, \sigma_N^2, \boldsymbol{\beta}, \sigma_N^2) = f(\mathbf{y}_s|\mathbf{B}_N, \sigma^2)$$

$$\propto (\frac{1}{\sigma_N^2})^{\frac{n}{2}}[1 - \frac{h(\mathbf{B}_N)}{\sigma_N^2}]^{\frac{N-m-n-2}{2}} I_{(h(B_N),\infty)}(\sigma_N^2).$$

Then,

$$g_N(\mathbf{B}_N, \sigma_N^2|\boldsymbol{\beta}, \sigma^2, \mathbf{y}_s) \propto (\sigma_N^2)^{\frac{N-m-n}{2}-1} \exp\{-\frac{1}{2\sigma^2}\{(N-m)\sigma_N^2 + Q(\mathbf{B}_N, \boldsymbol{\beta})\}$$

$$\times[1 - \frac{h(\mathbf{B}_N)}{\sigma_N^2}]^{\frac{N-m-n-2}{2}} I_{(h(\mathbf{B}_N),\infty)}(\sigma_N^2),$$

leading to

$$g_N(\sigma_N^2|\boldsymbol{\beta},\sigma^2,\mathbf{y}_s) \propto (\sigma_N^2)^{\frac{N-m-n}{2}-1}\exp\{-\frac{(N-m)}{2\sigma^2}\sigma_N^2\}.$$

Thus,

$$\sigma_N^2|\boldsymbol{\beta},\sigma^2,\mathbf{y}_s \sim Gamma(\frac{N-m-n}{2},\frac{N-m}{2\sigma^2}).$$

We can also obtain general results for the posterior distribution of σ_N^2which is considered next.

Corollary 5.2 *Under the assumptions of Proposition 5.1, we have that*

$$\pi(\sigma_N^2|\mathbf{y}_s) \propto (\sigma_N^2)^{\frac{N-m-n}{2}-1}\int(\frac{1}{\sigma^2})^{\frac{N-m-2}{2}}\exp\{-\frac{(N-m)}{2\sigma^2}\sigma_N^2\}\mu_2(d\sigma^2|\mathbf{y}_s),$$

where μ_2 is the marginal probability measure on $(0,\infty)$ induced by μ.

Proof: The marginal posterior distributions of σ_N^2 can be obtained from Proposition 5.1 by noticing that we can write

$$\pi(\sigma_N^2|\mathbf{y}_s) = \int\int g_N(\mathbf{B}_N,\sigma_N^2|\mathbf{y}_s,\boldsymbol{\beta},\sigma^2)\mu(d\boldsymbol{\beta},d\sigma^2|\mathbf{y}_s)d\mathbf{B}_N.$$

Interchanging the integrals in the above expression, it follows that

$$\pi(\sigma_N^2|\mathbf{y}_s) = \int g_N(\sigma_N^2|\mathbf{y}_s,\boldsymbol{\beta},\sigma^2)\mu(d\boldsymbol{\beta},d\sigma^2|\mathbf{y}_s).$$

Now, $g_N(\sigma_N^2|\mathbf{y}_s,\boldsymbol{\beta},\sigma^2)$ does not depend of $\boldsymbol{\beta}$; consequently

$$\pi(\sigma_N^2|\mathbf{y}_s) = \int \frac{(\frac{N-m}{2\sigma^2})^{\frac{N-m-2}{2}}}{\Gamma(\frac{N-m-n}{2})(\sigma_N^2)^{\frac{N-m-n}{2}}-1}\exp\{-\frac{(N-m)}{2\sigma^2}\sigma_N^2\}\mu_2(d\sigma^2|\mathbf{y}_s).$$

In a similar way we can obtain the posterior distribution of $\pi(\mathbf{B}_N|\mathbf{y}_s)$. It suffices to note that

$$\pi(\mathbf{B}_N|\mathbf{y}_s) = \int g_N(\mathbf{B}_N|\mathbf{y}_s,\boldsymbol{\beta},\sigma^2)\mu(d\boldsymbol{\beta},d\sigma^2|\mathbf{y}_s).$$

The term in the inner integral involving g_N can be obtained by using properties of the normal distribution.

Special cases can be studied by taking in (5.5) the natural prior with respect to a likelihood $N(\mathbf{X}\boldsymbol{\beta},\sigma^2)$, that is

$$\mu(d\boldsymbol{\beta},d\sigma^2) \propto (\sigma^2)^{-\frac{d_0+m+2}{2}}\exp\{(\boldsymbol{\beta}-\mathbf{b_0})^t B_0^{-1})\boldsymbol{\beta}-\mathbf{b_0}) + \frac{a_0}{\frac{1}{2\sigma^2}}\}d\boldsymbol{\beta}d\sigma$$

leading to

$$\mathbf{B_N}|\sigma_N^2 \sim t_m\left(\mathbf{b}_0, \frac{N-m+d_0}{(N-m)\sigma_N^2+a_0}(\mathbf{y}B_0+(\mathbf{X}^t\mathbf{X})^{-1}), N-m+d_0\right)$$

and

$$\sigma_N^2 \sim Gg\left(\frac{d_0+m}{2}, \frac{a_0}{N-m}, \frac{N-m}{2}\right)$$

where $Gg(\alpha,\beta,n)$ denotes the Gamma-Gamma distribution (Bernardo and Smith, 1994) with parameters $\alpha > 0, \beta > 0$ and $n > 0$.

By choosing in (5.11) $a_0 = d_0 = 0$ we obtain

$$\mathbf{B_N}|\sigma_N^2 \sim t_m\left(b_0, \frac{1}{\sigma_N^2}(\mathbf{y}B_0+(\mathbf{X}^t\mathbf{X})^{-1}), N-m\right)$$

and

$$\pi(\sigma_N^2) \propto \left(\frac{1}{\sigma_N^2}\right)^{\frac{m+2}{2}},$$

a type of noninformative prior for σ_N^2.

Straightforward algebraic manipulations allows to obtain the posterior distributions. Also, a class of priors such that $\mathbf{B_N}$ is independent of σ_N^2 can be obtained from (5.5) by taking

$$\mu(\boldsymbol{\beta}, d\sigma^2) = \mu_1(d\boldsymbol{\beta})\mu_2(d\sigma^2)$$

where μ_1 is a probability measure defined on the Borel sets of $\Re^p$ and μ_2 is degenerate on $\{\sigma_0^2\}$.

Acknowledgments

The authors thank Dr. Reinaldo Arellano-Valle and Dr. Nelson Tanaka for helpful discussions and suggestions.

References

1. Bernardo, J.M. and Smith, A.F.M, *Bayesian theory.* John Wiley (1994).
2. Barlow, R.E. and Mendel, M.B., The operational Bayesian approach in reliability theory. *Resenhas-IMEUSP*, **1**, 46–56 (1993).
3. Barlow, R.E. and Mendel, M.B., De Finetti-type representation for life distribution. *Journal of the American Statistical Association*, **87**, 1116–1122 (1992).

4. Basu, D., An essay on the logical foundations of survey sampling. Foundations of Statistical Inference. Edited by V.P. Godambe and D.A. Sprott. Hoet, Rinehart and Winston, Toronto, 203–242 (1971).
5. Basu, D., Statistical information and likelihood. *Sankhyā*, A, **37**, 1–71 (1975).
6. Bolfarine, H. and Zacks, S., *Prediction theory for finite populations. Springer Series in Statistics*, Springer-Verlag (1992).
7. Cassel, C.M., Sarndall, C.E. and Wretman, J.H., *Foundations of Inference in Sample Surveys.* Wiley, New York (1977).
8. Cochran, W.G., *Sampling Techniques.* Wiley, New York (1973).
9. Diaconis, P., Eaton, M.L. and Lauritzen, S.I., Finite de Finetti theorems in linear models and multivariate analysis. *Scandinavian Journal of Statistics*, **19**, 289–315 (1992).
10. Eaton, M., Group invariance applications in statistics. C.B.M.S. Regional Conference Series in Probability and Statistics, **v.1**, IMS, Hayward, California (1989).
11. Ericson, W.A., Subjective Bayesian models in sampling finite populations. *Journal of the Royal Statistical Society*, **B31**, 195–224 (1969).
12. Fang, K.T., Kotz, J. and Ng, K.W., *Symmetric multivariate and related distributions.* Chapman and Hall, N.Y. (1990)
13. Farrell, R.H., *Multivariate calculation. Use of the continuous groups.* Springer-Verlag, New York (1966).
14. Gasco, L.B., *Previsão e preditivismo em modelos lineares com e sem erros nas variáveis.* (In Portuguese.) Doctoral Thesis. Instituto de Matemática e Estatí stica da Universidade de São Paulo (1997).
15. Godambe, V.P., A new approach to sampling from finite populations, I, II. *Journal of the Royal Statistical Society*, **B28**, 310–328 (1966).
16. Iglesias, P.L., *Finite forms of de Finetti's theorem: a predictivistic approach to statistical inference in finite populations.* (In Portuguese.) Doctoral Thesis. Instituto de Matemática e Estatística da Universidade de São Paulo (1993).
17. Mendel, M.B., Bayesian parametric models for lifetimes. *Bayesian Statistics*, **4**, 697–705. (J.M. Bernardo, J.O. Berger, A.P. Dawid and A.F.M. Smith, eds.) Oxford University Press (1992).
18. Mendel, M.B., Operational parameters in Bayesian models. *Test*, **3**, 2, 195–206 (1994).
19. Mendel, M.B. and Kempthorne, P.J., Operational parameters for the use of Bayesian methods in engineering. *REBRAPE*, **10**, 1–13 (1996).
20. Nachbin, L., *The Haar integral.* Van Nostrand, Princeton, New York (1965).
21. Smith, A.M.F., On random sequences with centered spherical symmetry. *Journal of Royal Statistical Society*, **B43**, 208–305 (1981).

CHAPTER 23

BAYESIAN NONPARAMETRIC TESTING OF CONSTANT VERSUS NONDECREASING HAZARD RATES

Yu Hayakawa* and Jonathan Zukerman†
School of Mathematical and Computing Sciences
Victoria University of Wellington
P.O. Box 600, Wellington, New Zealand
*E-mail: *yu.hayakawa@mcs.vuw.ac.nz,*
E-mail: †jonathan.zukerman@mcs.vuw.ac.nz

Sue Paul
PA Consulting Group
P.O. Box 1659, Wellington, New Zealand
E-mail: Sue.Paul@paconsulting.com

Tony Vignaux
School of Mathematical and Computing Sciences
Victoria University of Wellington
P.O. Box 600, Wellington, New Zealand
E-mail: tony.vignaux@mcs.vuw.ac.nz

This chapter considers applications of a Bayesian nonparametric method for testing constant versus nondecreasing hazard rates. The weighted gamma process is selected as a prior on the space of nondecreasing hazard rates. Monte Carlo simulations of the weighted Chinese restaurant process provide an approximation of the Bayes factor. The method will be illustrated through examples.

1. Introduction

In reliability theory, the failure distribution is a probabilistic description of the length of time for which a device operates without failure. The exponential distribution is commonly used for this purpose. It is particularly characterised by the memoryless property and this implies a constant hazard rate. If the devices do not wear out with time, this is an appropriate choice for the failure distribution

Many real devices, however, deteriorate with time and may have a strictly increasing hazard rate function. The Weibull and gamma distributions, with some condition on their shape parameters, are commonly used as the failure distribution in these cases.

It is therefore of some practical interest to find methods to test whether a device has, on the one hand, a constant hazard rate or, on the other hand, an increasing hazard rate.

There are several approaches. For instance one can construct a test based on the total time on test statistics for a sample or normalised spacings (cf. Kotz and Johnson[14]). For more details the reader is referred to Barlow, et al.[11] and Doksum and Yandell[8]. In this Chapter, we use a fully Bayesian method to accomplish the task, i.e., to obtain the posterior probability of each hypothesis. Under H_o the gamma prior was used. For H_A, we chose the prior process for the set of nondecreasing hazard rate functions to be the weighted gamma process[3,10,13]. We simulated the weighted Chinese restaurant processes[4] which were needed to approximate the Bayes factor for the testing.

In the next section we describe the life testing model to be considered. In Section 3, the integrated likelihood function under the alternative hypothesis is simplified using the results in Lo and Weng[3], which include those of Dykstra and Laud[13] as a special case. Section 4 deals with approximation of the simplified function obtained in Section 3 via Monte Carlo simulations of the weighted Chinese restaurant process. In Section 5, numerical examples are used to illustrate the method. Finally, we state our conclusions and suggest further studies.

2. Life Testing Models and Hypothesis Testing of Constant versus Nondecreasing Hazard Rates

Consider a life testing situation in which n new items are tested. Their lifetimes are assumed to be independent and identically distributed. Let $N \equiv \{N(t) : t \in [0, \infty)\}$ be a counting process representing the number of operative units at time t and the testing is terminated at a prespecified time $T \in (0, \infty)$. Let $x_1, \cdots, x_k$ $(k \leq n)$ denote the failure times of n items on test, out of which k fail prior to or at T and the lifetimes of the remaining

$n-k$ items are censored. Let $F(\cdot)$ be the cumulative distribution function for each item with density $f(\cdot)$. Then the survival function corresponding to $F(\cdot)$ and its hazard rate function are respectively denoted by

$$\bar{F}(t) = 1 - F(t)$$
$$r(t) = \frac{f(t)}{\bar{F}(t)}$$

The likelihood function of $r(\cdot)$ given the data becomes

$$L\left(r(\cdot)\,|\,x_1,\cdots,x_k,Y(\cdot)\right) = \frac{n!}{(n-k)!}\left[\prod_{i=1}^{k} r(x_i)\right]\exp\left\{-\int I_{\{0\le s\le T\}}Y(s)\times r(s)ds\right\}$$

where $I_{\{A\}}$ is the indicator function for set A and $Y(t)$ denotes the number of remaining items on test just prior to t. We consider a model whereby the hazard rate $r(\cdot)$ can be represented as a mixture of known kernels with respect to some finite measure (See Lo and Weng[3]). Suppose that a nonnegative kernel $\kappa(t|\nu)$ on $([0,T]\times\Re, \mathcal{F}\times\mathcal{B})$ can be prespecified, where $\Re$ is the set of real numbers and $\mathcal{F}$ and $\mathcal{B}$ are Borel $\sigma-$fields of $[0,T]$ and $\Re$, respectively. Also assume that the following representation for $r(\cdot)$,

$$r(t\,|\,\mu) = \int_{\Re}\kappa(t|\nu)\mu(d\nu) \quad t\in[0,T]$$

where μ is a member of the space of finite measures on $(\Re,\mathcal{B})$ denoted by Θ.

Under this mixture model for the hazard rates, the likelihood function is given by

$$L\left(\mu(\cdot)\,|\,x_1,\cdots,x_k,Y(\cdot)\right) = \frac{n!}{(n-k)!}\left[\prod_{1\le i\le k}\int_{\Re}\kappa(x_i\,|\,\nu_i)\,\mu(d\nu_i)\right]\times \exp\left\{-\int_{\Re}\int_{\Re} I_{\{0\le s\le T\}}Y(s)\,\kappa(s|\nu)\;ds\;\mu(d\nu)\right\}.$$

Our main interest is to conduct the following testing.

H_0 : $X_1,\cdots,X_n$ have a constant hazard rate.
H_A : $X_1,\cdots,X_n$ have a nondecreasing hazard rate, but not constant.

Firstly, we assign prior probabilities to hypotheses denoted by $P(H_0)$ and $P(H_A)$. Secondly, Bayes theorem enables us to compute the posterior probability of each hypothesis. For instance, the posterior probability of the null hypothesis becomes

$$P(H_0 \mid x_1, \cdots, x_k, Y(\cdot)) = \frac{P(H_0)\int_0^\infty L(\lambda \mid \mathcal{D})\pi(d\lambda)}{P(H_0)\int_0^\infty L(\lambda \mid \mathcal{D})\pi(d\lambda) + P(H_A)\int_\Theta L(\mu(\cdot) \mid \mathcal{D})G(d\mu(\cdot))}$$

where

$$\mathcal{D} \equiv \{x_1, \cdots, x_k, Y(\cdot)\}$$
$$L(\lambda \mid \mathcal{D}) = \frac{n!}{(n-k)!}\lambda^k e^{-\lambda[\sum_{i=1}^k x_i + (n-k)T]}$$

and $\pi(d\lambda)$ and $G(d\mu(\cdot))$ denote priors for λ and $\mu(\cdot)$, respectively. Finally, one can make a decision by comparing $P(H_0 \mid x_1, \cdots, x_k, Y(\cdot))$ with $P(H_A \mid x_1, \cdots, x_k, Y(\cdot))$, namely, choose the hypothesis whose posterior probability is the largest. If a loss function is assessed, then the posterior expected loss with respect to each hypothesis needs to be evaluated.

As seen above, we need to compute

$$\int_0^\infty L(\lambda \mid x_1, \cdots, x_k, Y(\cdot))\pi(d\lambda) \tag{1}$$

$$\int_\Theta L(\mu(\cdot) \mid x_1, \cdots, x_k, Y(\cdot))G(d\mu(\cdot)). \tag{2}$$

In general, Expression (1) can be evaluated with no difficulty. In particular, if $\pi(d\lambda)$ is a gamma distribution, then Expression (1) is a Pareto density. On the other hand, evaluating Expression (2) directly may be quite difficult. We assume that a prior for $\mu(\cdot)$ is a weighted gamma process and we will approximate Expression (2) via a weighted Chinese restaurant process. Details are given in the following sections.

3. Prior Process for Hazard Rates and Predictive Distribution under the Alternative Hypothesis

Our choice of prior process for $\mu(\cdot)$ is a weighted (extended) gamma process defined below.

Definition 1: *Let $\alpha(s), s \geq 0$ be a nondecreasing, left-continuous, real-valued function with $\alpha(0) \equiv 0$. A continuous time stochastic process $Z \equiv \{Z(s), s \geq 0\}$ is referred to as a gamma process*[3,10,13] *with a shape function $\alpha(s), s \geq 0$, if the following properties are held.*

(1) $Z(0) \equiv 0$;
(2) Z *has independent increments;*
(3) for $t > s, (Z(t) - Z(s)) \sim gamma(\alpha(t) - \alpha(s), 1)$.

where $gamma(a, b)$ *denotes the gamma distribution with density*

$$f(x \mid a, b) = \frac{b^a}{\Gamma(a)} x^{a-1} e^{-bx} I_{\{0 \leq x < \infty\}}.$$

A generalisation of Z is defined next (cf. Dykstra and Laud[13], Lo and Weng[3], Singpurwalla[10]). Let $\beta(s)\,(s \geq 0)$ be a positive right continuous real-valued function, bounded away from 0, with left-hand limits existing. A continuous time stochastic process defined by

$$W(t) = \int_{[0,t)} \beta(s) dZ(s), \quad t \in [0, \infty)$$

is referred to as an extended (weighted) gamma process with shape function $\alpha(\cdot)$ and scale function $\beta(\cdot)$. The expected value and variance of $W(t)$ are given by

$$E\{W(t)\} = \int_{[0,t)} \beta(s) d\alpha(s)$$

$$Var\{W(t)\} = \int_{[0,t)} \beta^2(s) d\alpha(s)$$

Now we assume that $\{\mu(t), t \geq 0\}$ is a weighted gamma process with shape function $\alpha(\cdot)$ and scale function $\beta(\cdot)$ with its distribution denoted by $G(d\mu(\cdot) \mid \alpha(\cdot), \beta(\cdot))$. We are ready to evaluate the following integrated likelihood function.

$$\int_\Theta L\left(\mu(\cdot) \mid \mathcal{D}\right) G\left(d\mu(\cdot) | \alpha(\cdot), \beta(\cdot)\right) =$$

$$\int_\Theta \frac{n!}{(n-k)!} \left[\prod_{1 \leq i \leq k} \int_\Re \kappa(x_i \mid \nu_i)\, \mu(d\nu_i) \right] \times$$

$$\exp\left\{ - \int_\Re \int_\Re I_{\{0 \leq s \leq T\}} Y(s)\, \kappa(s|\nu)\; ds\; \mu(d\nu) \right\} G\left(d\mu(\cdot)|\alpha(\cdot), \beta(\cdot)\right) \quad (3)$$

Several existing theorems can be applied to simplify Expression (3). This is explained step by step below:

By Proposition 3.1 of Lo and Weng[3], Expression (3), which is also the Laplace transform of the prior for $\mu(\cdot)$, can be updated.

$$(3) = \exp\left\{-\int_{\Re} \log\left[1+\beta(\nu)f(\nu)\right]\alpha(d\nu)\right\} \times$$

$$\int_{\Theta} \frac{n!}{(n-k)!}\left[\prod_{1\le i\le k}\int_{\Re}\kappa(x_i\,|\,\nu_i)\,\mu(d\nu_i)\right] G\left(d\mu(\cdot)\,\middle|\,\alpha(\cdot), \frac{\beta(\cdot)}{1+\beta(\cdot)f(\cdot)}\right) \quad (4)$$

where $f(\nu) = \int_{\Re} I_{\{0\le s\le T\}} Y(s)\kappa(s|\nu)ds$. Then Lemma 3.1 of Lo and Weng[3] is repeatedly applied to interchange the order of integrals as follows.

$$(4) = C\int_{\Re}\int_{\Theta}\kappa(x_1|\nu_1)\left[\prod_{2\le i\le k}\int_{\Re}\kappa(x_i\,|\,\nu_i)\,\mu(d\nu_i)\right]$$

$$G\left(d\mu(\cdot)\,\middle|\,\alpha(\cdot)+\delta_{\nu_1}(\cdot), \frac{\beta(\cdot)}{1+\beta(\cdot)f(\cdot)}\right)\cdot\frac{\beta(\nu_1)}{1+\beta(\nu_1)f(\nu_1)}\alpha(d\nu_1)$$

$$= C\int_{\Re}\int_{\Re}\int_{\Theta}\kappa(x_1|\nu_1)\kappa(x_2|\nu_2)\left[\prod_{3\le i\le k}\int_{\Re}\kappa(x_i\,|\,\nu_i)\,\mu(d\nu_i)\right]$$

$$G\left(d\mu(\cdot)\,\middle|\,\alpha(\cdot)+\delta_{\nu_1}(\cdot)+\delta_{\nu_2}(\cdot), \frac{\beta(\cdot)}{1+\beta(\cdot)f(\cdot)}\right)\cdot$$

$$\frac{\beta(\nu_2)}{1+\beta(\nu_2)f(\nu_2)}(\alpha+\delta_{\nu_1})(d\nu_2)\cdot\frac{\beta(\nu_1)}{1+\beta(\nu_1)f(\nu_1)}\alpha(d\nu_1)$$

$$\vdots$$

$$\vdots$$

$$= C\int_{\Re^k}\int_{\Theta}\left[\prod_{1\le i\le k}\kappa(x_i\,|\,\nu_i)\cdot\frac{\beta(\nu_i)}{1+\beta(\nu_i)f(\nu_i)}\right]$$

$$G\left(d\mu(\cdot)\,\middle|\,\alpha(\cdot)+\sum_{1\le i\le k}\delta_{\nu_i}(\cdot), \frac{\beta(\cdot)}{1+\beta(\cdot)f(\cdot)}\right)$$

$$\prod_{k\ge i\ge 1}\left(\alpha+\sum_{1\le j\le i-1}\delta_{\nu_j}\right)(d\nu_i)$$

$$= C\int_{\Re^k}\left[\prod_{1\le i\le k}\kappa(x_i\,|\,\nu_i)\cdot\frac{\beta(\nu_i)}{1+\beta(\nu_i)f(\nu_i)}\right]$$

$$\prod_{k\geq i\geq 1}\left(\alpha+\sum_{1\leq j\leq i-1}\delta_{\nu_j}\right)(d\nu_i) \tag{5}$$

where $(\alpha+\sum_{1\leq j\leq i-1}\delta_{\nu_j})(d\nu_i)$ denotes $\alpha(d\nu_1)$ when $i=1$ and

$$C=\frac{n!}{(n-k)!}\exp\left\{-\int_{\Re}\log\left[1+\beta(\nu)f(\nu)\right]\alpha(d\nu)\right\}.$$

Expression (5) still has multiple integrals which make computation very difficult. However, through combinatorial techniques, Expression (5) can be converted into a function of uni-dimensional integrals. Some notation is introduced before presenting a simplification of Expression (5).

Consider a set of k elements, $\mathcal{S}=\{1,\ldots,k\}$. Let $\mathbf{p}$ and $n(\mathbf{p})$ be a partition of $\mathcal{S}$ and the number of cells of $\mathbf{p}$, respectively. C_i denotes the ith cell of $\mathbf{p}$ and e_i represents the number of elements in C_i. Applying Lemma 2 of Lo[2] , we obtain

$$(5)=C\sum_{\mathbf{p}}\prod_{i=1}^{n(\mathbf{p})}\left\{(e_i-1)!\int_{\Re}\prod_{l\in C_i}\left[\kappa(x_l\,|\,\nu)\cdot\frac{\beta(\nu)}{1+\beta(\nu)f(\nu)}\right]\alpha(d\nu)\right\} \tag{6}$$

where above terms are summed over all the possible partitions of $\{1,\ldots,k\}$. Since the case of nondecreasing hazard rates corresponds to $\kappa(t|\nu)=I_{\{\nu\leq t\}}$ (cf. Lo and Weng[3] and Dykstra and Laud[13]), we have

$$(6)=C\sum_{\mathbf{p}}\prod_{i=1}^{n(\mathbf{p})}\left\{(e_i-1)!\int_{\Re}\prod_{l\in C_i}\left[I_{\{\nu\leq x_l\}}\cdot\frac{\beta(\nu)}{1+\beta(\nu)f(\nu)}\right]\alpha(d\nu)\right\}$$

$$=C\sum_{\mathbf{p}}\prod_{i=1}^{n(\mathbf{p})}\left\{(e_i-1)!\int_{\Re}\left[\frac{\beta(\nu)}{1+\beta(\nu)f(\nu)}\right]^{e_i}I_{\{0\leq\nu\leq\min(i)\}}\alpha(d\nu)\right\} \tag{7}$$

where $\min(i)=\min\{x_l : l\in C_i\}$.

When $|\mathcal{S}|$ is large, a direct computation of Expression (7) is intractable, hence an approximation method is required to evaluate it. The next section describes a weighted Chinese restaurant process whereby we approximate Expression (7).

4. Monte Carlo Approximations to the Posterior Probabilities via the Chinese Restaurant Process

The Chinese restaurant process (termed by Jim Pitman) is a random mechanism to sequentially partition a group of "people" denoted by $\mathcal{S}=\{1,\ldots,n\}$

into "tables" denoted by $\{C_1, \ldots, C_{n(\mathbf{p})}\}$ where $n(\mathbf{p})$ denotes the number of tables, i.e., subgroups of $\mathcal{S}$ (see Aldous[5], Kuo[9]). An importance sampling variant of it called the weighted Chinese restaurant process is used to approximate Expression (7), which takes into account of similarities in failure times, x_js. The following description of the Chinese restaurant process is based on Lo, Brunner and Chan[4] to which the reader is referred for details.

Let $\rho(C)$ define the marginal weight for a table C by

$$\rho(C) = \int \prod_{j \in C} w_j(\nu)\alpha(d\nu)$$

where $w_j(u)$ is a nonnegative finite "likelihood" weighting function for $j \in \mathcal{S}$ and $\alpha(du)$ is a "prior" mixing measure. We also define a "predictive" weight of $r \notin C$ given C by the ratio

$$\rho(r|C) = \begin{cases} \rho(r, C)/\rho(C) & \text{if } \rho(C) \neq 0 \\ 0 & \text{if } \rho(C) = 0. \end{cases}$$

The weighted Chinese restaurant process proceeds as follows.

(1) `Step 1`: set $\lambda(0) = \rho(1)$ and assign 1 to C_1 with probability $\rho(1)/\lambda(0) = 1$.
(2) `Step r`: from Step $r-1$, we have tables $C_1, \ldots, C_{n(\mathbf{p})}$ with their respective sizes $e_1, \ldots, e_{n(\mathbf{p})}$ $(r = 2, \ldots, n)$.
   - Compute $\lambda(r-1) = \rho(r) + \sum_{1 \le i \le n(\mathbf{p})} e_i \rho(r|C_i)$.
   - Assign r to a new table $C_{n(\mathbf{p})+1}$ with probability $\rho(r)/\lambda(r-1)$; otherwise, assign r to C_i with probability $e_i\rho(r|C_i)/\lambda(r-1), i = 1, \ldots, n(\mathbf{p})$.
   - If r is assigned to a new table, $n(\mathbf{p}) \leftarrow n(\mathbf{p}) + 1$; otherwise $n(\mathbf{p})$ stays the same.

A density of the above algorithm is given by $q(\mathbf{p}|\alpha, \mathbf{w}) = \phi(\mathbf{p})/\Lambda_{n-1}(\mathbf{p})$ where

$$\phi(\mathbf{p}) = \prod_{1 \le i \le n(\mathbf{p})} (e_i - 1)!\rho(C_i)$$

$$\Lambda_{n-1}(\mathbf{p}) = \prod_{j=0}^{n-1} \lambda(j)$$

$$\mathbf{w} = (w_1(\cdot), \ldots, w_n(\cdot)).$$

If we let $w_j(\nu) = I_{\{0\leq\nu\leq x_j\}}$, then

$$\begin{aligned}\phi(\mathbf{p}) &= \prod_{1\leq i\leq n(\mathbf{p})} (e_i - 1)!\rho(C_i)\\ &= \prod_{1\leq i\leq n(\mathbf{p})} (e_i - 1)! \int \prod_{j\in C_i} w_j(\nu)\alpha(d\nu)\\ &= \prod_{1\leq i\leq n(\mathbf{p})} (e_i - 1)! \int \prod_{j\in C_i} I_{\{0\leq\nu\leq x_j\}}\alpha(d\nu)\\ &= \prod_{1\leq i\leq n(\mathbf{p})} (e_i - 1)! \int I_{\{0\leq\nu\leq \min(i)\}}\alpha(d\nu).\end{aligned}$$

Rewriting Expression (7) we get

$$C\sum_{\mathbf{p}}\left\{\prod_{i=1}^{n(\mathbf{p})}\left[(e_i-1)!\int_{\Re} I_{\{0\leq\nu\leq\min(i)\}}\alpha(d\nu)\right]\times \prod_{i=1}^{n(\mathbf{p})}\int_{\Re}\left[\frac{\beta(\nu)}{1+\beta(\nu)f(\nu)}\right]^{e_i}\alpha(d\nu\,|\,C_i)\right\} \tag{8}$$

where

$$\alpha(d\nu\,|\,C_i) = \frac{I_{\{0\leq\nu\leq\min(i)\}}\alpha(d\nu)}{\int_{\Re} I_{\{0\leq\nu\leq\min(i)\}}\alpha(d\nu)}.$$

Generate sufficiently large number of partitions $\mathbf{p}_1, \cdots, \mathbf{p}_M$ according to the density $q(\mathbf{p}|\alpha, \mathbf{w})$ of the weighted Chinese restaurant process. Then an approximation of Expression (8) is given by

$$C\cdot\frac{1}{M}\sum_{k=1}^{M} h(\mathbf{p}_k)\Lambda_{n-1}(\mathbf{p}_k)$$

where

$$h(\mathbf{p}) = \prod_{i=1}^{n(\mathbf{p})}\int_{\Re}\left[\frac{\beta(\nu)}{1+\beta(\nu)f(\nu)}\right]^{e_i}\alpha(d\nu\,|\,C_i).$$

5. Examples

In this section, the method is illustrated through examples.

5.1. *Assumptions and Related Issues*

Firstly, the assumptions made for the examples are described.

(1) Under H_0, the prior for the hazard rate denoted by θ of exponential density is a gamma distribution with density

$$\pi(d\theta) = \frac{\eta^\tau}{\Gamma(\tau)} \theta^{\tau-1} e^{-\eta\theta} d\theta$$

If the data only consist of complete failure times, then the predictive density becomes a Pareto density.

(2) Under H_A, $\beta(\nu) \equiv 1$ and $\alpha(\nu)$ is of the following form.

$$\alpha(\nu) = c\nu^d, \quad \nu \in [0, T]. \tag{9}$$

This can be considered as our prior guess about the hazard rate function of the distribution from which the data come.

One of the conditions on $\alpha(\cdot)$ assumed in Lo and Weng[3] suggests that $\lim_{\nu\to\infty} \alpha(\nu)$ be finite. However, the particular form of $\kappa(t|\nu) = I_{\{0\le\nu\le t\}}$ considered here implies that the integrated likelihood under H_A does not depend on $\alpha(\nu)$ for $\nu > T$.

According to Dykstra and Laud[13], in a very specialised case such that $\alpha(\nu)$ is constant save a jump at 0, i.e.,

$$\alpha(\nu) = \begin{cases} 0 & \text{if } \nu = 0 \\ c & \text{if } \nu > 0 \end{cases}$$

where c is a positive constant, the hazard rate function $r(t \mid \mu)$ is a constant function whose value is distributed as a gamma$(c, 1/\beta(0))$ random variable where $\beta(0)$ is the value of the scale function evaluated at 0. Hence, if we chose $\alpha(\cdot)$ to be of the special case above, H_A would be essentially equivalent to H_0 with the prior for θ being gamma$(c, 1/\beta(0))$. The gamma processes we have selected for the examples give no positive mass to the set of flat hazard rate functions.

5.2. *Data*

Two sets of data were simulated using the exponential and Weibull densities. Let $Exp(\theta)$ and $Weibull(\alpha, \lambda)$ denote the exponential and Weibull distributions with the respective densities:

$$f(t|\theta) = \theta e^{-\theta t} I_{\{t\ge 0\}}$$

and

$$f(t \mid a, \lambda) = a\lambda^a t^{a-1} e^{-(\lambda t)^\alpha} I_{\{t \geq 0\}}.$$

Note that the hazard rate function for $Weibull(a, \lambda)$ is given by

$$\begin{aligned} r(t \mid a, \lambda) &= (a\lambda)(\lambda t)^{a-1} \\ &= a\lambda^a t^{a-1} \end{aligned}$$

and is of the same form as $\alpha(\nu)$ given in Subsection 5.1. We now reparametrise Expression (9) as $\lambda(\nu) = a\lambda^a t^{a-1}$. The simulated data sets generated from $Exp(0.2)$ and $Weibull(2, 0.02)$ are tabulated below.

Table 1. **Data A** - $Exp(0.2)$

15.331131642	0.118787394	2.696430430	2.173715169	5.897698814
1.791502600	0.949234674	1.206629130	9.030635806	8.488133404
17.241880168	4.828624536	12.994042430	0.005868113	2.884614540
3.373834125	0.454258376	14.154665756	7.698991519	3.437420624

Table 2. **Data B** - $Weibull(2, 0.02)$

45.46250	33.23567	154.59818	79.37798	56.54835
77.87939	73.97001	101.16410	126.55424	46.03762
127.90861	88.25736	150.27799	98.31449	69.60570
54.33771	41.47340	65.58763	85.72739	86.92385

5.3. *Bayes Factor and Sensitivity Analysis*

A Bayesian approach to hypothesis testing was initiated by Jeffreys[6] using the Bayes factor, the ratio of the posterior odds to the prior odds. For a comprehensive review of the Bayes factor, see Kass and Raftery[12]. A definition of the Bayes factor follows. Let $\mathcal{D}$ denote the data. Then the posterior probability of H_0 given $\mathcal{D}$ is obtained by

$$P(H_0 \mid \mathcal{D}) = \frac{L(\mathcal{D} \mid H_0)P(H_0)}{L(\mathcal{D} \mid H_0)P(H_0) + L(\mathcal{D} \mid H_A)P(H_A)}$$

where $L(\mathcal{D} \mid H_0)$ and $L(\mathcal{D} \mid H_A)$ denote the integrated likelihoods under H_0 and H_A, respectively (e.g., Expressions (1) and (2)). Similarly, $P(H_A \mid \mathcal{D})$

is defined. Then it can be seen that

$$\frac{P(H_0 \,|\, \mathcal{D})}{P(H_A \,|\, \mathcal{D})} = \frac{L(\mathcal{D} \,|\, H_0)}{L(\mathcal{D} \,|\, H_A)} \cdot \frac{P(H_0)}{P(H_A)}.$$

In other words, the posterior odds ratio is obtained by multiplying the prior odds ratio by the integrated likelihood ratio which is called the Bayes factor. We denote the Bayes factor in favour of H_0 by

$$B_{0A} = \frac{L(\mathcal{D} \,|\, H_0)}{L(\mathcal{D} \,|\, H_A)}$$

and in favour of H_A by B_{A0}. Note that the Bayes factor becomes equal to the posterior odds ratio if $P(H_0) = P(H_A) = 1/2$.

It is essential to study how sensitive the Bayes factor is with respect to the choice of prior. Table 3 summarises a sensitivity analysis performed on B_{A0} given **Data A**. The prior mean and variance of θ are equal to τ/η and τ/η^2, respectively. Rows (1) - (4) and Rows (5) - (8) have the same prior mean of θ but the variance of the former is smaller than that of the latter. The same can be said about other 4 groups of rows. With fixed values of a, λ and $E(\theta)$, B_{A0} seems to be larger for the prior of θ with the larger $Var(\theta)$ than with the smaller $Var(\theta)$. Now for fixed values of τ and η, B_{A0} is more sensitive to the choice of a than of λ. However, the values of B_{A0} for each combination of priors we considered is very small compared with 1. This indicates that the data support H_0 very strongly.

Table 4 contains the results of the sensitivity analysis performed on B_{A0} given **Data B**. Contrary to Table 3, the values of B_{A0} are greater than 1 save the last two combinations of the priors. For the same pair of a, λ and $E(\theta)$, B_{A0} is larger for the prior of θ with the bigger $Var(\theta)$ than with the smaller $Var(\theta)$. The last two rows in Table 4 indicate that B_{A0} may be quite sensitive to the a value although λ is set to equal to the scale parameter value.

B_{0A} (B_{A0}) shows us a degree of how strongly H_0 (H_A) is supported by the data. Some suggestions for interpreting values of B_{0A} or B_{A0} are given in Kass and Raftery[12].

Table 3. **Sensitivity of B_{A0} - $Exp(0.2)$**

τ	η	a	λ	B_{A0}
4	20	2	0.2	0.000175884
4	20	2	0.1	0.000135817
4	20	3	0.2	1.07372e-10
4	20	3	0.1	1.29392e-07
0.4	2	2	0.2	0.000577912
0.4	2	2	0.1	0.000440917
0.4	2	3	0.2	3.95959e-10
0.4	2	3	0.1	4.29889e-07
1	10	2	0.2	0.000337272
1	10	2	0.1	0.000303696
1	10	3	0.2	2.56514e-10
1	10	3	0.1	2.98626e-07
0.1	1	2	0.2	0.00150293
0.1	1	2	0.1	0.00113446
0.1	1	3	0.2	9.78654e-10
0.1	1	3	0.1	1.312e-06
9	30	2	0.2	0.000276508
9	30	2	0.1	0.000222248
9	30	3	0.2	1.73186e-10
9	30	3	0.1	2.13353e-07
0.9	3	2	0.2	0.000355851
0.9	3	2	0.1	0.000301748
0.9	3	3	0.2	2.5229e-10
0.9	3	3	0.1	2.73134e-07

Table 4. **Sensitivity of B_{A0} - $Weibull(2, 0.02)$**

a	λ	τ	η	B_{A0}
2	0.01	1	100	46.6498
2	0.01	0.1	10	216.034
2	0.02	1	100	211.768
2	0.02	0.1	10	1076.67
2	0.005	1	100	9.78408
2	0.005	0.1	10	46.4756
4	0.01	1	100	36.9526
4	0.01	0.1	10	195.481
4	0.02	1	100	25.0599
4	0.02	0.1	10	124.529
4	0.005	1	100	1.83968
4	0.005	0.1	10	6.68633
1.5	0.01	1	100	28.0332
1.5	0.01	0.1	10	126.689
1.5	0.02	1	100	77.5004
1.5	0.02	0.1	10	361.359
1.5	0.005	1	100	9.73508
1.5	0.005	0.1	10	47.7305
10	0.01	1	100	0.000231274
10	0.01	0.1	10	0.000866646

6. Conclusions and Further Studies

We have demonstrated a fully Bayesian approach to the hypothesis testing of constant versus nondecreasing hazard rates. The gamma process was used as a prior for the set of nondecreasing hazard rate functions. The weighted Chinese restaurant processes were simulated to approximate the Bayes factor for the testing. We illustrated the method using examples and performed a sensitivity analysis with respect to the choice of priors.

For the null hypothesis, we used an informative prior for the hazard rate, θ, of the exponential density. When prior information about θ is not available, a noninformative prior such as Jeffreys prior could be used. Jeffreys priors are often improper and that for the exponential density: $1/\theta$, is no exception. This does cause difficulties with computing the Bayes factor. However, Dawid[1] has suggested an approach to circumvent this problem. For a good illustration of his method, see van Noortwijk and et al.[7]. It should be interesting to see how the numerical results obtained here would differ if we used a Jeffreys prior for H_0.

Acknowledgments

Studies related to Sections 2 to 4 were done while the first author was visiting Albert Y. Lo at the Department of Information and Systems Management, Hong Kong University of Science & Technology. The authors are grateful to him for his invaluable assistance. Also many thanks to Jan M. van Noortwijk and Ray Brownrigg for their help regarding Sections 5 and 6 which lead to an improvement of this Chapter.

References

1. A. P. Dawid. The trouble with Bayes factors. Technical Report Research Report No. 202, Department of Statistical Science, University College London, (1999).
2. A. Y. Lo. On a class of Bayesian nonparametric estimates: I. density estimates. *The Annals of Statistics*, **12**:351–357, (1984).
3. A. Y. Lo and C. Weng. On a class of Bayesian nonparametric estimates: II. hazard rate estimates. *Annals of the Institute of Statistical Mathematics*, **41**:227–245, (1989).
4. A. Y. Lo, L. J. Brunner and A. T. Chan. Weighted Chinese restaurant processes and Bayesian mixture models. Technical report, Department of Information and Systems Management, Hong Kong University of Science and Technology, (1998). Revision 1.1.
5. D. J. Aldous. Exchangeability and related topics. In *École d'Été de Probabilités de Saint-Flour XIII - 1983*, pages 1–198. Springer-Verlag, (1985).
6. H. Jeffreys. *Theory of Probability*. Oxford University Press, (1961).
7. J. M. van Noortwijk, H. J. Kalk, M. T. Duits and E. H. Chbab. The use of Bayes factors for model selection in structural reliability. In *8th International Conference on Structural Safety and Reliability (ICOSSAR), Newport Beach, California, U.S.A.*, June 17-21, (2001).
8. K. A. Doksum and B. S. Yandell. Tests for exponentiality. In *Handbook of Statistics*, volume 4, pages 579–611. Elsevier Science Publishers, (1984).
9. L. Kuo. Computaions of mixtures of Dirichlet processes. *SIAM Journal of Scientific and Statistical Computing*, **7**:60–71, (1986).

10. N. Singpurwalla. Gamma processes and their generalizations: an overview. In R. Cooke, M. Mendel and H. Vrijling, editors, *Engineering Probabilistic Design and Maintenance for Flood Protection*, pages 67–75. Kluwer Academic Publishers, (1997).
11. R. E. Barlow, D. J. Bartholomew, J. M. Bremner and H. D. Brunk. *Statistical Inference under Order Restrictions: The Theory and Applications of Isotonic Regression.* John Wiley & Sons, (1980).
12. R. E. Kass and A. E. Raftery. Bayes factors. *Journal of the American Statistical Association*, **90**:773–795, (1995).
13. R. L. Dykstra and P. Laud. A Bayesian nonparametric approach to reliability. *The Annals of Statistics*, **9**:356–367, (1981).
14. S. Kotz and N. L. Johnson, editor. *Encyclopedia of Statistical Sciences*, volume 8. John Wiley & Sons, (1988).

INDEX